CDM 2015 Questions and Answers

A practical approach

ice

Institution of Civil Engineers

publishing

CDM 2015 Questions and Answers

A practical approach

Pat Perry MCIEH, MIOSH, MIIRM

Published by ICE Publishing, One Great George Street, Westminster, London SW1P 3AA.

Full details of ICE Publishing sales representatives and distributors can be found at: www.icevirtuallibrary.com/info/printbooksales

First published 2015
Reprinted 2017

Other titles by ICE Publishing:

CDM Regulations 2015 Explained.
Raymond Joyce. ISBN 978-0-7277-6009-8
ICE Manual of Health and Safety in Construction, 2nd edition.
Ciaran McAleenan and David Oloke (eds). ISBN 978-0-7277-6010-4
Practical Guide to Using the CDM Regulations 2015.
Tony Putsman and Paul McArthur. ISBN 978-0-7277-5990-0

CDM Questions and Answers, published 1999.
CDM Questions and Answers, 2nd edition, published 2002.
CDM 2007: Questions and Answers, published 2008 (by Routledge).
This edition, *CDM 2015 Questions and Answers*, published 2015.

www.icebookshop.com

A catalogue record for this book is available from the British Library

ISBN 978-0-7277-6032-6

Commissioning Editor: Rachel Gerlis
Development Editor: Amber Thomas
Production Editor: Rebecca Taylor
Market Development Executive: Elizabeth Hobson

Typeset by Academic + Technical, Bristol
Index created by Indexing Specialists (UK) Ltd, Hove, East Sussex
Printed and bound by CPI Group (UK) Ltd, Croydon, CR0 4YY

Contents

Preface

The Construction (Design and Management) Regulations 2015 came into force in April 2015 and apply to virtually all construction works, from concept design through to project completion and beyond. They replaced the 2007 Regulations, which although considered to have been effective in improving health and safety in construction, were deemed inadequate by the European Commission (EC) because they excluded domestic projects.

The Regulations have become known as 'CDM' and were based on the EC Directive on temporary or mobile construction sites that was drawn up by EU member states in order to reduce the high rate of accidents and high incidences of ill health in the construction industry.

The CDM Regulations legally require all those involved in a construction project to consider health and safety issues, from the concept design stage through to project completion and building occupation.

The Regulations impose duties on clients, designers and contractors, and create the appointments of principal designer and principal contractor.

The Regulations now apply to domestic building projects and all types of construction project, and all projects where there is, or is likely to be, more than one contractor require a principal designer and a principal contractor to be appointed. Each has a duty to influence and manage health and safety matters relating to the design and construction of the building, and each must follow the 'principles of prevention' in respect of managing health and safety risk.

The objectives of the Regulations are to ensure better planning and design of construction projects, which, in turn, will lead to a reduction in accidents and incidents of ill-health within the industry. The application of the Regulations to domestic projects will reduce the number of accidents on these smaller sites, as the 2015 Regulations now require a fully thought through and written health and safety plan to be prepared before works start onsite.

This book is intended to provide a quick, practical reference guide to the 2015 Regulations, how all duty holders can comply with their legal responsibilities, and how the Regulations can be applied to the everyday problems encountered in a construction project.

The questions posed in this book have frequently been asked over the years by clients, architects, designers and contractors. Importantly, the questions have been answered by the author, allowing individuals to understand the CDM Regulations from a common-sense, practical point of view.

Rather than leave unrecorded the wealth of knowledge that has been built up in the fulfilment of the roles of CDM consultant and co-ordinator and environmental health consultant over many years, the author decided to collate all the questions and answers into book form in the anticipation that it would provide helpful, practical solutions to many of the queries raised regarding the CDM Regulations.

Acknowledgements

I would like to thank colleagues, clients, associates, contractors and all fellow professionals for sharing their thoughts on the CDM Regulations and for sharing experiences about health and safety in the construction industry.

I would also like to thank my family for all their support and patience during the numerous days of writing and typing the manuscript for the book.

About the author

Pat Perry MCIEH, MIOSH, MIIRM, qualified as an environmental health officer in 1978, and spent the early years of her career in local government enforcing environmental health regulations, particularly health and safety law, which became her passion. She has extensive knowledge of the sector and has served on various working parties on both health and safety and food safety issues. Pat also regularly contributes to professional journals as well as appearing frequently as a speaker at related conferences and seminars.

Pat set up her own environmental health consultancy in 1988, and fulfilled her vision of a creating a customer-focused provider of consultancy services to the commercial, hospitality and retail sectors.

The consultancy grew considerably over the years and provided advice to a wide range of high-profile clients across a variety of market sectors. In 2013 PSN joined the food supply chain risk management group Acoura. As an industry leader in its chosen field, Acoura has been able to provide support and expertise that has enabled PSN to realise improved growth while also providing customers with the broad range of services and solutions they require.

Hospitality, leisure and retail have become the consultancy's main sectors of specialisation, and PSN's team is particularly well versed in delivering CDM and environmental health services to clients operating in these areas. The team works on projects ranging from a few hundred thousand pounds to many millions, with new public house developments and major department store refits and refurbishments among the types of developments PSN assists with.

Now known as the specialist division of its parent company Acoura, PSN continues to build strong links with enforcing agencies, while also providing pragmatic, innovative solutions to compliance issues across a wide range of industries.

Glossary of terms in the 2015 CDM Regulations

In these Regulations—

"the 1974 Act" means the Health and Safety at Work etc. Act 1974;

"the 2007 Regulations" means the Construction (Design and Management) Regulations 2007;

"the Management Regulations" means the Management of Health and Safety at Work Regulations 1999;

"business" means a trade, business or other undertaking (whether for profit or not);

"client" means any person for whom a project is carried out;

"construction phase" means the period of time beginning when construction work in a project starts and ending when construction work in that project is completed;

"construction phase plan" means a plan drawn up under regulations 12 or 15;

"construction site" includes any place where construction work is being carried out or to which the workers have access, but does not include a workplace within the site which is set aside for purposes other than construction work;

"construction work" means the carrying out of any building, civil engineering or engineering construction work and includes—

(a) the construction, alteration, conversion, fitting out, commissioning, renovation, repair, upkeep, redecoration or other maintenance (including cleaning which involves the use of water or an abrasive at high pressure, or the use of corrosive or toxic substances), de-commissioning, demolition or dismantling of a structure;

(b) the preparation for an intended structure, including site clearance, exploration, investigation (but not site survey) and excavation (but not pre-construction archaeological investigations), and the clearance or preparation of the site or structure for use or occupation at its conclusion;

(c) the assembly on site of prefabricated elements to form a structure or the disassembly on site of the prefabricated elements which, immediately before such disassembly, formed a structure;

(d) the removal of a structure, or of any product or waste resulting from demolition or dismantling of a structure, or from disassembly of prefabricated elements which immediately before such disassembly formed such a structure;

(e) the installation, commissioning, maintenance, repair or removal of mechanical, electrical, gas, compressed air, hydraulic, telecommunications, computer or similar services which are normally fixed within or to a structure,

but does not include the exploration for, or extraction of, mineral resources, or preparatory activities carried out at a place where such exploration or extraction is carried out;

"contractor" means any person (including a non-domestic client) who, in the course or furtherance of a business, carries out, manages or controls construction work;

"design" includes drawings, design details, specifications and bills of quantities (including specification of articles or substances) relating to a structure, and calculations prepared for the purpose of a design;

"designer" means any person (including a client, contractor or other person referred to in these Regulations) who in the course or furtherance of a business—

(a) prepares or modifies a design; or

(b) arranges for, or instructs, any person under their control to do so, relating to a structure, or to a product or mechanical or electrical system intended for a particular structure, and a person is deemed to prepare a design where a design is prepared by a person under their control;

"domestic client" means a client for whom a project is being carried out which is not in the course or furtherance of a business of that client;

"excavation" includes any earthwork, trench, well, shaft, tunnel or underground working;

"the general principles of prevention" means the general principles of prevention specified in Schedule 1 to the Management Regulations;

"health and safety file" means a file prepared under regulation 12(5);

"inspector for the Executive" means an inspector within the meaning given in section 53(1) of the 1974 Act;

"loading bay" means any facility for loading or unloading;

"place of work" means any place which is used by any person at work for the purposes of construction work or for the purposes of any activity arising out of or in connection with construction work;

"pre-construction information" means information in the client's possession or which is reasonably obtainable by or on behalf of the client, which is relevant to the construction work and is of an appropriate level of detail and proportionate to the risks involved, including—

(a) information about—
  (i) the project;
  (ii) planning and management of the project;
  (iii) health and safety hazards, including design and construction hazards and how they will be addressed; and

(b) information in any existing health and safety file;

"pre-construction phase" means any period of time during which design or preparatory work is carried out for a project and may continue during the construction phase;

"principal contractor" means the contractor appointed under regulation 5(1)(b) to perform specified duties in regulations 12 to 14;

"principal designer" means the designer appointed under regulation 5(1)(a) to perform specified duties in regulations 11 and 12;

"project" means a project which includes or is intended to include construction work and includes all planning, design, management or other work involved in a project until the end of the construction phase;

"site rules" means rules which are drawn up for a particular construction site and are necessary for health or safety purposes;

"structure" means—

(a) any building, timber, masonry, metal or reinforced concrete structure, railway line or siding, tramway line, dock, harbour, inland navigation, tunnel, shaft, bridge, viaduct, waterworks, reservoir, pipe or pipeline, cable, aqueduct, sewer, sewage works, gasholder, road, airfield, sea defence works, river works, drainage works, earthworks, lagoon, dam, wall, caisson, mast, tower, pylon, underground tank, earth retaining structure or structure designed to preserve or alter any natural feature, and fixed plant;

(b) any structure similar to anything specified in paragraph (a);

(c) any formwork, falsework, scaffold or other structure designed or used to provide support or means of access during construction work,

and any reference to a structure includes part of a structure;

"traffic route" means a route for pedestrian traffic or for vehicles and includes any doorway, gateway, loading bay or ramp;

"vehicle" includes any mobile work equipment;

"work equipment" means any machinery, appliance, apparatus, tool or installation for use at work (whether exclusively or not);

"working day" means any day on which construction work takes place;

"workplace" means a workplace within the meaning of regulation 2(1) of the Workplace (Health, Safety and Welfare) Regulations 1992 other than a construction site.
(Extract from Construction (Design and Management) Regulations 2015.)

Throughout this book, references to the CDM Regulations are references to the Construction (Design and Management) Regulations 2015 or CDM 2015.

Part 1

The Construction (Design and Management) Regulations 2015, roles and duties

CDM 2015 Questions and Answers: A practical approach
ISBN 978-0-7277-6032-6

http://dx.doi.org/10.1680/cdmqa.60326.003

Chapter 1
The Construction (Design and Management) Regulations 2015 explained

What are the CDM Regulations and what is their purpose?

The CDM Regulations is the common name for the Construction (Design and Management) Regulations 2015, a set of regulations that came into force on 6 April 2015. They replace the Construction (Design and Management) Regulations 2007 (and amendments).

The purpose of the Regulations is to improve the health and safety record on construction sites by requiring all parties involved in a construction project to take responsibility for health and safety standards. By requiring better planning, design and management of a construction project it is believed that unacceptably high accident and fatality rates will be reduced both during the construction works and during ongoing maintenance and cleaning activities.

The Regulations are also intended to help reduce the high incident rate of occupational ill health that is common in the construction industry – chronic respiratory conditions, musculoskeletal conditions, industrial deafness, industrial dermatitis and so on.

Those who could create health and safety risks have been made responsible for considering and controlling them during all stages of the project – conception, design, planning, construction work, future maintenance and use of the building, including demolition.

Why have the 2007 CDM Regulations been replaced?

The CDM Regulations 2007 have been instrumental in improving the overall health and safety on construction sites but they have been relatively ineffective in reducing design risks and changing attitudes to the importance of planning and managing projects, and so have not brought the improvements in health and safety expected.

In particular, clients and designers have been slow to accept their duties, and have often failed to understand the intent of the Regulations regarding hazard elimination and risk management.

The CDM co-ordinator role introduced by the 2007 Regulations was considered by most parties involved in a construction project as bureaucratic and ineffective, and there was very little co-ordination of projects by them. Often there was confusion over the extent of the role, and clients were slow to appoint them, thus reducing the contribution that they could make to projects.

The Regulations have generated huge amounts of paperwork, and it seemed that the volume of paperwork created for a project would somehow indicate how effectively those involved in construction were complying with the legislation.

Often, duty holders under the 2007 Regulations would hide behind the paperwork, and little improvement was seen in relation to managing, communicating and co-ordinating risk on a construction project.

There were also concerns raised by the EU Commission that the UK had not adequately adopted the relevant EU directive, and improvements have been required to ensure that a consistent approach to the Regulations is adopted across all EU countries. In particular, the UK had to amend the 2007 Regulations to incorporate domestic projects, as the original EU directive included them but the UK government excluded them in the 2007 Regulations.

Do the CDM Regulations apply to all construction projects?

Yes. The 2015 CDM Regulations apply to all construction works irrespective of whether the construction project is for commercial or domestic clients.

Duties are placed on clients, designers and contractors, and additional duties apply when there is or is likely to be more than one contractor working on the project.

What is 'construction work' under the CDM Regulations?

Construction work means the carrying out of building, civil engineering or engineering construction work.

The definition includes

- the construction, alteration, conversion, fitting out, commissioning, renovation, repair, upkeep, redecoration or other maintenance (including cleaning that involves the use of water or an abrasive at high pressure or the use of substances classified as corrosive or toxic), de-commissioning, and demolition or dismantling of a structure
- the preparation for an intended structure, including site clearance, exploration, investigation (but not site survey), excavation (but not pre-construction

archaeological investigations), laying and installing the foundations of the structure, and the clearing or preparation of a site or structure for use or occupation at its conclusion

- the assembly on site of pre-fabricated elements to form a structure or the disassembly on site of pre-fabricated elements that, immediately before such disassembly, formed a structure
- the removal of a structure or part of a structure or any product or waste resulting from demolition or dismantling of a structure or from the disassembly of pre-fabricated elements that, immediately before disassembly, formed a structure
- the installation, commissioning, maintenance, repair or removal of mechanical, electrical, gas, compressed air, hydraulic, telecommunications, computer or similar services that are normally fixed within or to a structure.

What is not 'construction work' under the CDM Regulations?

Guidance issued by the Health and Safety Executive (HSE) on the 2015 CDM Regulations does not contain any information on works that would be classed as not construction, but based on previous guidance the following activities would generally not be classed as construction work

- general horticultural work and tree planting
- archaeological investigations
- erecting and dismantling of marquees
- erection and dismantling of lightweight partitions to divide open-plan offices
- creation of exhibition stands and displays
- erection of scaffolds for support or access for work activities that are not classed as construction works
- site survey works (e.g. taking levels, assessing soil types and examining structures)
- work to or on ships
- on-shore fabrication of elements for off-shore installations
- factory manufacture of items for use on construction sites.

The CDM Regulations refer to key appointments that must be made on a construction project. What does this mean?

The CDM Regulations identify key duty holders who have responsibilities for ensuring that health and safety matters are addressed during construction projects. They are

- the client
- the principal designer
- designers
- the principal contractor
- contractors.

The client is anyone for whom a construction project is carried out, and now includes anyone who commissions a domestic project – a major change in the 2015 Regulations.

The principal designer is appointed on projects where there is or is likely to be more than *one* contractor working on the project at any single time, and is the designer who is responsible for the pre-construction phase information.

Designers include anyone who prepares a design, whether it be architectural, building services, structural or civil engineering, landscape architecture or temporary works, and is anyone who carries on a trade, business or undertaking in connection with which they

- prepare a design or
- arrange for any person under their control to prepare a design, relating to a structure or part of a structure.

The principal contractor is appointed by the client when there is or is likely to be one or more contractors working on the project at any one time – and must be a contractor. They must take responsibility for all site-specific safety issues, including ensuring that contractors and sub-contractors are competent and have resources to carry out the work safely, and that a health and safety plan is developed. Principal contractors are also responsible for providing information, training and consultation with employees, including the self-employed.

Contractors are those who manage or carry out construction work.

The CDM Regulations refer to a 'notifiable project'. What is this?

CDM regulation 6 requires certain projects to be notified to the HSE by the client.

A construction project is notifiable to the HSE area office when

- it will, or is expected to, last more than *30 days* and involve more than *20 persons* working simultaneously at any point in the project or
- it will, or is expected to, involve more than *500 person days*.

The revised Regulations will reduce the number of projects notified, and this may enable the HSE to target inspections more effectively.

The HSE requires certain information that is outlined in Schedule 1 of the Regulations to be notified to it. As long as the relevant information is given, it can be supplied in any format, but in order to facilitate notification the HSE has produced form F10 (revised), which can be used for all projects.

When does the construction phase start, and are weekends and bank holidays counted?

The construction phase for the purposes of notification to the HSE is from the day 'construction works' start.

Remember, *site clearance* constitutes construction works and must be included in the calculation, although site survey does not constitute construction works.

If construction work is programmed to take place on Saturdays and Sundays and on any bank holiday, no matter how long the working shift will be, these days must be counted as 'construction days'.

What is a 'person day'?

A 'person day' is any day or part of a day (no matter how short) when someone is expected to carry out construction work. A person day relates to *one* individual, and includes site agents, foremen and supervisors.

Operatives do not actually have to be carrying out any physical work to be involved in 'construction work' – if they are managing the project they are included as a 'person day'.

Why does the HSE need to know about these projects?

The HSE enforces health and safety laws in the construction industry. Projects lasting over 30 days and involving more than 20 workers are considered to be substantial building or refurbishment projects where the risks to the health and safety of operatives and others can be high. The HSE has always been made aware of construction projects so that it can plan its inspection programme of enforcing the laws.

The HSE receive over 98 000 project notifications per year (2013–2014), and uses the notification process to help it target its on-site inspection priorities. In addition, it will review the F10 forms to try to influence the design process, enquiring about the provision for health and safety where appropriate.

What does the HSE do when it receives all these F10 forms?

The construction division of the local HSE office records all the forms received, and allocates the projects to individual inspectors. The HSE inspectors base their routine inspections on the type of projects notified, and will prioritise projects into perceived risk categories: for example, those where falls from height are likely, or those that involve excavations.

The construction inspector, or the administration officer, will look at key dates on the F10 form. They are interested to note when the form is signed by the client (or on their behalf) and received by them, and when works are proposed to start on site.

The time allowed for preparation and planning on the project will be of critical importance to the HSE.

If there are only a few days between the HSE receiving the form and works starting on site, it will investigate because the essence of the CDM Regulations is to involve key duty holders in the planning process of a project so that health and safety is considered and to allow the principal contractor or the contractor adequate time to plan and prepare for the start on site.

The HSE will want to establish whether

- the client appointed a principal designer and principal contractor early enough in the project (i.e. as soon as they had information about the project and the construction work involved)
- the client notified the project to the HSE as soon as was practicable after their appointment.

A client who fails to appoint the relevant duty holders early enough in the project is guilty of an offence.

Case study

A developer was undertaking 'core and shell' works for new retail units, and had notified the HSE in the usual way. The shell unit went under offer to a restaurant company that commissioned its designers to fit out the shell. The developers' work over ran the timescales, and legal complications delayed the client's purchase. The client appointed the principal designer as soon as the legal work had been completed and the shell building was handed over. The client wanted works of fitting out to start as soon as possible. The client requested the principal designer to notify the HSE on form F10, as required under CDM regulation 6, indicating that works were due to start on site in approximately 2 weeks.

The HSE wrote to the principal designer to enquire at what date they were appointed and why notification had only just taken place. It asked to see letters of appointment for the principal designer, with a view to determining whether the client had failed in their duty to appoint the principal designer 'as soon as practicable after information about the project becomes available'. Also, the inspector wanted to know what time had been allowed for preparation and planning by the principal contractor – 2 weeks seemed inadequate to him for the size of the project.

If the client had complied with their legal duty of an early appointment, the HSE wanted to establish whether the principal designer was dilatory in making the notification.

In this instance, the HSE advised that, in its view, the client had appointed the principal designer too late in the design process and that, in future, the client should appoint the principal designer as soon as an intended purchase had been agreed.

What happens if a project was originally going to last less than 30 days and have less than 20 workers (or 500 person days) and not be notifiable but due to unforeseen circumstances it will now take longer?

First, you must review *why* you thought it would be lasting less than 30 days and would involve less than 20 workers in the first place. The CDM Regulations state that where it is reasonable to assume that a project may take longer than 30 days and will or is likely to involve more than 20 workers on site simultaneously (or 500 person days), the HSE must be notified. The HSE will view seriously any intent to evade the Regulations by avoiding notification.

If the project will last longer than anticipated, the client (advised by the principal designer) should notify the HSE office on form F10, and include a covering letter explaining the reasons for the project over-run and the late notification.

Is demolition work covered by the CDM Regulations?

Yes. Demolition works are construction works, and the CDM Regulations apply.

If the project involves more than one contractor working on the project at any one time, the client must appoint a principal designer and a principal contractor.

If demolition works are expected to last for more than 30 days and involve more than 20 workers working simultaneously at any stage of the project or involve more than 500 person days, then the project must be notified to the HSE.

Specifically, under the Regulations, demolition works are required to have a written plan showing how danger is to be avoided. This plan is to be drafted by those in control of the works (CDM regulation 20).

What does demolition and dismantling work include?

Demolition is taken to mean the deliberate pulling down, destruction or taking apart of a structure, or a substantial part of the structure.

Dismantling is the taking down or taking apart of all, or a substantial part, of a structure, and includes situations where the structure is carefully taken down for re-use.

A 'structure' is defined as

(a) any building, timber, masonry, metal or reinforced concrete structure, railway line or siding, tramway line, dock, harbour, inland navigation, tunnel, shaft, bridge, viaduct, waterworks, reservoir, pipe or pipeline, cable, aqueduct, sewer, sewage works, gasholder, road, airfield, sea defence works, river works, drainage works, earthworks, lagoon, dam, wall, caisson, mast, tower, pylon, underground tank, earth retaining structure or structure designed to preserve or alter any natural feature, and fixed plant,
(b) any structure similar to anything specified in paragraph (a),
(c) any formwork, falsework, scaffold or other structure designed or used to provide support or means of access during construction work,

and any reference to a structure includes part of a structure.

Is there a difference between 'demolition' and 'dismantling' works?

Demolition is the deliberate pulling down, destruction or taking apart of all, or a substantial part, of the structure.

Dismantling is the taking down or taking apart of all, or a substantial part, of a structure.

Dismantling for re-erection or re-use will be demolition for the purposes of the CDM Regulations.

The formation of openings for windows, doors and services are not in themselves demolition works.

The removal of cladding, roof tiles or scaffolding is not, in itself, demolition or dismantling works unless included in or combined with other building operations.

> **Case study**
> A small retailer is having a brick outhouse demolished to make way for additional car parking. The building is single-storey, single-skin brickwork under a flat roof. The local builder can undertake the work in 2 days with three men on site.

The works involved fall within the definition of construction work, and irrespective of their duration and number of operatives on site the works come under the control of the CDM Regulations.

The client (the shop owner) and the contractor have to comply with all the requirements of the Regulations.

The contractor will be the only contractor on the project, and so the client does not have to appoint a principal designer or principal contractor. The client takes on some responsibilities, and discusses with the contractor the use of the rear yard, the previous use of the outhouse and passes to the contractor any information as to how it was constructed.

The contractor makes notes that the rear yard is used for access to the shop by delivery vehicles but it is not used as staff access, nor does anyone else have rights of access. The information on the building to be demolished is brick, single skin and flat felt roof. No services are connected to the out-building, no asbestos is present and the structure to be demolished is straightforward. To comply with CDM regulation 20, a short written plan of works or method statement is produced by the contractor and agreed with by the client.

The contractor stipulates that the rear yard will be cordoned off, that the shop owner will re-schedule deliveries for later in the week, after demolition, that the skip will be positioned away from the rear emergency exit door, that his operatives will be able to use the shop WC and tea-making facilities, and that waste will be removed by a licensed disposal company. In addition, the builder prepares a brief method statement covering the sequence of demolition, identifying hazards and risks (dust inhalation, falling from height, manual handling, etc.).

The shop owner is given a copy of the health and safety plan, and it is agreed before works start.

How do the CDM Regulations apply to term maintenance contracts?

If the works are 'construction works', then the CDM Regulations apply to those aspects of the project. Therefore, any painting, renewals, repairs, redecoration, maintenance, improvements and so on are likely to be construction works – it is a wide-ranging definition.

If the works are 'construction works' and last more than 30 days and involve more than 20 workers working simultaneously at any point on the project or involve more than 500 person days then they will be notifiable, and the CDM Regulations will apply.

Where some of the works are outside of the Regulations due to the definition of construction works and some are within it, it would be advisable to apply the Regulations to all of the works, although any management arrangements should clearly outline how both aspects of the project would be managed.

Projects involving term maintenance must be reviewed individually.

Do the CDM Regulations apply to emergency works?

In any emergency, the first priority is to make sure that the premises or structures are safe and without imminent risk to the health and safety of members of the public and others.

Once the making safe has been carried out, then the CDM Regulations should be applied: that is, the client should make the appointment of a principal designer and a principal contractor as soon as is practicable.

Duty holders, once appointed, must fulfil their responsibilities under CDM regulations 9, 11, 13, 14 and 15.

The construction phase health and safety plan should be developed as soon as is practicable, and if time does not allow a written plan to be completed before the emergency works take place, verbal discussions and agreement should be reached regarding key health and safety issues. If possible, the key issues should be written down, and the body of the plan can then be developed as the works progress.

Emergency works that are likely to be substantial will inevitably fall under the Regulations. It would be wise to consult the local HSE office about any plans and so on, even if there is not time to prepare the paperwork.

Who are domestic clients?

Domestic clients are clients who have work done that does not relate to any trade, business or other undertaking. This is usually someone who commissions work on their own home, or the home of a family member.

Duties are placed on domestic clients by the 2015 CDM Regulations, which is a major change from the 2007 Regulations. Those working on a project for a domestic client will assume the duties allocated to clients.

Do the CDM Regulations apply to domestic house building or repairs?

Yes. A major change in the 2015 CDM Regulations is that they will apply to domestic building projects. If you commission a builder to build you a house for your own occupation, then the Regulations will apply. If you have a new conservatory, extension, loft conversion or similar, then the Regulations will apply.

CDM regulation 7 sets out the application of the Regulations for domestic projects.

A principal designer and a principal contractor will have to be appointed if there is or likely to be more than one contractor involved in the project, and the principal designer will have to ensure that health and safety requirements are met.

The client may appoint the principal designer and principal contractor, but the legal duty to do so under regulation 4 does not apply.

Where no appointments are made, the first designer appointed and the first contractor appointed will automatically assume responsibility for the duty holder positions.

If you commission a house to be built to sell it to someone else, then you are likely to be a developer, and the Regulations will apply, including the duties imposed on clients.

Notification of projects to the HSE will be required if the client is a domestic client and when the requirements of regulation 6 is met: that is, a project will last more than 30 days and involve more than 20 workers or the project will involve more than 500 person days. If notification is not made by the domestic client, it must be made by either of the other duty holders (i.e. the designer or the contractor).

What are the duties in respect of the CDM Regulations when work is done for a domestic client?

Domestic clients have duties under the CDM Regulations, but by way of regulation 7 they are automatically transferred to the contractors. Where more than one designer or contractor will be working on the project, the principal contractor will take on the client's duties, and designers will work with them as if they were the client.

Contractors have to comply with all the duties placed on them as laid out in Part 4 and Schedule 2 of the Regulations.

Designers have duties under regulation 9: that is, they must have adequate regard to health and safety when preparing their designs and must provide health and safety information to the domestic client and contractor as necessary to ensure that health and safety standards are met. Where the principal contractor or sole contractor is assuming the client's duties, the designer must provide the information to the contractor as if they were the client.

Domestic clients can assume their full responsibilities, and appoint both a principal contractor and a principal designer, and may appoint the designer, in writing, to assume responsibility for all client duties and to manage the project.

What constitutes a 'developer' under the CDM Regulations?

A developer is someone who carries on a trade, business or undertaking (whether for profit or not) in connection with which

- land or an interest in land is granted or transferred to the client
- the developer undertakes that construction work will be carried out.

In effect, a 'developer' is a commercial developer who sells domestic premises before a project is complete and arranges for construction work to be carried out.

Developers in this category include housing associations, local councils, self-build companies and other such bodies, whether they are profit-making or non-profit-making national house builders.

What are the consequences of failing to comply with the CDM Regulations?

All 'duty holders' have legal responsibilities under the CDM Regulations.

Health and safety legislation is triable 'either way' (i.e. either on summary conviction or on indictment). This means that prosecutions can be heard in the magistrates' court or the Crown Court.

If prosecutions are brought in the magistrates' court for contraventions of the Regulations, then the maximum fine per offence is unlimited. If prosecutions are brought in the Crown Court, then fines are unlimited and custodial sentences are possible.

Any of the following can be prosecuted if they fail to discharge their legal duties:

- the client
- the principal designer
- the principal contractor
- the designers
- the contractors
- the workers.

Where an investigation in to potential breaches of the Regulations is carried out by the HSE, a 'fee for intervention' charge could be levied, as there may well be a 'material breach' of the regulations.

Case studies

Prosecution of a principal contractor by the HSE

Following an investigation into two accidents, one a fatality, which happened on a construction site in London, the HSE brought a prosecution against the principal contractor for failing to have an adequate construction phase health and safety plan in operation over the period of the two accidents.

The HSE also prosecuted the ground works contractor for failing to take all reasonable steps to ensure that an excavation did not collapse accidentally, and for failing to take suitable and sufficient measures to prevent vehicles overrunning the edge of an earthwork.

The construction phase health and safety plan failed to consider the hazards and risks of the ground works, and it did not include risk assessments and suitable method statements.

The prosecutions were heard in the Crown Court due to the severity of the accidents, and the principal contractor was fined £20 000 plus £3000 costs. The ground works contractor was fined a total of £20 000.

Prosecution of a client, designer and principal contractor by the HSE

A client, designer and principal contractor, as well as the contractors undertaking the work, were prosecuted in the Crown Court following the fatal accident to a scaffolder who died after falling through a fragile roof light while working on an extension to a warehouse.

The principal inspector for the HSE said

> This was a tragic incident in which a man lost his life in circumstances that could so easily have been prevented. The dangers presented by fragile roof materials have been well known for many years, yet were not taken into consideration in this case. Considerable time had been spent planning the job and any one of those involved had the opportunity to realise that somebody could fall through the warehouse roof. Simple steps could then have been taken to prevent this death. If this had been done, the deceased would still be alive today. Everyone has responsibility to consider the hazards and risks of the site and to share information. The designers knew that the roof material was fragile and they should have done more to highlight this. The client did not obtain up-to-date information on the existing building structure as he wanted to save costs. The contractors generally, failed to work safely.

What civil liability extends to the CDM Regulations?

The CDM Regulations constitute criminal law – it is a criminal act to contravene them. Anyone who does so incurs a criminal record.

Civil law applies to situations where an individual can sue another person (or corporate body) for damages due to their negligence (i.e. failing in their common law duty of care).

Being prosecuted for criminal offences under the Regulations does not infer an automatic right to bring civil proceedings. A successful prosecution does not necessarily imply a failure in the duty of care under civil law, and a successful civil case will depend on the facts of the case.

Claimants will have to show negligence by either their employer or some other person in order to bring a civil claim, and the court hearing will determine the evidence. The fact that the employer or other person contravened the law will not automatically lead to a guilty verdict.

The law on strict liability was amended as a result of the review into health and safety legislation, leading to fewer automatic civil claims for compensation.

Will the CDM Regulations be enforced only by the HSE?

No, not necessarily. Generally, most construction sites come under the jurisdiction of the HSE for health and safety enforcement, and therefore enforcement of the CDM Regulations.

But the changes introduced in the new 2015 Regulations mean that there will be the opportunity for local authorities to enforce the Regulations in those premises for which they have statutory inspection duties (e.g. retail premises, offices, entertainment venues and sports facilities). The Health and Safety (Enforcing Authority) Regulations 1998 set out which enforcement body is responsible for enforcing the law in various types of workplace.

Where works are construction works and generally, where they are not notifiable, the local environmental health officer could be responsible for ensuring that the Regulations are being followed – especially in relation to site safety matters (working at heights, use of equipment, provision of welfare facilities, etc.).

The HSE is responsible for enforcing health and safety – and fire safety – on construction sites, and if the area where the work is being carried out can be described as 'a place set aside for construction work', then the HSE will enforce the Regulations.

Projects that are notifiable will be enforced by the HSE.

Do the CDM Regulations apply to all forms of building procurement such as the private finance initiative and public–private partnerships?

Yes – because, irrespective of how the project is funded or acquired, the works will be construction and design work will take place.

Project originators should take on the duties of clients and ensure that a principal designer is appointed early on into the project and that HSE notification is made in a timely manner.

The role of the client can be assigned to other parties during the course of the project, but there must always be someone clearly identified in that role.

Where multiple clients are discussing and agreeing a new project, one of them must be appointed as the client under the CDM Regulations, in order to comply with the legislation. It would be sensible to record this as part of the management process, and communicate to all other parties.

If there are several clients involved in a project, do they all have duties under the CDM Regulations?

Not necessarily. Regulation 4 of the CDM Regulations allows for multiple clients to elect either one or more of them to be classed as the client under the Regulations.

All clients have to agree in writing that such an election has taken place, and, once this has been done, no other client will be required to fulfil the client duties, except that they will all be required to provide any information to other parties as appropriate.

If there is no agreement in writing or no one client wants to take on the full responsibility for everyone else, then all clients will remain as clients under the Regulations, and their various principal designers will have to ensure good co-operation and co-ordination of information during the course of the project.

It would be sensible to start the initial project meetings with a review of responsibilities under the CDM Regulations and to minute the discussion and decisions made regarding the appointment of duty holders and so on.

What are the general duties imposed on everyone under the CDM Regulations?

The CDM Regulations emphasise the need for duty holders to the competent to do the jobs they are appointed to and for everyone involved in the construction project to receive information, instruction and training and/or to be properly supervised.

The Regulations contain general duties on

- co-operation
- co-ordination
- competency, knowledge and training
- provision of information.

The Regulations also contain the general duty that all persons involved in planning, designing or carrying out construction work follow the principles of prevention in respect of health and safety.

No person who is offered an appointment under the Regulations should take on the role unless they themselves are satisfied that they are competent to undertake the role.

No person may instruct a worker to carry out or manage construction work unless they have received information, instruction or training or are under the supervision of a competent person.

This means that anyone who wishes to appoint or engage someone to do a construction activity, or design a new building and so on, must be satisfied that the person they want to appoint has sufficient knowledge and experience to carry out the tasks.

Often, accidents and major incidents occur because the people carrying out potentially complex tasks have been appointed because they are the cheapest and can do the job the quickest. There have been many incidents involving the collapse of buildings because the contractors involved have not understood the correct demolition sequences or have not been experienced enough in undertaking the job, thereby being ignorant of the hazards and risks involved.

All persons have the duty to appoint competent persons, so if a designer needs to engage the services of, say, a structural engineer, they must ensure that the person has the necessary skills, experience, knowledge and training, and must not just appoint the cheapest company.

A client should ultimately be satisfied that the whole design, procurement and implementation team is experienced and suitably trained and informed, and should therefore ensure that if members of the project team appoint sub-contractors and others that they have rigorous procedures in place for assessing competencies.

Everyone involved in a project (including workers) has a duty to report instances where they or others are working in a way that puts them or others in danger. Any such instances must be reported to the person in control of the work. Those in control of the construction works should encourage workers to stop work and report dangerous occurrences or conditions when they see them.

What are the duties in respect of co-operation under the CDM Regulations?

Regulation 8 of CDM Regulations states that every person concerned in a project, on whom a duty is placed by the CDM Regulations should

- seek the co-operation of any other person concerned in any project involving construction work at the same or an adjoining site so far as is necessary in order that any person with a duty under the Regulations can fulfil that duty
- co-operate with any other person concerned in any project involving construction work at the same or an adjoining site so far as is necessary to enable that person to perform any duty under the Regulations.

Any person who is working on a project under the control of another person must inform that person if they believe there is any issue that could affect their own or others' health and safety.

Regulation 8 requires everyone to co-operate with each other, whether part of the project team for the construction project or involved in a separate project that may have some influence on the main project.

Co-operation will be required, for instance, in areas with shared access. A retail store may be having an extension built and the rear service yard may be used by other retailers. The principal contractor may need to co-ordinate their materials deliveries with the other premises stock deliveries. The other retailers must then co-operate with the principal contractor so as to enable the principal contractor to operate safely.

HSE sees the improvement in co-operation between duty holders as a significant step in improving the incidents of poor health and safety: for example, often contractors will have to handball materials long distances, increasing musculoskeletal risks, because others have blocked the vehicle entry to the delivery area.

Are there any duties in respect of co-ordination under the CDM Regulations?

The CDM Regulations do not contain any specific duties in respect of co-ordination, although co-ordination is implied throughout the Regulations, especially in respect of information transfer among duty holders.

All persons concerned with a project on whom duties are placed by the Regulations should co-ordinate their activities with one another in a manner that ensures, so far as is reasonably practicable, the health and safety of persons

- carrying out the construction work and
- affected by the construction work.

Co-ordination means to 'bring together and cause to work together efficiently'. If all parties involved in a project co-operated with and co-ordinated their work with each other, better planning would result, and conflicts involving health and safety would be significantly reduced.

How many times do builders work and mechanical and electrical installations clash on site, often creating unsafe working conditions when one or other of the parties has to 'make do' to work on site, usually ending up working in an unsafe way to get the job done?

Do the CDM Regulations apply to Scottish projects?

Yes. The CDM Regulations, in their entirety, apply in Scotland.

Do the CDM Regulations apply in Wales?

Yes. The Regulations, in their entirety, apply in Wales.

Do the CDM Regulations apply in Northern Ireland?

Yes. The Regulations, in their entirety, apply in Northern Ireland – although the Regulations have the addition of 'Northern Ireland' in the title.

Do the CDM Regulations apply to term contracts?

The CDM Regulations do not apply to the 'term contract' but may apply to the individual projects undertaken as part of the term contract.

The HSE's general view regarding term contracts is that any form F10 notification will be project specific within the general term of the contract.

Notification for a term contract for general work that may not take place is not of benefit.

The HSE is interested in works of construction that will last more than 30 days and involve more than 20 persons on site at any one time or take more than 500 person days. These projects will be notifiable whenever they occur in a term contract.

The Regulations apply to all construction work, so the only consideration will be to decide whether the work package is notifiable.

If there is, or is likely to be, more than one contractor on site, then the client must appoint a principal designer and a principal contractor.

It may be sensible to address the roles and responsibilities for the Regulations, and outline the procedural approach for compliance within the term contract.

A practical approach to a term contract in which minor works are carried out to a range of buildings by the same team is to appoint someone to a similar role to that of the principal designer: that is, someone to co-ordinate all the health and safety issues and any design issues across all the contractor teams.

What are some of the key definitions included in the CDM Regulations 2015?

In these Regulations—

"the 1974 Act" means the Health and Safety at Work etc. Act 1974;

"the 2007 Regulations" means the Construction (Design and Management) Regulations 2007;

"the Management Regulations" means the Management of Health and Safety at Work Regulations 1999;

"business" means a trade, business or other undertaking (whether for profit or not);

"client" means any person for whom a project is carried out;

"construction phase" means the period of time beginning when construction work in a project starts and ending when construction work in that project is completed;

"construction phase plan" means a plan drawn up under regulations 12 or 15;

"construction site" includes any place where construction work is being carried out or to which the workers have access, but does not include a workplace within the site which is set aside for purposes other than construction work;

"construction work" means the carrying out of any building, civil engineering or engineering construction work and includes—

(a) the construction, alteration, conversion, fitting out, commissioning, renovation, repair, upkeep, redecoration or other maintenance (including cleaning which involves the use of water or an abrasive at high pressure, or the use of corrosive or toxic substances), de-commissioning, demolition or dismantling of a structure;

(b) the preparation for an intended structure, including site clearance, exploration, investigation (but not site survey) and excavation (but not pre-construction archaeological investigations), and the clearance or preparation of the site or structure for use or occupation at its conclusion;

(c) the assembly on site of prefabricated elements to form a structure or the disassembly on site of the prefabricated elements which, immediately before such disassembly, formed a structure;

(d) the removal of a structure, or of any product or waste resulting from demolition or dismantling of a structure, or from disassembly of prefabricated elements which immediately before such disassembly formed such a structure;

(e) the installation, commissioning, maintenance, repair or removal of mechanical, electrical, gas, compressed air, hydraulic, telecommunications, computer or similar services which are normally fixed within or to a structure,

but does not include the exploration for, or extraction of, mineral resources, or preparatory activities carried out at a place where such exploration or extraction is carried out;

"contractor" means any person (including a non-domestic client) who, in the course or furtherance of a business, carries out, manages or controls construction work;

"design" includes drawings, design details, specifications and bills of quantities (including specification of articles or substances) relating to a structure, and calculations prepared for the purpose of a design;

"designer" means any person (including a client, contractor or other person referred to in these Regulations) who in the course or furtherance of a business—

(a) prepares or modifies a design; or

(b) arranges for, or instructs, any person under their control to do so,

relating to a structure, or to a product or mechanical or electrical system intended for a particular structure, and a person is deemed to prepare a design where a design is prepared by a person under their control;

"domestic client" means a client for whom a project is being carried out which is not in the course or furtherance of a business of that client;

"excavation" includes any earthwork, trench, well, shaft, tunnel or underground working;

"the general principles of prevention" means the general principles of prevention specified in Schedule 1 to the Management Regulations;

"health and safety file" means a file prepared under regulation 12(5);

"inspector for the Executive" means an inspector within the meaning given in section 53(1) of the 1974 Act;

"loading bay" means any facility for loading or unloading;

"place of work" means any place which is used by any person at work for the purposes of construction work or for the purposes of any activity arising out of or in connection with construction work;

"pre-construction information" means information in the client's possession or which is reasonably obtainable by or on behalf of the client, which is relevant to the construction work and is of an appropriate level of detail and proportionate to the risks involved, including—

(a) information about—
    (i) the project;
    (ii) planning and management of the project;
    (iii) health and safety hazards, including design and construction hazards and how they will be addressed; and

(b) information in any existing health and safety file;

"pre-construction phase" means any period of time during which design or preparatory work is carried out for a project and may continue during the construction phase;

"principal contractor" means the contractor appointed under regulation 5(1)(b) to perform specified duties in regulations 12 to 14;

"principal designer" means the designer appointed under regulation 5(1)(a) to perform specified duties in regulations 11 and 12;

"project" means a project which includes or is intended to include construction work and includes all planning, design, management or other work involved in a project until the end of the construction phase;

"site rules" means rules which are drawn up for a particular construction site and are necessary for health or safety purposes;

"structure" means—

(a) any building, timber, masonry, metal or reinforced concrete structure, railway line or siding, tramway line, dock, harbour, inland navigation, tunnel, shaft, bridge, viaduct, waterworks, reservoir, pipe or pipeline, cable, aqueduct, sewer, sewage works, gasholder, road, airfield, sea defence works, river works, drainage works, earthworks, lagoon, dam, wall, caisson, mast, tower, pylon, underground tank, earth retaining structure or structure designed to preserve or alter any natural feature, and fixed plant;
(b) any structure similar to anything specified in paragraph (a);
(c) any formwork, falsework, scaffold or other structure designed or used to provide support or means of access during construction work,

and any reference to a structure includes part of a structure;

"traffic route" means a route for pedestrian traffic or for vehicles and includes any doorway, gateway, loading bay or ramp;

"vehicle" includes any mobile work equipment;

"work equipment" means any machinery, appliance, apparatus, tool or installation for use at work (whether exclusively or not);

"working day" means any day on which construction work takes place;

"workplace" means a workplace within the meaning of regulation 2(1) of the Workplace (Health, Safety and Welfare) Regulations 1992 other than a construction site.

(2) Any reference in these Regulations to a plan, rule, document, report or copy includes a copy or electronic version which is —

(a) capable of being retrieved or reproduced when required; and

(b) secure from loss or unauthorised interference.

(Extract from Construction (Design and Management) Regulations 2015.)

CDM 2015 Questions and Answers: A practical approach
ISBN 978-0-7277-6032-6

http://dx.doi.org/10.1680/cdmqa.60326.025

Chapter 2
The client

Who is a client?

Regulation 2 of the CDM Regulations defines a client as 'any person for whom a project is carried out'.

Regulation 2 further defines a project as 'a project which includes or is intended to include construction work and includes all planning, design, management or other work involved in a project until the end of the construction phase'.

It is not necessary for the construction works to actually take place for someone to fall within the definition of a client under the Regulations: the term refers to any person who first thinks about having a structure built, repaired, refurbished, demolished or maintained.

Concept and feasibility schemes are commissioned by a 'client' under the Regulations, and even though actual construction work may not take place, the principles of the Regulations must be applied.

Clients can include

- businesses
- school governors
- charities
- insurance companies
- private finance initiative consortia
- developers
- social clubs or voluntary organisations
- private members' clubs
- individuals who are undertaking a business (e.g. pub landlords and restauranteurs)
- residential property owners
- domestic clients (i.e. those for whom a project is carried out that is not in the course or furtherance of a business of that client).

Why do the CDM Regulations have to apply to clients?

Research has shown the Health and Safety Executive (HSE) over the years that clients have tremendous influence on the planning, organisation and implementation of a construction project. If the client has a cavalier attitude to health and safety, then it is likely that everyone on the project will develop a similar approach. If standards of health and safety on construction sites is to be improved, clients must 'lead from the top' and must ensure that adequate provision, including time and financial resources, is made for health and safety on every project.

Clients also often hold considerable amounts of information about their building or the project to be undertaken, and by ensuring that this information is made available, designers and contractors can plan for known hazards and so on. Hence the duty placed on clients to provide information.

What are the duties of a client under the CDM Regulations?

Under CDM regulation 4 the client has duties for managing projects. In particular, the client has a duty to

- ensure that suitable arrangements are made to manage the project safely
- ensure that sufficient time and resources are allocated to the project
- ensure that construction work is carried out so far as is reasonably practicable without risk to the health and safety of any person
- ensure that the requirements of Schedule 2 of the Regulations are complied with
- ensure that designers and contractors are promptly supplied with information relevant to their purposes
- ensure that any arrangements made for managing the project safely are maintained and reviewed throughout the project
- ensure that the principal designer complies with the duties set out in CDM regulations 9 and 11
- ensure that the principal contractor complies with the duties set out in CDM regulations 12 and 13
- ensure that before construction works start there is a suitable construction phase plan drawn up either by the principal contractor or a contractor as appropriate under the Regulations
- ensure that the principal designer prepares an appropriate health and safety file for the project that must include any information relevant to asbestos-containing materials.

Under CDM regulation 5, the client must appoint a principal designer and a principal contractor where there is, or is likely to be, more than one contractor appointed to the project. Such an appointment must be in writing.

The duty to appoint a principal designer and a principal contractor does not apply to a domestic client because the duties fall automatically to the first appointed designer and contractor. Should they wish, the domestic client could make the appointments or appoint the designer to manage the project rather than the duties passing to the principal contractor.

What duties, if any, does a client have regarding notification of projects?

CDM regulation 6 states that a project is notifiable to the HSE if the construction work on the construction site is scheduled to

- last longer than 30 working days and have more than 20 workers on site working simultaneously at any point in the project or
- exceeds 500 person days.

The client is responsible for making the notification, and must ensure that the information set out in Schedule 1 of the Regulations is provided.

The notification form or document must be displayed on the site where it can be read easily by any worker.

Form F10 will be available for notifications, but clients are at liberty to use their own format provided the requisite information is provided.

What are a client's specific duties regarding the arrangements for managing construction projects?

Every client has to take reasonable steps to ensure that arrangements are in place, or made, for managing a project, including the allocation of sufficient time and resources and that they are suitable to ensure that 'the construction work can be carried out, so far as is reasonably practicable, without risk to the health and safety of any persons' (CDM regulation 4).

The client must take reasonable steps to ensure that the arrangements made are maintained and reviewed throughout the project.

The client is expected to ensure that things are done, not necessarily to do them themselves.

Reasonable steps have to be taken to review the procedures and arrangements adopted by all duty holders, not just contractors.

Clients will need to ensure that all duty holders have good health and safety knowledge, have access to competent advice, know what their legal responsibilities are, have suitable trained and experienced people available to work on the project and so on.

Some of the arrangements will be covered in any general competency enquiries before the contractor is appointed (health and safety arrangements, accident records, etc.), and these may not need to be reproduced for every project.

It is important that there is proper clarity of roles and responsibilities for the project, and information can be included in the pre-construction information pack.

Clients must influence the time available for the specific stages of the project, and while some clients are sceptical about scheduling ('if you give a contractor an inch, they'll take a mile') and therefore prolong the construction phase unduly, they must now demonstrate that they have taken the time needed for the project and must show that they have not enforced unreasonable timescales.

One of the greatest concerns of both major and smaller contractors is that clients impose unrealistic timescales on them, forcing them to accelerate works and take shortcuts, thus compromising health and safety on site.

Designers often complain that clients force them to cut costs and omit preferred safety solutions (e.g. roof edge protection), thereby not allowing them adequate financial resources to design the project safely.

Clients must ask the project team how they are proposing to communicate, co-operate and co-ordinate the information for the project, and will need to be satisfied that the project meetings, design meetings and progress meetings are being scheduled.

The circulation of meeting minutes may be reasonable evidence that the project is being managed effectively.

Can another party act on behalf of the client in ensuring that management arrangements are in place?

No. The client cannot delegate their legal responsibilities for ensuring that the management arrangements are in place.

The client can ask the competent person appointed to advise them on health and safety – as required under the Management of Health and Safety at Work Regulations 1999, to advise and assist them.

Clients are not expected to be experts in managing construction projects, but they are expected to understand that they cannot ignore the requirement for good management and health and safety on a project. They are expected to ask relevant questions and to query less than satisfactory answers.

Clients are expected to make a judgement call on the appropriateness and adequacy of management arrangements – doing nothing is not an option.

Where there is more than one client on a project (e.g. the development of a shopping centre), one or more of the clients may elect, in writing, to be treated for the purposes of the Regulations as the only client or clients. The duty to provide information will apply to all clients even if agreements are made to share the client duties.

What are a client's duties regarding the provision of welfare facilities on a construction site?

A client has to ensure that provision has been made for adequate welfare facilities during the course of construction works in accordance with Schedule 2 of the CDM Regulations.

The client does *not* have to provide welfare facilities, but if they have them available (e.g. a hotel is being refurbished and the existing staff toilets can be used), they are expected to co-operate with the contractor and make them available if at all possible.

The client must assist the contractor in achieving the provision of a good standard of welfare facilities, especially sanitary accommodation. If, for instance, the client can facilitate early connections into mains drainage, this would enable the contractor to install flushing toilets, thus avoiding the use of inferior chemical toilets.

A client does not have to tell the principal contractor or the contractor (whichever is appropriate) what facilities to provide, as the contractor is responsible for ensuring that adequate facilities are available and need not stipulate numbers of WCs, urinals or wash hand basins and so forth. However, a client should be able to assess the general adequacy of the facilities and should encourage the contractor to provide flushable WCs, running water and mains drainage or drainage to suitable tanks.

Chemical closets, inadequate water supplies and so on are not acceptable, and generally imply to the workforce that their welfare is not important.

Does information only have to be provided for notifiable projects?

No. The duty to provide information is placed on the client for *all* projects.

The client must ensure that pre-construction information is provided as soon as is practicable to each designer involved in the design of a structure and to each contractor who is or might be engaged by the client in relation to a project.

Where the project is undertaken for a domestic client, the duty to provide information sits with the first designer appointed and the first contractor appointed.

The best approach is to ask the designers and other members of the team, and the contractor if appointed early enough, what information they would hope to have and what information they believe to be essential in order for them to do their job safely.

The information need only be proportionate to the job in hand – reams of paper and piles of manuals are not required.

The arrangements for managing the project could set out what information is available from whom, and list who is responsible for providing it and when.

Do detailed surveys need to be commissioned in order to provide this information?

The more information available at the planning stage of the project regarding potential health and safety issues, the more likely it will be that the project can be programmed efficiently and safely.

Finding out about the presence of hazardous substances (e.g. asbestos) once the construction phase has started means downtime, costs and potential health risks to all operatives and others. Breaches of legislation are inevitable.

It makes sound business sense to obtain as much information about the site before progressing the design phase. Asking a structural engineer to advise on load-bearing walls and the necessary temporary works will prevent a major on-site collapse, which could cause major and fatal injuries to those on site.

Commissioning the following would not be unreasonable

- a building survey, including services
- a structural survey
- an asbestos and other hazardous substances survey
- a pest survey
- a contaminated land survey
- an environmental noise survey
- a land use survey, including adjacent premises and so on
- an overhead and underground power lines/cables survey.

The CDM Regulations are concerned with reducing hazards and risks to health and safety: that is, being prepared to deal with what is there because of being forewarned.

If asbestos is present, it poses a *serious* health hazard. The HSE expects all clients to have identified the existence of asbestos on site *before* construction works start and to have included proposals to deal with it in the pre-construction health and safety pack, such as remove it completely or encapsulate it, label it and manage the risks. (The preferred option is always to consider the removal of asbestos from a site following the legal requirements of the Control of Asbestos Regulations 2012 unless removal is considered more hazardous than leaving it in situ.)

Each project needs to be assessed on its merits – if the building is a new 'developers' shell', then detailed surveys will not be necessary as the developer will have dealt with these issues. If the building is post-1980 it is unlikely (though not guaranteed) that it will have sprayed asbestos coating and lagging as the insulation and fire protection material – but it may have asbestos boards or asbestos cement tiles.

A high-street retail premises that has been in existence for decades may not need a contaminated land survey unless a review of previous planning permissions indicates an industrial factory or petrol station had been present on the site and extensive ground works are proposed.

What are a client's duties regarding the provision of information on a project?

Every client has a duty to provide information to designers and contractors prior to the commencement of construction that is relevant to them to enable them to carry out their duties under the CDM Regulations.

All available information must be collated by the principal designer in the pre-construction information pack.

The amount of information provided need only be relevant to the project in hand, but must include everything that could be considered relevant, such as

- information on the site and existing buildings
- any survey information
- the future use of the building
- in-house health and safety procedures.

Clients perhaps unwittingly can withhold information about a site or a building, and designers and contractors can be exposed to unnecessary risks to their health and safety. Typical examples would be not advising the project team and contractors about asbestos-containing materials, or fragile roof surfaces, rotten flooring, overhead power lines and so on.

What are the duties of a client before giving the go ahead for a construction project to start on site?

The client must ensure that they have complied with the requirements CDM Regulations to appoint a principal designer and principal contractor where there is or is likely to be more than one contractor working on the project at any one time. Each of these functions must also have adequate resources to carry out the tasks required.

An assessment of competency, including a review of information, instruction and training and the resources required for the project, needs to have been carried out, and this process should be ongoing throughout the duration of the project, although the client has no legal duty to monitor the standards of their appointees during the project.

The only specific legal duty on the client prior to construction work commencing on a project is to ensure that the construction phase health and safety plan has been prepared and that adequate welfare facilities will be provided. The client does *not* have to approve this plan, but must ensure that there is one relevant for the project.

The client also has a duty to ensure that the principal contractor or contractor, where only one is appointed to the project, has sufficient *time* to prepare the construction phase health and safety plan prior to the commencement of works, and they must provide information to the principal contractor on the amount of time that they have for preparation and planning for the project.

CDM 2015 requires clients to make appointments in writing. What does this mean?

Clients have a duty to appoint a principal designer and a principal contractor on all projects where there will be or is likely to be more than one contractor working on the project at any one time.

The appointments must be in writing.

It is sensible to record appointments in writing so that evidence is available to show that each party was aware of their responsibilities and knew what was expected of them.

Both the principal designer and principal contractor need to know that they have been given specific legal duties and the client must set them out in writing.

Example letters of appointment are included in the appendix to this chapter.

When does a client need to appoint a principal designer?

A client must appoint a principal designer when there is or is likely to be more than one contractor working on site at any one time.

The appointment must be made as soon as practicable and always before construction works start.

If the client fails to make the appointment he assumes the duties of the principal diesigner.

The client has to ensure that arrangements are in place for managing health and safety on a construction project. How can this be done?

The best way for a client to be confident that the requirements of CDM have been met before construction works start is to instigate a formal documented procedure which clearly sets out the actions taken to comply with the various statutory duties.

Such a document, called a Permit to Proceed, is included in the appendix to this chapter. The various stages of the project are recorded and actions confirmed.

The permit is signed by those with accountabilities and once every question is completed and everyone is happy works can start on site.

When must the client appoint a principal contractor?

A client must appoint a principal contractor as soon as is practicable after the client has such information about the project that it will, or is likely to, involve more than one contractor working on the project at any one time.

It is essential that the principal contractor is appointed in sufficient time to develop the health and safety plan before construction works start.

If a contract is to be negotiated with a preferred contractor, it is essential that they are formally advised that they are deemed to be the principal contractor for the project under the CDM Regulations. A formal letter of appointment will be required. An example is in the appendix to this chapter.

If a construction contract is to be awarded by competitive tender, it would be useful to include an introductory paragraph in the pre-construction information pack, which is sent to all tenderers, that states that the successful tenderer for the project will be deemed to be the principal contractor as defined in the Regulations.

Information and/or clarification on who will be appointed the principal contractor can be included in the bill of quantities or the employer's requirements.

Are there any specific requirements regarding the appointment of the principal contractor that must be considered under the CDM Regulations?

No. The CDM Regulations do not stipulate any specific criteria for a principal contractor other than they must meet the criteria for a contractor. A contractor is defined in CDM regulation 2 as any person who, in the course or furtherance of a business, carries out, manages or controls construction work.

What does the principal contractor have to do, and will I have to pay additional fees?

The principal contractor is responsible for site safety, assessing the competency and resources of contractors, co-ordinating work activities, organising site training, setting up communication between contractors and monitoring site safety.

The principal contractor must develop any pre-construction health and safety information into the *construction* phase health and safety plan. This plan is a critical document that sets out the rules for safety on site, how the site will be managed, arrangements for training and communication, welfare facilities and so on.

In addition, the principal contractor must appoint competent contractors, carry out induction training, provide protective equipment for general use, carry out general risk assessments, and co-ordinate the health and safety of all trades working on the project.

The principal contractor's duties involve co-ordinating and planning site safety. Therefore, it takes time and resources, and these must be paid for. If the project is tendered on a bill of quantities or is negotiated, then an allowance should be included for undertaking the role of principal contractor.

The costs will depend on the value and duration of the project.

The client must allow the principal contractor the resources to do the job. This means ensuring that adequate financial provision is made to do the job properly. The client may need to allow for a full-time safety manager on site – this should be included in the pre-construction information pack (if appropriate), and an item included in the preliminaries of any bills of quantities for costing purposes. Also, if a full-time safety manager is required, this could be included in the employer's requirements on how the project is to be managed.

If the principal contractor is effective at managing, co-ordinating, planning and monitoring site safety, it should have beneficial effects on the project in both financial and time-scale terms. Good site organisation could reduce site accidents and down time, thereby allowing the project to progress in accordance with the construction programme.

Poorly managed sites create 'accidents waiting to happen'. Accidents cost all parties money to a greater or lesser extent, including the client. Therefore, additional fees in the 'prelims' may be recouped many times over by having fewer accidents or near misses on site.

If awarded the construction contract on a project, is a contractor automatically appointed as the principal contractor?

Not necessarily so. The client must ensure that they properly appoint the contractor as the principal contractor in writing, as this is legally required and will clarify responsibilities.

The pre-construction information pack may include a statement to the effect that the successful tenderer will be appointed as the principal contractor and must allow for such responsibilities in their tender.

A section may be included in the preliminaries to the bill of quantities or the specification for works.

The client cannot proceed with a construction project where there is one or more contractors working on it without appointing a principal contractor. If the client fails to appoint a principal contractor, they assume the role themselves. Conflict and, more importantly, legal contraventions of the CDM Regulations could occur if the client thinks that they have appointed a principal contractor but the successful tenderer has not assumed the responsibility and other contractors are unaware of the roles required.

Can there be more than one principal contractor on a project to which the CDM Regulations apply?

There can be only one principal contractor at any one time on a project, although there may be many main contractors undertaking the works.

The principal contractor is a specific post required under the CDM Regulations with the responsibility for managing and co-ordinating all of the construction phase health and safety issues. The principal contractor must be given the overall responsibility for co-ordinating construction phases, and, as this is such an important function, there can only be one such appointment.

Notwithstanding the above, at the same location there could be one or more different projects being carried out for different clients, and, in these circumstances, one or more principal contractors could be appointed. Projects have to be distinct from each other and must not rely on one another for their viability and completion. If projects share common entrances or site access, need communal lifting equipment, share services to site and so on, then it is preferable to appoint one principal contractor with the overall site management and co-ordinating responsibility so that these 'common resources' can be managed to the benefit of the site.

Each construction project could still have a main contractor responsible to their client, but a representative should liaise with the principal contractor. Everyone needs to co-operate so that they can relay information to the respective contractors regarding site and design hazards and risks generated by other parts of the project.

Any person who is responsible for appointing a contractor has to ensure that they have received the necessary information, instruction and training and has appropriate supervision. What does this mean?

Anyone who appoints a contractor must ensure that they are competent to do the job and that they have received either information, instruction and training about construction safety and any specific information about the project.

Clients and others must take the responsibility for ensuring that only suitably trained and competent contractors are appointed.

The Health and Safety at Work etc. Act 1974 and the Management of Health and Safety at Work Regulations 1999 contain specific requirements for employers to ensure that their employees receive information, instruction and training. Individual experience and training are important, as it is essential that operatives know how to work safely and how to evaluate hazards and risks.

Clients and others will need to have procedures in place to test how contractors train their workforce, including any sub-contractors appointed.

The need for detailed competency assessments has been removed in the 2015 CDM Regulations. Although there is no legal requirement for competency schemes, they can still be a good indicator that a contractor has provided suitable information, instruction and training to their employees.

The HSE suggests that compliance with industry standard schemes such as PAS 91 – the British Standards Institute document that sets standards for procurement in construction – would be a good indicator for the client or other person to satisfy themselves that the people they are appointing are properly trained and competent.

Appendix to Chapter 2

CDM Regulations – client duties

We, [*Client*], confirm that we are aware of and understand, the following duties placed on Clients in the Construction (Design and Management) (CDM) Regulations 2015:

- Ensure co-operation and co-ordination of all persons involved in the project so far as it relates to our duties as the Client.
- Ensure that there are suitable management arrangements in place for the project, in particular in relation to
  - construction work to be undertaken without risk to health and safety of any person
  - compliance with Schedule 2 of the CDM Regulations (welfare facilities)
  - allowing sufficient time and resources for planning and preparation of all stages of the project
  - providing pre-construction information
  - providing information for the Health and Safety File to the Principal Designer
  - retaining and providing access to the Health and Safety File
  - ensuring that arrangements made for managing the project are maintained and reviewed throughout the project.
- Appointing a Principal Designer for projects where there is or will be more than one contractor and ensuring that they comply with their duties.
- Appointing a Principal Contractor for projects where there is or will be more than one contractor and ensuring that they comply with their duties.

In addition to the above, we acknowledge that it is the duty of the Client to sign form F10 sent to the Health and Safety Executive for all notifiable projects or to authorise another to sign it on our behalf provided they are confident that we understand our duties.

[*Client*] hereby authorises [*Principal Designer*] to sign all F10 forms that relate to construction works, new build works, refurbishment and repair works to our premises that operate under the names of [*Client*].

Signed: ______________________________

Company and position

Date: ______________________________

Appointment letter – principal designer

We, [*Client*], do hereby appoint [*Principal Designer*], of [*Principal Designer's address*], to the position of Principal Designer as required by the Construction (Design and Management) (CDM) Regulations 2015.

The Principal Designer will supply the services as listed in the Specification of Duties attached to this statement of appointment.

The position of Principal Designer is assumed until the final handover of a project from the Principal Contractor to [*Client*].

In making this appointment, [*Client*] confirms that they are fully aware of the duties imposed on them by the CDM Regulations 2015 and understand the role that the Principal Designer must perform.

Signed: ______________________	Signed: ______________________
For [*Client*]	For [*Principal Designer*]
Date: ______________________	Date: ______________________

Appointment letter – principal contractor

We, [*Client*], do hereby appoint [*Principal Contractor*], of [*Principal Contractor's address*], to the position of Principal Contractor as required by the Construction (Design and Management) (CDM) Regulations 2015.

The Principal Contractor will supply the services as listed in the Specification of Duties attached to this statement of appointment.

The position of Principal Contractor is assumed until completion of all construction activity and the final handover of the project to [*Client*].

In making this appointment, [*Client*] confirms that they are fully aware of the duties imposed on them by the CDM Regulations 2015 and understand the role that the Principal Contractor must perform.

The name of the Principal Designer as appointed under CDM 2015 is

____

As the Client we will monitor the arrangements put in place for managing health and safety on the project either directly or via a third-party independent CDM adviser.

Signed: ____________________	Signed: ____________________
For [*Client*]	For [*Principal Contractor*]
Date: ____________________	Date: ____________________

Permit to proceed

Name of Project: ______________________________

Address of Project: ______________________________

Name of Designers: ______________________________

Name of Principal Contractor: ______________________________

Name of Principal Designer: ______________________________

The following information is confirmed as being available and, where necessary, checked and approved by the appropriate person.

		YES	NO	N/A
1.	Asbestos survey, asbestos removal and issue of clear air certificates Signed:			
2.	Party wall agreements Signed:			
3.	All necessary planning permissions and Building Control applications Signed:			
4.	Fire Safety Risk Assessments/Plans for existing buildings Signed:			
5.	Access/facilities for those with disabilities considered – new building and major refurbishments Signed:			

6.	Welfare facilities as specified in the pre-construction information: No. of operatives expected on site: No. of WCs: No. of urinals: No. of wash hand basins: Canteen facilities:			
		YES	NO	N/A
7.	F10 form completed and sent to the Health and Safety Executive			
8.	Management arrangements for health and safety on site			
	Name of the site supervisor/agent:			
9.	Satisfactory completion of the Construction Phase Health and Safety Plan			
10.	Successful design co-operation and agreement by all parties that best practice has been achieved in respect of health and safety issues relevant to the designs of all elements of the project, e.g. construction, monitoring and evaluation, services, end users, maintenance and direct suppliers			
11.	Confirmation by designers that their designs meet the requirements of the principles of prevention Signed: (Designer)			
12.	Comment on any aspect above not completed or confirmed and state whether such omissions should delay the start on site date:			
13.	Confirmation that adequate time and resources have been allowed for planning and preparation before start on site date, i.e. indicate the time allowed: __________ weeks			

14.	Subject to agreement by all parties, this project is signed off to commence on site on: ____

Signed: ____ Principal Designer

____ Principal Contractor

____ Principal/Lead Designer

Permit to proceed – small projects

Name of Project: ______________________________

Description of works: ______________________________

Date works due to commence: ______________________________

Duration: ______________________________

(NB: must be 30 days or less and involve less than twenty workers, or involve 500 person days or less and involve only one contractor)

Requirements for the Construction (Design and Management) Regulations 2015	**Yes No N/A**	**Comments**
Approved competent contractors appointed		
Appropriate pre-construction information about the project, procedures, etc., issued to operatives, e.g. asbestos location, opening hours, use of staff accommodation and residential hazards		
Workers aware of safety procedures, e.g. permit to work schemes and evacuation procedures		
The contractor has produced a written health and safety construction plan commensurate with the risks on the project and has or will communicate to all other workers		
Adequate management provided on the site Named site agent:		
Adequate arrangements made for welfare facilities, e.g. WCs, wash hand basins, eating of meals and drinking water. Please describe:		
Appropriate co-operation between designers has taken place and all are aware of and agree with the works to be undertaken		

Design Risk Assessments have been completed and provided to those who will need them		

Preparation and planning

The project has been properly planned and resourced and the following timescales have been used:

Design, planning and preparation: ____________________ days/weeks

Pre-start preparation, i.e. the time between the appointment of the contractor and the start on site: ____________________ days/weeks

All agreed works can start on site: Yes/No/N/A

Date: ________________________________

Signed: ________________________________ On behalf of the Client

________________________________ On behalf of the Contractor

________________________________ On behalf the Designer

CDM 2015 Questions and Answers: A practical approach
ISBN 978-0-7277-6032-6

http://dx.doi.org/10.1680/cdmqa.60326.045

Chapter 3
The principal designer

Who is the principal designer in respect of the CDM Regulations?

The principal designer is an appointment that the client must make under regulation 5 of the CDM Regulations to ensure that health and safety is co-ordinated during the design and planning phase of any project that has or is likely to have more than one contractor, working on the project at any one time.

The principal designer is a member of the design team, as it is felt that many health and safety issues relating to construction projects, and subsequent maintenance tasks, can be 'designed out' in the design and planning process. The principal designer can help to ensure that these issues are discussed and resolved early.

The principal designer is expected to have influence over the design team (and client) and to highlight health and safety issues and to initiate discussions regarding ways to implement the hierarchy of risk control or the principles of prevention.

The principal designer is required to put together the pre-construction information for the project.

The principal designer must co-ordinate information relating to health and safety among the design team and facilitate the sharing of information so that a holistic approach to health and safety on the project is achieved. They must ensure good communication, co-operation and co-ordination between all duty holders.

What is the purpose of the principal designer?

The role of the principal designer is to provide the client with a key project advisor in respect of health and safety risk management matters.

The principal designer's role is seen as pivotal in helping to improve the overall standards of health and safety on projects. In particular, they are expected to be able to guide, advise and influence clients into making good, sound decisions about health and safety on projects.

Clients may not be too familiar with the design and construction process, and sometimes it is easy for them to get carried along by professionals who have their own interests at heart and not those of their clients. A designer may, for instance, decide to change their minds about a design solution and consider little the consequences on the overall programme, budget or buildability. Ultimately, the client will pay for the changes, but a good principal designer will be able to step into debates about the need for the change, and bring an independent, objective view on the health and safety consequences.

The principal designer is seen as having a pivotal role in ensuring that designers understand their roles under the CDM Regulations, and follow the correct procedures and thought processes to address the hierarchy of risk control and the principles of prevention in relation to health and safety.

Designers may have excellent design creativity but sometimes lack a practical approach to buildability and future maintenance.

It is not acceptable to leave the contractors or future occupiers with problems such as 'How do I build it?' and 'How do I get at it to clean it or replace defective parts?'

The principal designer can ask these questions very early on in the design process, and hopefully act as a catalyst to 'joined up thinking' by the whole design team.

The principal designer is seen as the conduit down which information flows, and will ensure 'the right information to the right people at the right time'.

What are the key duties of a principal designer?

The principal designer has the overall responsibility for co-ordinating the health and safety aspects of the design and planning phase and for the pre-construction information pack and for producing the health and safety file.

The principal designer must ensure that designers have given adequate regard to health and safety within their design, in particular that they have

- avoided foreseeable risks to the health and safety of any person at work carrying out construction work or cleaning work in or on the structure at any time, or of any person who may be affected by the work of such a person at work
- combated at source the risks to the health and safety of any person at work carrying out construction work or cleaning work in or on the structure at any time, or of any person who may be affected by the work of such a person at work
- given priority to measures that will protect all persons at work who may carry out construction work or cleaning work at any time and all persons who may be

affected by the work of such a person at work over measures that only protect each person carrying out such work.

The principal designer has to ensure that the designer has followed the 'hierarchy of risk control' or has implemented the 'principles of prevention'.

The principal designer must ensure that designers co-operate with one another, as far as it is necessary for each of them to comply with the requirements of the CDM Regulations (i.e. promote good design risk management).

A health and safety file must be produced for the project, and the principal designer must ensure that this is done and handed over to the client. In addition, the principal designer must review, attend or add to the health and safety file as necessary so as to ensure it contains all relevant information when handed over to the client.

The principal designer is expected to advise the client on the time and resources needed to effectively discharge their responsibilities in respect of health and safety on the project, as a key duty is to assist the client in the preparation of the pre-construction information, which should include preparation and planning time as this is extremely critical to managing projects safely.

Is there any recognised qualification and set of competencies that the principal designer must have?

No. However, designers must obviously be competent to do their job and have suitable experience and expertise.

The client has to ensure that the principal designer complies with their duties under regulations 9, 10 and 11 of the CDM Regulations, and so the client should look for evidence that the designer will be able to comply with those duties before appointing them.

A principal designer is described as the designer in control of the pre-construction phase appointed under CDM regulation 5(1)(a) to perform the functions in regulation 11.

A designer is described in regulation 2 as

> any person (including a client, contractor or other person referred to in these Regulations) who in the course or furtherance of a business—
>
> (a) prepares or modifies a design; or
>
> (b) arranges for or instructs any person under their control to do so,

relating to a structure or to a product or mechanical or electrical system intended for a particular structure, and a person is deemed to prepare a design where a design is prepared by a person under their control.

A formal qualification in respect of health and safety would be helpful but is not essential when appointing a principal designer. They must, however, have design qualifications appropriate for the designs they are undertaking.

Can a client appoint themselves as the principal designer?

Yes, provided they can demonstrate that they have the competency and resources to undertake the function.

The client must be familiar with the requirements of regulation 11 in the CDM Regulations, as this lays down the duties that are applicable to the principal designer.

The client must be able to demonstrate knowledge of the construction process, design function, and health and safety (including fire safety), and be fully conversant with construction and related practices.

A client that is a company may appoint an employee as the principal designer under the Regulations (e.g. a facilities manager or a project manager). They must have the resources, including time, to do the job and must not be placed in conflicting situations that are detrimental to the health and safety of the project.

If the client fails to formally appoint a principal designer where one should be appointed (i.e. on projects where there is or is likely to be more than one contractor working simultaneously on the project at any one time), they automatically take on that role themselves. Clients will be vulnerable to prosecution if any investigation shows that they did not have the competencies to carry out the role.

Can the client appoint the principal contractor as the principal designer?

Yes, provided the client is satisfied that the principal contractor is competent and has adequate resources to do the job.

If the principal contractor has their own health and safety person, then they may be in a good position to discharge the duties of the principal designer. Also, design and build contractors are likely to be able to fulfil both roles competently.

The principal designer has to have the ability to communicate with the design team in order to discuss health and safety issues. A contractor's safety officer could be an ideal person to communicate with the architect, building services consultant or structural engineer because they are familiar with construction processes and could advise the design team that the proposed sequence of construction or programming is not the preferred one from a health and safety point of view.

The principal contractor must be assessed in the same way that anyone else would be in respect of the position of principal designer.

Does the principal designer have to be independent of the other members of the design team?

No. Provided the person appointed meets the requirements for competency, a member of the project team could be appointed.

Architects, quantity surveyors, project managers, engineers and similar can all be appointed as the principal designer to a project even if they are already on the design team. They may have a separate 'in house' division that can carry out the function, or specialist health and safety teams.

Independent principal designers are available, and such an appointment can bring objectivity to a project in respect of health and safety matters. It may be easier for an independent principal designer to question a designer's design risk assessment than for the designer to objectively do it themselves. The principal designer should act as the go-between to all parties – the client and the designer, the client and the principal contractor, the designer and the contractor, and the designer and the designer should ensure that there is a good flow of information regarding health and safety of the project.

How soon does the principal designer have to be appointed to the project?

As soon as the client believes that a project is likely to happen (e.g. at the feasibility stage), or when outline design/planning applications are submitted and there is evidence that the project will involve more than one contractor.

The Health and Safety Executive (HSE) believes that the principal designer has a pivotal role in the adoption of health and safety principles throughout the project. It wants to ensure that the principal designer has been involved in the planning stage of the project and that the appointment is not superficial. The HSE will want to be satisfied that the principal designer will have had the opportunity to develop information for the pre-construction information pack prior to the tendering or negotiating phase of the contract.

The principal designer needs to be in a position to be able to co-ordinate the health and safety aspects of the design work and advise on the suitability and compatibility of designs.

The principal designer should therefore be appointed before significant detailed design work begins.

Can the principal designer be changed during the project?

Yes. The principal designer's role is best carried out by people who have relevant experience, and this may vary from stage to stage in a project.

The CDM Regulations do not specifically address the changes to the principal designer role, but do allow for the principal designer to be 'terminated', 'changed' or 'renewed' as necessary as long as the position remains filled at all times until the end of the construction phase.

The client must not allow a construction project to proceed at any stage without the suitable appointment of a principal designer (unless only one contractor will be working on the project). If they do, they take on the responsibility of principal designer by default.

A project may start at the feasibility stage, and the architect or project manager may be appointed as the principal designer. During the first phase of the project the principal designer assumes responsibilities for co-ordinating the design, gathering information on the site (e.g. contaminated land surveys), advising the client and so on. The next phase of the project may be agreed as a design and build project. The role of principal designer could be passed from the architect/project manager to the design and build contractor, provided they can demonstrate competency, as they would be best placed to co-ordinate health and safety issues in the design and build phase of the project.

Case study

The client commissioned a feasibility and concept design project for a new block of residential flats. They appointed an independent principal designer to the design team as they believed that this would give an objective view to the siting of the flats, access to the site and the overall design (flat roof versus pitched roof, window cleaning access, etc.).

Following receipt of planning permission and lease agreements, the principal designer put together an initial pre-construction information pack dealing with the concept design of the building. The client's preferred procurement route was design and build for the detailed design and construction of the project.

The principal designer for the concept stage advised the client that the design and build contractor would be best placed to assume the principal designer's role for the remainder of the project, having satisfied themselves on behalf of the client of the design and build contractor's competency and resources.

The design and build contractor was duly appointed as the principal designer, and the HSE was informed by 'revised information' on form F10.

Does a client have to take the advice of the principal designer?

There is nothing in the CDM Regulations that states the client must take the advice of the principal designer. The principal designer must be in a position to give adequate advice to the client to enable them to comply with the Regulations.

There is little point in appointing a principal designer to a project if the client does not value their opinion. A principal designer should have the client's interest as a priority and should offer advice in respect of the health and safety aspects of the project in an objective and unbiased way. The post of principal designer should be seen as the 'client's friend' – implying that a good working relationship is essential.

Should a client fail to take the advice of the principal designer, they could be in breach of regulation 7 of the Management of Health and Safety at Work Regulations 1999 by ignoring the advice of a 'competent' person in relation to health and safety.

Does the principal designer have to visit the site during the design stage?

The role of the principal designer is to co-ordinate information in respect of health and safety for the project, and to ensure that the pre-construction information pack is completed and available to contractors and designers as necessary.

The principal designer needs to understand the site, design proposals, hazards and risks of the project, and how all of these things will inter-relate to the existing environment.

The principal designer should visit site to carry out their own survey of health and safety issues and to familiarise themselves with the area (e.g. busy roads, pedestrian flow and access restrictions).

There is, though, nothing in the CDM Regulations that requires a principal designer to visit the site. It could be perfectly acceptable to assimilate vital information for compiling the pre-construction information pack and for issuing advice on hazards and risks in respect of the project from information supplied by way of surveys, reports, photographs and so on.

However, a diligent principal designer will always want to visit the site at the beginning of a project, and they should be expected to do so.

Does the principal designer have to carry out site safety inspections once the project has started on site?

No. There is no duty on the principal designer to undertake any site safety activity in respect of construction works. Therefore, the client does not need to pay for this service, as it does not need to be carried out under the CDM Regulations.

The client may, of course, appoint the principal designer to carry out this duty for them in addition to their legal responsibilities. It has useful benefits for ensuring standards of safety are maintained and that the principal contractor complies with the construction safety plan. The principal designer is also in a good position to give advice to the client after site safety visits. Site visits during the construction phase do, however, allow the principal designer to review ongoing design issues and gather information for the health and safety file.

If the client requires the principal designer to assume the responsibilities of a safety adviser, then they should make sure that there is a specific agreement to cover the duties, and agree the fee in advance.

There is a role for an independent health and safety professional on many larger projects to take on the responsibilities of checking on site safety for the client and for ensuring that health and safety is being managed during the course of the works.

Does the principal designer have to attend every site meeting?

No. There is no requirement for a principal designer to do this, as there is no legal duty placed on the role to be responsible for site safety.

The principal designer does have a duty to ensure co-operation between designers so far as it is necessary to enable each designer to comply with the requirements of regulation 9 of the CDM Regulations, and this may be best achieved by attendance at site meetings so that the principal designer can ensure health and safety information is freely shared. However, sitting in at lengthy site meetings may be an expensive way to achieve this duty when a few regular phone calls can be undertaken to each designer and the principal contractor to ensure that they have all the information they need.

The principal designer role is to be seen as an integral part of the project team, and the consideration of health and safety should be a natural part of the project. The principal designer role should be seen not as a separate position but as one that is integrated into the team.

It would be sensible to have the principal designer attend a site meeting towards the end of the project when they will be able to issue requests for information to all parties for inclusion in the health and safety file and check the information that has already been collected.

The principal designer may need to attend a specific design team meeting if fundamental design changes have been made to the project, as these may have health and safety consequences. The role is to help co-ordinate issues and ensure the co-operation of all duty holders.

The client should ensure that the principal designer is sent copies of site meeting minutes and issued instructions to *read* them and take whatever appropriate action to ensure that they are fully up to speed with the project.

What fee should be paid for the services of a principal designer?

Fees can be either on a percentage of project or contract costs or can be a 'fixed fee' for the project, and may be included in the general design fees.

If an independent appointment is made to work alongside the design team, then fees could be a percentage of project costs.

A fixed fee should be based on an hourly or daily rate and the anticipated number of hours/days needed to complete the statutory duties placed on the principal designer.

The client should have a clear brief of what they want the principal designer to do. The client's requirements should be based on the CDM Regulations (i.e. the duties placed on the principal designer). If the client requires *other* services (e.g. site safety audits), they should make sure that this is specified as additional services and indicate whether fees are to be included in the percentage or fixed fee or to be invoiced separately on a time-charged basis.

It is easy for a client to pay a lot of money and to receive poor value when appointing a principal designer. The position should be beneficial to the client and their design team and be a source of advice and information on health and safety issues.

The client should ask the principal designer the following questions in order to obtain an indication of how much time they propose to devote to the project:

- How do you intend to gather information regarding the project?
- Will you be carrying out an initial site visit?
- How will you co-ordinate designers' responsibilities under Regulations?
- How do you propose to check design risk assessments and other documentation?
- How long will it take you to prepare the pre-construction information pack?
- How do you propose to communicate with the principal contractor during the construction phase to ensure that design issues, including temporary works, are co-ordinated?
- How often do you intend to visit the construction site, and for what purpose?
- How do you propose to co-ordinate the mechanical and electrical designers and contractors?
- How will you gather information for the health and safety file, and how long will it take you to prepare the health and safety file.

- How will you communicate your advice to me regarding the adequacy of management arrangements for the project?

As the principal designer, my client wants me to give advice regarding the knowledge and resources of the design team. What do I need to consider?

Although the CDM Regulations do not specifically address competency, a client must not appoint any designer who is not competent in respect of health and safety matters relating to their design nor must they appoint anyone who does not have the resources to comply with the requirements of the Regulations.

Many clients will delegate the assessment of the design team to their principal designer. If this is the case, it is important that the principal designer has a written procedure in place to demonstrate the steps taken.

Step 1 will be for you as the principal designer to issue a designers' questionnaire that seeks general information about the designer, partnership, practice or company. Information on the types of projects they normally deal with will be relevant – a design company experienced in domestic dwellings may not necessarily be competent to design industrial buildings.

Details of individual qualifications, membership of professional bodies and so on will be relevant.

Details of the designer's health and safety policies, procedures and practices will be needed.

Details of how the designer prepares and considers design risk assessments would be helpful.

Has the designer had experience of the client's type of project before? You should ask for references and go to see similar work done by them. You should also ask to see examples of design risk assessments.

How does the designer keep their employees up to date about health and safety issues, design innovations and so on? Self-development through reading is fine, but what about their commitment to the subject of health and safety by spending money on courses, for example?

Remember that the design team includes

- architects

- quantity surveyors
- building services consultants
- structural engineers
- civil engineers
- project managers (if they are allowed to *specify*)
- interior designers
- landscape architects.

All of them should be subjected to competency and resources checks.

Having gathered all the relevant information, devise a system where you can quantitatively score their responses to their responsibilities under the Regulations.

Information on any formal or informal action by statutory authorities (e.g. the HSE) will be relevant, as will details of any accidents that have happened to their own personnel or on sites for which they have held some responsibility.

Nominated individuals responsible for health and safety and issues relating to the Regulations should be included, and examples of safety policy documents, CDM procedures and protocols and so on should be included with any responses.

Obtaining information is relatively easy. The tricky bit is assessing it objectively so that you can conclude that they are 'competent'. It is easy to believe that all 'professionals' are competent otherwise they would not be in business. Unfortunately this is not true!

You should review the information against a 'benchmark' of acceptable answers. If an answer seems inconsistent or lacks details, you should request more information.

Just because a designer is honest enough to admit to having being served with a statutory notice does not mean that they should automatically be excluded. You should seek information about what they learned from the experience, what procedures they reviewed and how they improved. Persistent contraventions of laws indicate an unwillingness to accept responsibilities and usually implies poor standards and attitudes from management downwards. These designers would be best avoided.

Approvals can be given for certain types of projects, projects up to *X* value and so on.

You should encourage the client to develop an 'approved list' of designers that indicates their strengths and weaknesses, approval status and so forth.

The list should be reviewed annually, or more frequently if necessary.

The client wants to pay minimum attention to the CDM Regulations and to do things on a shoestring. What should I do?

Refusing to accept the commission as the principal designer would be a good starting point. Clients who refuse to accept their legal responsibilities will probably be difficult to work with in other areas, and professional integrity is as important as fee income. Even more importantly, a criminal record for failing to discharge the duties of the principal designer will stay on your records long after the project has finished.

The client has to allow the principal designer adequate resources to do the job and has to provide information regarding the construction project. Failure to do so could be a breach of the client's statutory duties.

The principal designer's role relies heavily on good communication and inter-personal skills because it is largely an 'ensuring' and 'influencing' role. It would be sensible for you to discuss with the client the reasons for their attitude to CDM and why they wish to avoid their responsibilities. It may be because of fees – a question of 'what do I get for my money'.

A professional approach by the principal designer would be to explain that there may be no need to appoint a separate, independent person to the role, and that perhaps another member of the design team could fulfil the duties, saving on fee expenses if the client feels that the lead designer's fees are too expensive

You should advise the client what in your view needs to be undertaken in order to apply the CDM Regulations to the project (e.g. what information should be made available), and confirm such matters in writing.

Also, persuade the client of the benefits of applying the CDM Regulations (i.e. better planning and design reduces site accidents, which in turn reduces site delays, etc.) and reiterate the legal duties that fall to the client.

You should discuss the type of health and safety file that the client would prefer – perhaps the majority of it could be collected by the principal contractor, reducing the time needed by the principal designer. The principal designer has to prepared the health and safety file, but there is no reason why the information cannot be provided by others in a form that requires little input from the principal designer.

CDM is an approach to planning and managing health and safety on construction sites, and afterwards, when the building is maintained. It should be seen as an integral part of the design and construction process, and need not cost considerable sums of money.

Review your own fees! Are you misleading your client about the complexities of the principal designer function and demanding fees for unnecessary activities (e.g. site safety visits)?

The client refuses to accept my advice as the principal designer. What should I do?

Disagreements with the client may damage the relationship with the principal designer, and, when and if communication has irrevocably broken down, you may have no choice but to resign the role.

If you as the principal designer feel that you are being prevented from carrying out their legal duties, and legal compliance is in jeopardy, then a resignation of the role will be essential.

But perhaps the situation hasn't got to that stage. You should then work diplomatically to advise the client of the reasons behind your advice, the implications for complying and not complying with it, the project-wide benefits and so on.

Always consider alternative approaches if possible – especially if the client's reluctance is based on perceived costs associated with your advice.

Ultimately, the client will be legally responsible for ensuring that they comply with the client duties under the CDM Regulations, and if you can show that you have performed your duties diligently, you may not necessarily be implicated if there is any subsequent prosecution for non-compliance with the Regulations.

The Management of Health and Safety at Work Regulations 1999 require an employer to appoint a competent person to advise and assist them in undertaking the measures they need to take to comply with the requirements and prohibitions imposed upon them by or under the relevant statutory provisions. The CDM Regulations were made under the enabling provisions of the Health and Safety at Work etc. Act 1974, and are therefore 'relevant statutory provisions'.

A principal designer could be classed as a 'competent person' for giving health and safety advice, and the client would be in breach of duty if they failed to consider the advice given by you.

It is imperative to have written systems in place as a principal designer that demonstrate what advice has been given to the client and when.

If you do resign from the project, it will be important to put the reasons in writing to the client and clearly state the termination date. Advise the client that they will assume the principal designer's role until they appoint a replacement.

Provided you have exercised your statutory duties as the principal designer competently as required by CDM Regulations, there will be no liability to prosecution because others have failed their duties.

Can the principal designer be appointed by verbal agreement?

Regulation 5 of the CDM Regulations states that any references to appointments made for projects that have more than one contractor working on site must be in writing.

So, although clients can verbally agree to appoint someone to the role of principal designer, either they, or the principal designer, must issue a formal appointment letter in writing.

The letter or contract of appointment should set out as a minimum the statutory duties expected of the principal designer and the scope of services to be provided.

Copies should be retained by both parties.

Where the principal designer is appointed to numerous projects, the contract or letter of appointment should clearly state this and either list all of the projects or provide a general statement the appointment is to all projects undertaken by the client for a specific time period.

The letter of appointment should also state under what circumstances the principal designer acts on behalf of the client: for example, by signing the client's declaration on form F10 sent to the HSE.

As the principal designer, can I sign form F10 on behalf of the client?

Guidance issued in regulation 6 (notification) of the CDM Regulations states that the client can request someone else to make the notification on their behalf.

If the principal designer signs form F10 on behalf of the client, they must make sure that they have clearly advised the client what their duties are under the Regulations.

It would be good practice for the principal designer to obtain an authorisation from the client that they may sign form F10 and a confirmation that the client is aware of their responsibilities.

The HSE expects the client to take full responsibility for their duties under the Regulations, and therefore expects the client to sign form F10, as this is a clear statement that they are aware of their legal duties. If someone else signs form F10, the client could easily avoid their duties by saying that they were unaware of them because they did not sign form F10.

What steps or procedures should the principal designer follow in respect of co-ordination and co-operation when appointed to a project?

Once the principal designer is appointed, they should immediately become part of the project team. If all members of the team have worked together before, this will be straightforward, but where the principal designer is new to the team they will need to form relationships with all the key duty holders (i.e. designers, the client and contractors).

The client should give the principal designer a list of the members of the project team. Good relations will be achieved if the principal designer arranges to meet with the project team and establishes the roles and responsibilities of the team members in the project. A visit to appropriate offices might be beneficial, especially if the principal designer is to advise the client on the competency, skills and experiences of the team members.

All designers have responsibilities to ensure that they follow the principles of prevention in respect of their designs, and the principal designer should establish that each designer has procedures in place to demonstrate how they achieve compliance with this duty.

The role of any lead designer should be established.

Mechanical and electrical designers and contractors need to be included, together with engineers, specialist designers and so on.

The principal designer should establish whether each designer has their own procedure for identifying risks on the projects, and should enquire as to how this information may be shared among the team.

The principal designer may decide that, in the interests of co-ordination, a project-specific risk register or pro forma will be used, and will then raise this issue with the design team.

The principal designer should bring real benefits to the project team, and must not be seen as someone who over-complicates issues or creates mountains of paperwork.

As an example, design risk assessments that are in writing are not actually required under the CDM Regulations, but the principal designer has to ensure that designers consider the hazards and risks in their designs. As long as they are satisfied that such concerns are being addressed, then they need not impose requirements for meaningless design risk assessments.

The principal designer may need to attend a number of design meetings in the initial stages of the project, as this is where they can bring most benefit in ensuring that all designers and other duty holders co-operate with one another. Integrating mechanical and electrical design solutions into the overall design of the project will be invaluable and, hopefully, will enable hazards to be eliminated.

The principal designer is not responsible for checking the designs and the risk assessments of designers, but they are expected to raise any concerns or to give feedback if they believe health and safety is being compromised – both for the construction activity and for the future use, maintenance or cleaning of the building.

The principal designer should question decisions made on the principles of prevention, design solutions chosen and so on but support sensible and well-reasoned arguments that provide alternative health and safety solutions.

The principal designer should be able to call for a design review meeting if they believe that proper co-ordination on the project is lacking.

One way to ensure that design co-ordination has been addressed is for the principal designer to chair a review meeting prior to the project going out to tender or starting on site.

The principal designer should ask all of the designers, and the contractors, whether they have all the information they need to progress their roles safely, and ask them to sign a project declaration.

If co-operation and co-ordination is not forthcoming and once the principal designer has tried various options to improve the situation, they must advise the client that co-operation and co-ordination is not happening and that health and safety may be at risk. The client should then step in to remind all the duty holders of both their legal duties and those of the client.

Once the principal contractor is appointed, the principal designer should arrange to meet with them to go through all of the information they have been given and to discuss how they will manage design changes on site, provide information for the health and safety file and so on.

The client is looking to me, as the principal designer, to advise them that all legal duties have been met before a project starts on site. What steps or procedures should I take?

Clients, especially those not too familiar with the construction process, will often require the principal designer to act as their eyes and ears and to advise on the adequacy of arrangements regarding health and safety.

A number of duties have to be met before works can start on site, and the principal designer must be able to advise the client that they have been fulfilled.

A simple procedure that could be adopted is one called a 'permit to proceed'.

A specific permit to proceed document is designed for the project by the principal designer, and is completed and signed, as necessary, by the appropriate duty holders before works start on site.

Such a document will be able to give a clear indication that all duties have been completed or that something is missing.

If duty holders (e.g. designers and the principal contractor) have to sign that they are satisfied with the information provided to them, then there will be a clear record should disputes arise in the future about, for example, lack of information on drawings or inadequate risk assessments.

The client has to be satisfied that adequate arrangements have been made for managing the construction project and that welfare facilities provided meet the requirements of Schedule 2 of the CDM Regulations, and the detail of these can be included in the permit to proceed (an example is included in the appendix to Chapter 2).

Is there anything that a principal designer does not have to do under the CDM Regulations?

Principal designers do not have to

- approve the appointment of other duty holders, although they are expected to give advice to the client on the appointments if requested
- approve or check designs, although they do have to be satisfied that designers have addressed the hierarchy of risk control and the principles of prevention
- approve or supervise the principal contractor's construction phase health and safety plan
- supervise or monitor works on site
- approve the principal contractor's risk assessments and method statements
- advise on the detail included in any management arrangements for the project.

However, any of the above tasks can be carried out by a competent principal designer if the client has authorised them to do so.

Principal designers should not take on the responsibilities of being the principal contractor's safety advisors, although if so requested by the client they can offer advice and guidance to the contractor so as to ensure the efficiency and effectiveness of the project.

The principal designer has to prepare or review and update a record containing information relating to the project that is likely to be needed during any subsequent construction work. What does this mean?

The principal designer has to prepare (where none exists) or review and update a document known as the 'health and safety file' for the project.

The health and safety file is a record of the information needed to allow future construction work, including cleaning, maintenance, alterations refurbishment and demolition, to be carried out safely.

Information in the file should alert building users, owners or occupiers of any residual hazards associated with the structure and should alert any workers of any potential health and safety hazards, especially maintenance and refurbishment workers.

The health and safety file must be prepared by the principal designer.

The structure and content of the health and safety file should be agreed between the client and the principal designer at the beginning of the project.

The principal designer should issue guidance to the design team and contractors on the type and detail of the information they will be requiring.

The health and safety file should focus on health and safety information, and does not need to replicate the building or operations manuals.

What key duties should the principal designer have completed before a project starts on site?

The majority of the principal designer's duties are undertaken before works commence on site.

The principal designer has to ensure the following:

- clients are aware of their legal responsibilities under the CDM Regulations
- all parties who carry out design work on a project collaborate with each other and pay attention to reducing risks to health and safety wherever possible
- the design is being progressed to avoid foreseeable risks or, where this is not possible, to combat risks at source (i.e. that the hierarchy of risk control and/or the principles of prevention are followed)
- all designers co-operate with each other and provide information to each other in order that health and safety risks can be assessed

- a pre-construction information pack has been produced and that it has been instrumental in developing the construction phase health and safety plan prior to commencement of construction works
- advice has been given to the client, if requested, in respect of the suitability of the construction phase health and safety plan and the management of the project.

The principal designer should have a major influence over the design team in respect of considering health and safety issues from the design and the construction methods envisaged for the construction phase.

The role of co-ordinating the design team is an important one for the principal designer, as it means that all the different designers (architects, interior designers, building services consultants, quantity surveyors, etc.) should interact with one another to ensure that each of their respective designs do not conflict or create foreseeable health and safety risks.

Do the CDM Regulations specifically require design risk assessments to be generated on every construction project?

No. The CDM Regulations do not specifically state that design risk assessments have to be produced. The Regulations do state that designers must manage design risks by avoiding foreseeable risks and reducing residual risks so far as is reasonably practicable: that is, they must follow the principles of prevention.

A designer must take reasonable steps to provide with their designs sufficient information about aspects of the design of the structure or its construction or maintenance (including cleaning) as will adequately assist

- clients
- other designers
- contractors

to comply with their duties under the Regulations.

The information that designers have to provide to other duty holders can be in any format, but it should be brief, clear, precise and project specific.

Information can be notes on drawings, written information provided with the design drawings, formal risk assessment forms and project-specific instructions.

HSE inspectors are not looking for meaningless paperwork – they want to see evidence that designers have thought about health and safety issues and consequences during the development of the project, and that they have done everything that they can to assist in reducing on-site accidents and ill health and that they have also contributed to a reduction in accidents for maintenance and cleaning workers together with building users.

What are the 'principles of prevention'?

The 'principles of prevention' are set out in Schedule 1 of the Management of Health and Safety at Work Regulations 1999 and are referenced in Appendix 1 of the HSE's guidance document on the CDM Regulations.

The requirements are

(*a*) avoiding risks
(*b*) evaluating the unavoidable risks
(*c*) combating the risks at source
(*d*) adapting the work to the individual
(*e*) adapting to technical progress
(*f*) replacing the dangerous with the non- or less dangerous
(*g*) developing a prevention policy
(*h*) giving collective/individual protection
(*i*) giving appropriate instruction.

The CDM Regulations require the principal designer to 'ensure' that various duties are carried out. What does this mean?

Principal designers must do all that they can to fulfil their duties under regulation 11 of the CDM Regulations but the duty is qualified by 'as far as is reasonably practicable'. Ensuring that something happens is not the same as actually doing it, but the principal designer must have mechanisms in place that could demonstrate that they did everything they could to ensure that others fulfilled their duties or that they provided available information. Principal designers are seen as having an important role in influencing how the risks to health and safety should be managed and incorporated into the wider management of the project. The principal designer's role involves co-ordinating the role of others in the project team to ensure that significant and foreseeable risks are managed throughout the design process.

The HSE would expect to see evidence that the principal designer had set up and was chairing design team meetings, and that information was freely flowing among the teams and specialists.

Leaving the designers within the project to their own devices and never reviewing how they are approaching the CDM Regulations will not comply with the duties within regulation 11, and principal designers would be criticised by the HSE. In addition, a fee for intervention may apply, as there could be a material breach of the Regulations.

CDM 2015 Questions and Answers: A practical approach
ISBN 978-0-7277-6032-6

http://dx.doi.org/10.1680/cdmqa.60326.065

Chapter 4
Designers

The CDM Regulations place responsibilities on designers. Who are designers?

Under the CDM Regulations, designers are all those who have some input into design issues in respect of a project. These include

- architects and engineers contributing to, or having overall responsibility for, the design
- building services engineers designing details of fixed plant
- surveyors specifying articles or substances or drawing up specifications for remedial works
- contractors carrying out design work as part of a design and build project
- anyone with the authority to specify or alter the specification of designs to be used for the structure, including the client
- temporary works engineers designing formwork and false work
- interior designers, shop fitters and landscape architects.

The above includes architects, quantity surveyors, structural engineers, building services engineers, interior designers, project managers (if they can change or issue specifications), landscape architects/designers, temporary works engineers designing propping systems, and so on.

The 'designer' must be carrying on a trade, business or other undertaking in which they prepare or modify a design for a structure or arrange or instruct any person under their control to do so.

Anyone who also prepares or modifies a design to a product or mechanical or electrical system intended for a particular structure is also a designer.

Where clients specify certain sequences of work or the use of specific materials or design features, they will be designers themselves under the Regulations.

Who decides who is a designer?

Ultimately, only the courts can make definitive interpretation of the CDM Regulations and determine whether an individual or company is liable for the duties imposed on them.

In actuality, an individual must decide for themselves whether they fall within the definition given in regulation 2 of the CDM Regulations. A designer is any person (including the client, contractor or other person referred to in the Regulations) who in the course or furtherance of a business prepares or modifies a design or arranges for or instructs any person under their control to do so in relation to a structure or to a product or mechanical or electrical system intended for a particular structure, and a person is deemed to prepare a design when a design is prepared by a person under their control.

The client can decide that they expect all members of the design team to assume 'designer responsibilities', and may conduct competency and resources assessments on the members of the team. By doing this, the client will be able to show that they took steps to show that they were 'reasonably satisfied' regarding the designers' competency and resources.

It would be prudent for a designer to discuss with professional indemnity insurers their definition of 'designer' under the Regulations and to ensure that they have appropriate insurance cover for the role they will undertake.

What are the responsibilities of a designer?

Designers from all disciplines have a contribution to make in avoiding and reducing health and safety risks that are inherent in the construction process and subsequent work (e.g. maintenance).

The most important contribution a designer can make to improve health and safety will often be at the concept and feasibility stage, where various options can be considered so as to avoid potential health and safety issues.

Designers must therefore give due regard to health and safety in their design work.

Designers must provide adequate information about health and safety risks of the design to those who need it (e.g. proposed roof access routes and use of fragile materials).

Designers must co-operate with the principal designer and other designers on the project and ensure that information is freely available regarding health and safety issues and that they consider the implications of their designs with other aspects of the design (e.g. structural works in relation to building services).

Designers must advise clients of their duties under the CDM Regulations, as specified in regulation 9. Designers must take reasonable steps to advise their clients of the existence of the Regulations and their duties within them, the existence of the CDM guidance document, good health and safety management, and the benefit of making early appointments.

Are designers permitted to specify fragile materials?

Fragile materials (i.e. materials that give way on impact, point loading, etc.) are a major cause of both construction accidents and building maintenance accidents. Many of these accidents end as fatalities, or with disabling injuries.

Fragile materials constitute a safety hazard, and a design risk assessment must be completed that outlines the consideration given in the design process to the use of fragile materials, substitute options and so on.

The risks associated with fragile materials include falling through them, injury, death and collapse causing injury to those below.

As the first responsibility of a designer is to *eliminate* known hazards, the implication is that fragile materials should not be designed or specified into a project.

If the design scheme demands, for example, a glazed atrium roof, then there are inherent hazards with the design. A solid non-fragile material cannot be specified because the glazing is needed to allow for natural lighting. Where such a scenario exists, the designer must specify safety features for the glazed atrium (safety railings, guardrails, a running rail and safety harness fittings, gantry access, etc.) and also reduce risks with the use of safety glass.

A netting could be designed underneath the glazed atrium to prevent falls should the material give way. Access for cleaning of the glazing, maintenance of the paintwork of the frame and so on needs to be considered during the design.

If a fragile material is used and design considerations allow a *residual* risk, then the designer must specify information that must be included in the health and safety file. The designer may recommend that a 'permit to work/enter' system is adopted by the premises occupier, and that additional safety precautions are needed. The designer must also consider the safety precautions that the principal contractor should adopt so as to ensure safe erection or construction of the fragile material.

Can the principal designer require changes to designs?

The principal designer has to ensure that designers have had adequate regard to their responsibilities for considering health and safety during the project design process, and where they believe the designer has been reticent they are duty bound to raise their concerns with the designer and also to advise the client.

The principal designer should be an expert source of advice about applying the hierarchy of risk control and the principles of prevention, and should be consulted by the designers whenever necessary.

The principal designer must use their best endeavours to advise the designer if they have concerns about safety issues, and must look for alternatives, compromises and so on. If the designer refuses to listen, then the principal designer will need to advise the client of a potential conflict, particularly if the principal designer believes there will be implications for future occupancy (e.g. specifying fragile roof coverings unnecessarily).

If the designer refuses to listen to advice offered by the principal designer regarding health and safety in design issues, then the designer may be in a vulnerable position should future accidents occur as a result of the design principles chosen.

A prosecution of the designer will be possible under regulation 9 of the CDM Regulations, and the prosecution case would be strengthened if evidence was available that the designer refused to listen to professional advice.

If the principal designer's advice is ignored, it may provide the principal designer with sufficient evidence to question the competency of the designer, and they may advise the client accordingly, leading to the designer being dismissed from any approved list and probably the project.

What can designers be prosecuted for under the CDM Regulations?

Any designer can be prosecuted for not complying with their statutory duties.

If a designer fails to advise the client about the latter's duties under the CDM Regulations, the designer can be prosecuted. More likely, any designer who fails to advise the client is likely to receive a written informal or formal notification from the Health and Safety Executive (HSE), and this will incur a fee for intervention because the HSE would cite a material breach of the law. If the designer *does* advise the client about the requirement to apply the Regulations to a project and the client ignores the advice, provided the designer has written records that they did everything reasonably practicable to advise and inform the client of their duties, they may have a defence against prosecution – the HSE preferring to bring the prosecution against the client.

However, the designer will need to consider carefully whether they should be working for a client whom they know to be blatantly ignoring the law. Professional ethics may preclude taking such an appointment.

If the client has not appointed a principal designer, nor a principal contractor in relation to projects where such appointments are required nor provides information, then the designer will have great difficulty in carrying out their own statutory duties arising from the Regulations.

Designers can also be prosecuted for failing to comply with their duties to design any construction works safely and with a view to foreseeable risks.

If an accident were to happen on site and the subsequent investigation by the HSE concluded that the designer had specified a not commonly known fragile material as a roof covering and that they had not given any information on the hazardous nature of the material to the principal designer or the principal contractor, then it could be argued that the accident was due to the designer's negligence because they had failed to specify safety precautions or supply relevant information to allow others to specify safety precautions to be taken.

If a designer is insistent on specifying heavy weight blocks or materials for use on the project, and adequate provision has not been made for handling them on site (e.g. mechanical aides provided), then they could be liable for any accident or claims.

Equally, a designer could be liable to a prosecution if they specify a substance that is severely harmful (e.g. causes cancer) when there are suitable alternatives on the market.

Are the CDM Regulations designed to stifle all design creativity?

No. The CDM Regulations impose responsibilities on designers to consider health and safety issues in respect of their designs.

Many accidents happen on construction sites because designers give no consideration to the material they specify, the buildability process, or the time and resources it takes to do the job. Research by the European Community found that designers can have a tremendous positive influence over the number of site accidents and industry ill health by giving more thought to health and safety in their design.

The Regulations do not intend to turn all building design into grey boring boxes; rather, the expectation is for buildings with design flair to be safe to build and for future use.

Designers can still do adventurous things but must consider the practicalities and safety of their designs. Glazed atriums are ideal for some buildings and are perfectly acceptable, if consideration is given to safety issues such as

- protection from falling through the fragile material
- access for cleaning

- access for maintenance
- lighting features.

Provided solutions are given and the design incorporates these, there is nothing legally that can prevent a designer from designing a glazed atrium.

Buildings can be innovative in design but they need to be practical and safe for all future occupiers. If designers consider the 'hierarchy of risk control' for all design intentions, they will comply with the law.

Case study

Designers of a new concept retailing outlet sourced a unique mesh-type ceiling material that was to be installed in one single sheet to give a ripple effect to the ceiling.

Design considerations included

- How will it be delivered to the site?
- Will it be difficult to handle?
- Will the metal mesh have jagged edges?
- How will it be fixed to the ceiling frame?
- How will lighting be fixed?
- How will sprinklers be fixed?
- Will it be a fragile surface?
- How will maintenance personnel walk above it to access fittings?
- How will it be cleaned?

All of the above, and many more questions, formed part of the design risk assessment. The principal designer was asked to comment on the information available and to offer advice.

The new ceiling concept was installed, and created the innovative design that the designers and client had hoped for. Everyone was satisfied because it was easily cleaned and maintained and created no specific residual health and safety risks.

Do duties as a designer under the CDM Regulations only apply when the project is notifiable?

No. The duties of a designer under regulation 9 of the CDM Regulations apply to all projects where there is design. There does not have to be a client, principal designer or principal contractor.

Therefore, for all design commissions designers must have a procedure in place for documenting how they comply with the legal responsibilities imposed on them, as contained in the Regulations.

Can the health and safety considerations of designs be left to the principal designer?

No. The legal duty to consider health and safety matters in design rests with *designers.*

The principal designer can be used for advice on how to apply the hierarchy of risk control and the principles of prevention, and may have experience of what has worked in similar design scenarios.

Designers must provide information to the principal designer where appointed, when asked to do so. The principal designer has to ensure that designers co-operate and co-ordinate with each other – sharing information that could have health and safety implications.

The principal designer must ensure that a designer includes among the design considerations, adequate regard for health and safety issues, and therefore has a duty to ask relevant questions to establish that this has been the case. Designers must therefore expect the principal designer to want to see their design risk assessments or any other records kept by the individual or practice and to discuss these with them, making recommendations if appropriate.

As a designer, I advised the client of their responsibilities under the CDM Regulations but they refuse to make any of the statutory appointments or address the Regulations. Am I liable to be prosecuted if the project goes ahead?

No. Provided you have fulfilled your responsibility to advise the client about the CDM Regulations and how, in your opinion, they will be applicable to the project, you will not be liable to a prosecution (unless of course you fail in any other duties imposed on designers under the Regulations).

It would be advisable to have a system of written notification to a client that the Regulations apply to the project. This would demonstrate that you had taken your responsibilities as a designer seriously.

A client who was properly briefed about the Regulations by the designer but who refused to appoint a principal designer or principal contractor on a project, or made the appointment too late, will be in breach of the Regulations and will be liable to prosecution by the HSE.

What are a designer's key responsibilities before a construction project commences on site?

The duties of designers are itemised in regulations 9 and 11 of the CDM Regulations. The predominant duty of a designer is to manage the process of hazard and risk in respect of health and safety, as they have a responsibility to reduce hazards within their designs and also in respect of the materials, methods and processes that they specify for the construction phase.

Designers must consider the consequences of their designs in relation to future maintenance and cleaning.

All aspects of hazard and risk in relation to the design must have been considered before works commence on site.

The acceptable way to consider hazards and risks is to produce design risk assessments. These should have been produced before construction commencement and passed to the principal designer or contractor for inclusion in the pre-construction information pack.

The principles of regulation 9 must be followed in respect of risk management, namely

- eliminate the hazards
- tackle hazards at source by designs that reduce the risks to an acceptable level
- develop designs that protect all people exposed to the hazard and do not rely on control measures that just protect an individual.

As a design practice, are we legally responsible for the actions of our employees?

Yes. All employers have vicarious liability for their employees, and the duties placed on designers apply to the employer of all employed designers. If designers are self-employed, then they are responsible individually for ensuring that they comply with the law.

Employers must ensure that their employees comply with the CDM Regulations.

As designers, must we provide information on all hazards and risks associated with the project?

No. The CDM Regulations require designers to take account of significant risks that a competent contractor might *not* be aware of in respect of their designs.

All competent contractors, for instance, should be aware of the hazards and risks of working at heights, and the designer does not need to give the principal contractor chapter and verse on the safety precautions for working at heights. However, the designer should consider how to *reduce* the need to work at height, and if a particular sequence of construction is envisaged, then the designer must provide this information to the principal contractor by way of the principal designer on notifiable projects, and direct to the contractor on other projects.

Providing too much information on all hazards and risks of construction obscures the important detail about significant risks.

Contractors need to know about any specific materials specified or construction sequences planned in order to achieve the desired design effect of the building that they may not be overly familiar with (e.g. the installation of glazed atria).

The principal designer insists that we co-operate with other designers. How can we best achieve this?

Designers have to co-operate with the principal designer and other designers and co-ordinate their work so that the principles of the CDM Regulations for reducing hazards and risks in construction are met.

The reason that co-operation is needed between designers is to ensure that hazards due to incompatibilities between different design functions are identified and addressed.

Co-operation and co-ordination can be achieved by

- regular meetings of the design team and others
- agreement on the common principles of hazard and risk assessments and the overriding requirement to reduce risks from design and construction
- regular reviews and ongoing design team meetings to address changes in designs, developments in design and so on
- co-ordinated design drawings that incorporate all design disciplines (e.g. architects, structural engineers, mechanical and electrical services, and interior designers)
- initial site survey meetings and co-ordinated site visits.

Designers need to ensure that their clients are informed of their responsibilities under the Regulations and, in particular, that they must ensure that designers have adequate resources to undertake their responsibilities. This means allowing enough time during the project design process for such co-ordination meetings to take place.

As a designer, what do I not need to do under the CDM Regulations?

Designers are not required to

- take account of risks that were not reasonably foreseeable at the time the design was prepared
- provide information about unforeseeable risks
- provide information about insignificant risks

- specify construction methods unless the design proposed is unusual and a competent contractor may need information (e.g. a new construction technique or material used abroad and being introduced into this country)
- monitor, manage, review or provide any health and safety management function over contractors (other than that required under general health and safety laws or professional ethics)
- review and report on contractors' health and safety performance
- keep copious notes and documents on all aspects of design risk assessment other than conclusions about particular hazards and reasons for their design solutions.

Building services consultants (or mechanical and electrical design and build contractors) advise that they do not have to produce design risk assessments. Is this true?

Building services consultants, mechanical and electrical design and build contractors, public health engineers and so on are all classed as designers under the CDM Regulations, and therefore the duties imposed under CDM regulation 9 applies to them. They must therefore consider health and safety in respect of their designs.

Building services installations are often some of the most hazardous activities on site, and it is essential that due consideration is given to the hazards and risks of such processes.

Services designers must provide information to the principal designer where one is appointed or directly to the contractor regarding their proposed design solutions for the installation of plant and equipment, excavations for services and so on, and the principal designer must ensure that this information is co-ordinated throughout the design team.

The best method for building services designers to provide this information is by way of design risk assessments.

The true intent of the Regulations can be seen with the co-ordination of building services information into the general design, as this will allow any conflicting processes to be considered and resolved prior to construction work commencing: for example, rather than build solid walls and then have to chase out conduits for cabling, the design solution could incorporate hollow walls where the cabling could be run in the void, thus preventing the need for chasing concrete and the subsequent health hazards of cement dust, noise and vibration and so on.

Building services design risk assessments are required to be provided to the principal contractor so that they can address the identified hazards and risks in the construction phase health and safety plan.

Specialist contractors are likely to be appointed to install building services, and it is vital that the principal contractor has all available information so that they can ensure that hazards and risks that will affect the whole site and all operatives are considered (delivery of large and heavy plant and equipment to the site, proposed lifting procedures, etc.).

Do the CDM Regulations specifically require design risk assessments to be generated on every construction project?

No. The CDM Regulations do not specifically state that design risk assessments have to be produced. The Regulations do state that designers must manage design risks by avoiding foreseeable risks and reducing residual risks so far as is reasonably practicable by following the principles of prevention.

Designers must take reasonable steps to provide with their designs sufficient information about aspects of the design of the structure or its construction or maintenance (including cleaning) as will adequately assist

- clients
- other designers
- contractors

to comply with their duties under the Regulations.

The information that designers have to provide to other duty holders can be in any format, but it should be brief, clear, precise and project specific.

Information can be notes on drawings, written information provided with the design drawings, formal risk assessment forms and project-specific instructions.

HSE inspectors are not looking for meaningless paperwork – they want to see evidence that designers have thought about health and safety issues and consequences during the development of the project, and that they have done everything that they can to assist in reducing on-site accidents and ill health and that they have also contributed to a reduction in accidents for maintenance and cleaning workers together with building users.

Designers should not start work on a project where there will be one or more contractors until the principal designer has been appointed. Why?

The principal designer is seen as having a pivotal role in influencing health and safety issues during the design process, and, in the past, designers have progressed their designs without anyone raising objective questions about safety. Often, projects would have progressed so far down the design process that even if health and safety improvements to the design were suggested it would be too late to make meaningful changes: a case of 'Oh well, we've designed it this way now and it's awkward to change so we'll stick with it'.

The designer can in fact undertake 'initial' design work on a project before the principal designer is appointed, but this is a fairly limited activity.

The principal designer should be the client's 'best friend' and, one hopes, the designer's too, and this should allow non-confrontational dialogue about designing out hazards and risks on the project that the principal designer may see but the designer does not.

By being appointed early on into the design process, the principal designer can influence decisions and will be able to ensure that designers are fulfilling their legal duties, in particular that they are applying the principles of prevention to a project.

Principles of prevention

The 'principles of prevention' are set out in Schedule 1 of the Management of Health and Safety at Work Regulations 1999 and are referenced in Appendix 1 of the HSE's guidance document on the CDM Regulations

(*a*) avoiding risks
(*b*) evaluating the unavoidable risks
(*c*) combating the risks at source
(*d*) adapting the work to the individual
(*e*) adapting to technical progress
(*f*) replacing the dangerous with non- or less dangerous
(*g*) developing a prevention policy
(*h*) giving collective/individual protection
(*i*) giving appropriate instruction.

What is initial design work?

Initial design is generally taken to be

- appraisal of project needs
- setting of project objectives
- feasibility in relation to costs
- possible constraints on the project
- possible desktop studies for contaminated land and remedial works
- likely procurement methods

- a strategic brief
- confirmation of key project team members (positions rather than names).

The Royal Institute of British Architects (RIBA) suggests that stages 1 and 2 in its Plan of Work (2013) would be initial design work (see ribaplanofwork.com).

Once drawings and specifications are being drawn up and once planning permission is being sought, the design has moved into outline design, and the principal designer should definitely have been appointed.

Designers' best practice tips

- Have a clear policy endorsed at the board or senior partner level on the management of health and safety, including how it will be addressed during the design stages of a project.
- Establish a programme of health and safety training and continual professional development.
- Establish a system to demonstrate that design staff have a good understanding of the construction process and a good working knowledge of health and safety legislation and guidance and knowledge of the Workplace (Health, Safety and Welfare) Regulations 1992.
- Ensure that hazard and risk information is in the practice library concerning products and materials regularly specified and used.
- Establish systems for design reviews, including procedures for communicating with principal designers, other designers and so on.
- Draw up lists of key hazards and risks associated with designs that are to be targeted on every project (e.g. red, amber and green lists of products/processes to be encouraged/discouraged).

Design best practice – things designers should do

- Ensure adequate access for construction vehicles, to minimise reversing requirements (one-way systems and turning radii).
- Ensure provision of adequate access and headroom for maintenance in plant rooms, and adequate provision for replacing heavy components.
- Ensure thoughtful location of mechanical/electrical equipment, light fittings, security devices and so on, to facilitate access and away from crowded areas.
- Specify concrete products with pre-cast fixings, to avoid drilling.
- Specify half-board sizes for plasterboard sheets, to make handling easier.
- Ensure early installation of permanent means of access, and prefabricated staircases with hand rails.
- Ensure provision of edge protection at permanent works where there is a foreseeable risk of falls after handover.

- Ensure practical and safe methods of window cleaning (e.g. from the inside).
- Appoint a temporary work co-ordinator (BS 5975, 2008: Code of Practice for temporary works procedures and permissable stress design of falsework).
- Specify off-site timber treatment if hazardous preservatives are used (boron or copper salts can be used for cut ends on site).

Design best practice – things designers should avoid whenever possible

- Internal manholes in circulation areas.
- External manholes in heavily used vehicle access zones.
- 'Lip' details (i.e. trip hazards) at the tops of pre-case concrete staircases.
- Shallow steps (i.e. risers) in external paved areas.
- Heavy building blocks (i.e. those weighing >20 kg).
- Large and heavy glass panels.
- The chasing out of concrete/brick/blockwork walls or floors for the installation of services.
- Heavy lintels (the use of slim metal or concrete lintels is preferred).
- Solvent-based paints and thinners, or isocyanates, particularly for use in confined area.
- Curtain wall or panel systems without provision for the tying of scaffolds.
- Blockwork walls >3.5 m high and retarded mortar mixes.

Design best practice – things designers should never do

- Issuing the pre-construction information pack before detailed structural surveys, asbestos surveys and so on have been completed.
- Specify the scabbling of concrete ('stop ends', etc.).
- Specify the demolition by hand-held breakers of the top sections of concrete piles (pile cropping techniques are available).
- Specify fragile roof lights and roofing assemblies.
- Specify processes giving rise to large quantities of dust (dry cutting, blasting, etc.).
- Specify the on-site spraying of harmful particulates.
- Specify structural steelwork that is not purposely designed to accommodate safety nets.
- Design roof-mounted services requiring access (for maintenance, etc.), without provision for safe access (e.g. barriers).

HSE inspector's assessment checklist for designers

Design stage	What to look for	Basis for intervention (i.e. what action to take)
Prior to concept selection (at or before project sanction)	■ Policy for safety in design ■ Criteria for concept selection (e.g. life cycle aspects considered) ■ Roles, responsibilities and competence of relevant personnel ■ Adequacy of health and safety advice	Sector specific – generally advice
Detailed design	As above, and: ■ Application of relevant and current good practice ■ Effective approval processes, including interaction with risk assessment ■ Application of formal and structured risk assessments and their assessments and their effectiveness in reducing risks ■ Effective change control procedures ■ Development of appropriate information to enable safe operation, maintenance and repair	Sector-specific legislation. Also, codes, standards and other good practice
Construction	As above, and: ■ Effective material control ■ Effective quality control (e.g. leak testing, welding procedures, etc.) ■ Conformity to design	Sector-specific and general legislation (e.g. the CDM Regulations). Also, codes, standards and other good practice

Source: HSE (2003), *Policy and Guidance on Reducing Risks as Low as Reasonably Practicable in Design*

CDM 2015 Questions and Answers: A practical approach
ISBN 978-0-7277-6032-6

http://dx.doi.org/10.1680/cdmqa.60326.081

Chapter 5
Designing for safety and design risk assessments

Designers have duties under the CDM Regulations to avoid foreseeable risks to the health and safety of people constructing their design, affected by the construction works or using or resorting to the building or structure once built. What does this mean?

Designers must consider the health and safety aspects of their designs and must avoid foreseeable risks: that is, those risks that their experience, knowledge and competence tells them that they should know about.

Many accidents occur in buildings because the end design is not user friendly, or the materials looked good when first built but after some wear and tear they become hazardous, or the access to plant and equipment is poor and maintenance personnel have to undertake hazardous tasks to do their jobs and so on.

Designers have great influence on the future safe use of the building and the Regulations require them to recognise this fact by considering health and safety.

Designers can also have great influence on the safety of contractors by paying attention to health and safety when they choose materials, the building location, the site set out, access, construction sequences and so on.

Designers can do more to ensure that contractors and their operatives work more safely on construction sites by ensuring that they specify more appropriate materials, equipment and so on: for example, by paying greater attention to size and weight.

If the risk is foreseeable (i.e. you could see or know about it beforehand), then the Regulations require the designer to avoid it.

Risks may be foreseeable but they may not always be avoidable. What happens then?

Risks will always be with us, and it will be impossible to eliminate all risks. Health and safety legislation is qualified by the term 'as far as is reasonably practicable', and this means that a judgement can take place.

If a risk cannot be eliminated completely, it may be able to be replaced by a lesser risk: for example, a substance known to cause cancer could be replaced with one known to cause skin irritation. It would still be a risk but the consequences of use are less and perhaps more acceptable.

Health and safety law refers to

- the hierarchy of risk control
- the principles of prevention.

Designers thus have to demonstrate that whenever they are unable to eliminate a hazard and risk they have ensured that they have chosen a design with the minimum risk or that they have protected against the hazard and risk.

What is safe design?

Safe design is the integration of hazard identification and risk assessment methods to eliminate or minimise the risks of injury throughout the life of a product or structure.

The safe design approach begins with an emphasis on making choices about design, materials and methods of manufacture or construction to enhance the safety of the finished product.

Safe design is good business because it improves safety and reduces costs by

- improving the risk management of workplace health and safety issues during the construction phase
- protecting constructors from injury
- reducing the need for redesign and retrofitting.

It should begin at the design phase of a project and not wait until construction starts.

Designers therefore have the greatest opportunity to influence the safety of the structure and the process of construction

Is there a recognised format for design risk assessments?

No. The CDM Regulations do not specify an exact format that has to be used for design risk assessments; in fact, they do not actually mention the term 'design risk assessment'. They require merely that hazards and risks in the design are considered and, where appropriate, recorded so that the information can be passed to other members of the design team, the principal designer and then on to the principal contractor and all contractors, as appropriate.

Suitable information on hazards and risks can be annotated on drawings if this will give clear guidance to whoever will be looking at the drawings. Information can equally be collated in pro-forma design risk assessment forms – although the Health and Safety Executive (HSE) does state that complicated design risk assessments are not expected.

If drawings are designed on computer-aided design systems, text can be annotated on drawings to convey hazard and risks and the proposed control measures.

Information on hazards and risks can also be recorded using building information modelling (BIM) if this is being used on the project.

Information on hazards and risks associated with the design of the structure, its construction and subsequent use and its maintenance must be freely available to all members of the project team.

No one has a monopoly on the safest way to do things, and previous experiences of other team members may add valuable contribution to the debate on health and safety.

Some clients and principal designers advocate that a project risk register is collated – usually by the principal designer – and that all hazards, risks and solutions are recorded, with dates of actions taken, changes made and so on. Common and expected hazards are not expected to be laboriously recorded – competent contractors should be aware of the dangers of using ladders to work at height, for example. What is more important in the design risk assessment is the thought process the designer has undertaken to see if works can be undertaken at ground level thus negating the need for ladders and eliminating a site risk. Could staircases be installed early on, thereby reducing the use of ladders as a means of access to upper levels? The designer can influence this process, and it forms part of the design risk assessment.

What is initial design work?

Initial design is generally taken to be

- appraisal of project needs
- setting of project objectives
- feasibility in relation to costs

- the possible constraints on the project
- possible desktop studies for contaminated land and remedial works
- likely procurement methods
- a strategic brief
- confirmation of key project team members (positions rather than names).

The Royal Institute of British Architects suggests that stages 1 and 2 in its Plan of Work (2013) would be initial design work.

Once drawings and specifications are being drawn up and once planning permission is being sought the design has moved into outline design and the Principal Designer should definitely have been appointed.

What are some of the considerations I need to think about regarding safety in design?

All designers should have procedures in place to help them review and record the processes they follow to ensure that their designs follow the principles of prevention in respect of heaslth and safety. The following ideas can be adopted by designers on a wide range of projects.

- Electrical safety:
  - the earthing of electrical installations
  - the location of underground and overhead power cables
  - the location of leads/cables within structures
  - the number and location of power points so as to avoid the use of extension leads and overloading sockets.
- Movement of people and materials:
  - safe access and egress, including for people with disability to the building when in use
  - traffic management during construction and for future use (the separation of vehicles from pedestrians, etc.)
  - loading bays and ramps both during construction and future use (in retail areas, goods-in areas, etc.)
  - safe crossings across vehicle routes, staff car parks, customer car parks and so on
  - exclusion zones around any chemical bunds and other high-risk areas (gas storage areas, etc.).
- Working environment:
  - ventilation for thermal comfort and general air quality and specific ventilation requirements for the work to be performed on the premises
  - temperature within work areas and compliance with the Workplace (Health, Safety and Welfare) Regulations 1992

  - lighting including that of plant rooms, minimising glare, and localised lighting for focused and intricate tasks
  - acoustic properties and noise control (noise isolation, insulation and absorption, the location of plant and equipment, the provision of sound enclosures, etc.)
  - seating requirements (whether fixed or movable)
  - floor surfaces to prevent slips and trips (the suitability of floor coverings, slip co-efficient/slip resistance values, the avoidance of changes in levels, etc.)
  - space for occupants so as to prevent overcrowding and for ease of use of the working environment.
- Plant and equipment:
  - tower crane locations, and loading and unloading during construction and any future maintenance
  - mobile crane loads on ground surfaces
  - plant and machinery installed in a building or structure (location, ease of access, route for replacement parts, etc.)
  - materials handling of plant and equipment during construction
  - maintenance access to plant and equipment (the location of access hatches, provision of access ladders, fixed access routes, etc.)
  - the guarding of plant and machinery (adequate space around equipment and plant, etc.)
  - lift installations (access to motor rooms, access to lift shafts, future maintenance, etc.)
  - avoiding confined spaces.
- Earthworks:
  - excavations (risks from earth collapsing or engulfment, ease of access, guarding, shuttering, etc.)
  - the location of underground services.
- Structural safety:
  - the erection of steelwork or concrete frameworks (weight loads, ease of access, prefabrication of site, reduction in on site welding, etc.)
  - load-bearing requirements (provision of information, lifting eyes, etc.)
  - stability and integrity of the structure.
- Manual tasks:
  - methods of material handling during construction, specifying lower weights, lightweight materials and so on
  - accessibility of material handling both during construction and for future operation (getting plant to higher levels in the building, etc.)
  - loading docks and storage facilities
  - workplace space and layout to prevent musculoskeletal disorders, including facilitating use of mechanical aids

  - assembly and disassembly of prefabricated fixtures and fittings.
- Substances:
  - exposure to hazardous substances and materials, including insulation and decorative materials
  - exposure to volatile organic compounds during construction and any future maintenance
  - exposure to irritant dust and fumes during construction and subsequent dust or fume release during refurbishment works
  - storage and use of hazardous chemicals, including cleaning products (ensure adequate storage facilities and well ventilated spaces).
- Falls prevention:
  - guard rails to all drops, leading edges, lift shafts and so on
  - window heights and cleaning (internal cleaning where possible, access for mobile elevating platforms, storage for ladders, etc.)
  - anchorage points for building maintenance and cleaning
  - access to working spaces for construction, cleaning, maintenance and repairs (considering safe access)
  - scaffolding (ease of erection, access for maintenance, etc.)
  - temporary work platforms
  - roofing materials and surface characteristics such as fragility, slip resistance and pitch (avoiding fragile materials, using long-life or maintenance-free materials, etc.).
- Specific risks:
  - exposure to radiation (electromagnetic radiation, etc.)
  - exposure to biological hazards
  - fatigue caused by poor environment, lack of ventilation, lack of rest spaces and so on
  - working alone (consider rescue facilities and methods, communication routes, etc.)
  - confined spaces (plant rooms, internal rooms without ventilation, those prone to gas and fume build-up, etc.)
  - over- and underwater work, including diving and work in caissons with compressed air supply.
- Fire and emergencies:
  - fire risks during construction (use of timber framed components, etc.)
  - fire detection and fire-fighting both during construction and future use of the building
  - emergency routes and exits (adequacy, availability, suitability, etc.)
  - access for fire tenders in and around the building both during construction and future use
  - other emergency facilities (likelihood of gas escapes, fuel containment, building collapse, etc.).

Appendix to Chapter 5

Design risk assessments

This form can be used by all Designers to provide health and safety information to the Principal Designer in respect of specific hazards and risks associated with their design.

Project	*[Name and address of project]*
Designer	*[Name and address of Designer]*
Discipline	[*Architect, structural engineer, building services consultant, specialist designer, etc.*]

(a) Design/project/works activity Describe the design concept under consideration or the project or works activity (creation of a glazed atrium, installation of mechanical and electrical plant on the roof, use of pre-stressed concrete, erection of structural steel work, etc.)
(b) Common major hazards Describe common hazards (falls from height, impact, collapse, manual handling, access, health hazards, etc.)
(c) Site-specific hazards Describe any site restrictions (confined spaces, awkward access, occupied buildings, known hazardous materials (e.g. asbestos), live services, transport restrictions, etc.)

(d) Hazard considerations in design Describe what considerations have been given to 'designing out' the hazard (why use pre-stressed concrete, what options have been considered for lifting steelwork into position, have environmental factors been taken into consideration, etc.?)
(e) What design options have been adopted to avoid or mitigate hazards identified? Describe what actions have been taken to reduce hazard and risk (plant to be specified in component parts for assembly on site to prevent the need for lifting plant into position thereby reducing the need for cranes, concrete to be post-tensioned, lifting eyes specified for steel beams, etc.)
(f) What safety hazards are envisaged regarding maintenance, cleaning, etc.? Describe how maintenance and cleaning activities will be carried out. Will there be falls from height, how will plant and equipment be accessed, will there by sufficient room to undertake repairs, are components accessible, etc.?
(g) What design considerations have been taken to eliminate or reduce the operational hazards? Describe how steps will be taken to mitigate the hazard (light fittings will not be above 4 m high or else they will be lowered to ground level, access to escalator lighting will be from within safe ceiling void access, safety railing will be specified, atrium services will be accessible via a movable gantry, etc.)
(h) What information does the Client require to ensure that either staff or contractors can operate, maintain or clean equipment safely? Outline the information needed (maintenance or operating manuals, specialist contractor activity, etc.)

(i) Has this assessment been forwarded to the Principal Designer for inclusion in the Health and Safety File?
YES/NO
If yes – by whom and when?
If no – who will supply the information and when?
(j) Have specific considerations been given to the following?
■ Following the principles of prevention ■ Fire risk ■ Health hazards (dust, noise, fumes, etc.) ■ Ease of access on to the site ■ Ease of access for future maintenance ■ Procedure for cleaning (what tools and equipment will be used, etc.) ■ Customer safety ■ Staff access to areas ■ Buildability ■ Specification of low-risk materials (no MDF, lightweight blocks, laminated glass panels, etc.) ■ Disability discrimination in relation to access and wayfaring
Describe any specific measures taken to address any of the above
(k) Please describe any other relevant information

Signed: ____________________ Dated: ____________________

Print name: ____________________

Example design risk register

Project hazard	Risk	Control measures required	Actioned		Comments
			Yes	No	
Fire safety					
Lack of fire strategy	■ Non-compliance with the Regulatory Reform (Fire Safety) Order 2005 (RRO)	■ Fire safety consultants to advise. ■ Fire protection standards required (e.g. 2 or 4 hours)			
Lack of Fire Risk Assessments (FRAs) ■ Operational ■ Construction	■ Contravention of the RRO ■ Life safety ■ Building safety	■ —— to produce an FRA for the loading bay and the sub-basement ■ Principal Contractor to produce an FRA for the demised area			
Lack of fire safety plan from contractor	■ Contravention of the RRO ■ Poor co-ordination ■ Life safety ■ Building safety	■ Principal Contractor to include a comprehensive Fire Safety Plan as part of the Construction Phase Health and Safety Plan			
Impairment of: ■ Sprinklers ■ Fire alarms ■ Emergency lighting Note: not only in demised areas but also in adjacent areas	■ Contravention of the RRO ■ Contravention of insurance requirements ■ Life safety ■ Building safety	■ Rigid compliance to the isolation permit procedure operated by —— ■ Review of all areas affected by impairment – discussion with the FM and the operations teams ■ Principal Contractor to manage			
Hot works	■ Life safety ■ Building safety	■ Permit procedures to be followed: —— Contractor's own in demised area —— 's in non-demised area			

Compromised means of escape from loading bay and sub-basement areas	■ Contravention of the RRO ■ Life safety	■ Thorough planning of all construction activity ■ Routes to be walked ■ FRAs to be completed ■ Fire safety consultant to advise on escape routes and distance of travel ■ Use the BS_9999 standard			
Storage of combustible materials required on the project	■ Life safety ■ Building safety ■ Insurance breach ■ Contravention of the RRO	■ Contractor to include details of materials and storage methods in the Construction Phase Health and Safety Plan			
Emergency procedures: ■ Other escape routes ■ Other					
Access/egress					
Access route to demised area	■ Interface with ——'s operations				
Access route for materials	■ Injury ■ Obstructions				
Operative safety and vehicle access in the loading bay	■ Injury ■ Fatality ■ Disruption to deliveries/ store operations				

Project hazard	Risk	Control measures required	Actioned		Comments
			Yes	No	
Access/egress from demised areas	■ Injury	■ Dedicated route to be planned by the Principal Contractor ■ Drawings to be marked up			
Other					
Asbestos					
Finding asbestos materials Disturbing asbestos materials already known Unexpected asbestos materials being disturbed Fibre release	■ Legal contravention ■ Health hazard to operatives and ——'s staff and others	■ Review the Asbestos Register ■ Consult the asbestos consultant ■ Carry out a refurbishment/demolition survey ■ Remove asbestos materials prior to work in the area ■ Manage asbestos materials in adjacent areas not disturbed by works			
Other					
Demolitions					
Structural stability	■ Collapse of a structure ■ Injury ■ Fatality	■ Proper planning of all works ■ Design review by the Structural Engineer ■ Plan of work and Risk Assessments provided by the Principal Contractor			

Removing the wrong thing	■ Collapse of a structure ■ Injury ■ Fatality	As above			
Dust	■ Respiratory illness ■ Stock contamination	■ Safe systems of work ■ Damping down ■ Dust control measures			
Removal of waste materials	■ Obstructions ■ Manual handling ■ Dust	■ Proper planning ■ Use of correct receptacles ■ Use of movement aids ■ Risk Assessments ■ Daily checks			
Temporary works – propping	■ Collapse of a structure ■ Injury ■ Fatality	■ Proper planning of all works ■ Design review by the Structural Engineer			
Area not ready or cleared by ——	■ Stock contamination ■ Project delays	■ Co-ordination of works ■ Proper planning ■ Communication between the Principal Contractor and ——			
Lift installations					
Structural works Demolitions Access/egress	■ See previous sections				
Installation of steelwork	■ Manual handling ■ Collapse of a structure	■ Design Risk Assessments ■ Choice of suitably sized steels, fixed together on site			

Project hazard	Risk	Control measures required	Actioned		Comments
			Yes	No	
Installation of lift cars	■ Manual handling ■ Collapse of a structure ■ Obstructions ■ Conflict with operations	■ Installation of lifting beam ■ Proper planning ■ Method statements and Risk Assessments ■ Clearing the area ■ Suspend operation activities			
Operations in a confined space	■ Injuries ■ Short-cuts to get work done	■ Proper planning ■ Method statements and Risk Assessments			
Electrical services	■ Electric shock ■ Isolation of the service ■ Fatality ■ Business interruption	■ Detailed design drawings ■ Method statements and risk assessments ■ Planning and preparation ■ On-site attendance Services Consultants ■ Isolation of live services – follow Permit procedures			
Installation of dock levellers and scissor lifts					
See section above for lift installations	■ See section above for lift installations				
Creation of trip hazards	■ Injury ■ Non-compliance	■ Check specification ■ Check installation standard			

Non-compliance with health and safety legislation	■ Contravention – action by the local authority environmental health officer	■ Review specification issued by the lift company ■ Check against the code of practice/approved guidance			
Access for future maintenance and cleaning	■ Injury ■ Poor maintenance standards	■ Review access to all areas of the equipment (e.g. motors and moving parts) ■ Determine access strategy with manufacturers/installers			
Location of switches, emergency cut off switches	■ Inaccessible locations will lead to unsafe systems of work ■ Injury	■ Review drawings and specification to ensure all controls are easily accessible in normal operation and in emergencies			
Lighting					
Erection of lighting at heights above 4m	■ Falls from height ■ Fatality ■ Injury ■ Non-compliance with legal requirements	■ Review designs to ensure lighting is at the optimum height ■ Position light fittings so they can be accessed from a cherry picker ■ Use long-life bulbs and low-maintenance fittings			
Access for future maintenance and cleaning	■ Falls from height ■ Fatality ■ Injury ■ Non-compliance with legal requirements	■ Position light fittings so they can be accessed from a cherry picker ■ Design Risk Assessment ■ Low-maintenance fittings			
Lighting levels not adequate	■ Accidents ■ Non-compliance	■ Review specification to ensure meets the Workplace (Health, Safety and Welfare) Regulations 1992			

Project hazard	Risk	Control measures required	Actioned		Comments
			Yes	No	
Maintenance					
Inadequate access to all plant and equipment	■ Falls from height ■ Inadequate maintenance ■ Unsafe working practices ■ Injury ■ Fatality	■ Proper planning for all equipment and plant ■ Access Strategy Statement to be completed for all plant and equipment ■ Provision of Design Risk Assessments ■ Provision of information for the Health and Safety File			
Environmental health issues					
Failure to address critical areas raised by the local authority environmental health officer (EHO) Failure to comply with the Workplace (Health, Safety and Welfare) Regulations 1992	■ Prosecution due to contravention ■ Service of statutory notices ■ Brand reputation	■ Review EHO letter/correspondence ■ Check specifications of all works against legal requirements and approved code of practice ■ Pay particular attention to – ventilation – lighting – safe access routes – vehicle and pedestrian separation – flooring repairs – edge protection – manual handling – stock movement – lifting devices			
Communication, co-ordination and co-operation between contractors and store operations					

Store personnel placed at risk as not familiar with building works ■ Overlap of ——'s operations with the contractor's area ■ Lack of forward planning – ——'s operations teams not aware of proposed works and their impacts	■ Injury ■ Unsafe systems of work ■ Contravention of legal requirements				
Clashes with deliveries – contractors' materials and ——' stock	■ Injuries ■ Obstructions to access/ egress routes ■ Unsafe working conditions				
Personnel gaining access to hazardous areas	■ Injury ■ Breach of the CDM Regulations	■ Proper hoarding of area – demised site ■ Security systems to prevent unauthorised access			
Movement of building materials around the loading bay and the sub-basement	■ Injury ■ Obstructions to access/ egress routes				
Conflict of operational needs	■ Injury ■ Disruption	■ Preparation and planning ■ Communication ■ Good programming			
Lack of clear accountability and responsibility for areas	■ Injury ■ Contravention of the CDM Regulations	■ Preparation and planning ■ Good organisation chart and information ■ Who's who			
Poorly informed sub-contractors creating hazards for store operations	■ Injury to store staff ■ Injury to operatives	■ Principal Contractor to properly induct sub-contractors ■ Regular meetings with sub-contractors ■ Communication			

Project hazard	Risk	Control measures required	Actioned		Comments
			Yes	No	
Conflict of live services and isolations, etc., affecting store trading ■ Wrong services isolated, causing disruption	■ Injuries ■ Fatalities ■ Contravention of health and safety laws ■ Brand reputation ■ Press/publicity ■ Disruption to trade	■ Thorough planning of all works involving services ■ Detailed drawings indicating isolations, etc. ■ Follow permit procedures ■ Consult with services consultants ■ Communications and briefing with the operations team			
Discomfort to ——'s staff due to poor ventilation, isolation of services	■ Contravention of legal requirements	■ Minimise disruption through good planning ■ Inform staff that isolations are temporary ■ Install mobile ventilation or other equipment			
Staff and others put at risk due to life safety system isolated – lack of proper communication to inform of the increased risk	■ Injury ■ Fatality ■ Fire	■ Proper planning – Fire Safety Plan ■ Use radio-controlled portable fire alarm system ■ Communication with the operations teams in a timely manner			

CDM 2015 Questions and Answers: A practical approach
ISBN 978-0-7277-6032-6

http://dx.doi.org/10.1680/cdmqa.60326.099

Chapter 6
Notification of projects

When is a project notifiable under the CDM Regulations?

Regulation 6 of the CDM Regulations requires certain projects to be notified to the Health and Safety Executive (HSE) by the client.

A construction project is notifiable to the HSE area office when

- it will, or is expected to, last more than *30 days* and involve more than *20 persons* working simultaneously at any point in the project *or*
- it will, or is expected to, involve more than *500 person days*.

The revised 2015 Regulations will reduce the number of projects notified, and this may enable the HSE to target inspections more effectively.

The HSE requires certain information that is outlined in Schedule 1 of the Regulations to be notified to it. As long as the relevant information is given, it can be supplied in any format, but in order to facilitate notification the HSE has produced form F10 (revised), which can be used for all projects.

An example of a pro forma based on form F10 (revised) is provided in the appendix to this chapter.

Does the principal designer have to notify the HSE about a construction project?

No. The duty to notify a project that will last more than 30 days and involve more than 20 persons on site at any one time or last more than 500 persons days rests with the client.

As soon as the principal designer is appointed to the project and determines that the project is notifiable, they should advise the client about the need to notify the project.

All the relevant information needed for the notification form may not be known at the time of the principal designer's appointment, but this need not delay the client from making the notification – tick the initial notification box on form F10.

As soon as further information is available (e.g. the confirmed start-on-site date or the name of principal contractor), tick the 'additional notification' box on form F10 and send the more detailed form to the HSE.

Make sure that either the 'initial' or 'additional' box is clearly indicated. The HSE uses the F10 forms to assess project complexity, and may believe that a notification is late if it is unclear whether it is providing initial or additional information.

The client can request that another person submits the notification on their behalf. The important duty is that the information about the project is submitted to the HSE – who does it is irrelevant, although the duty to ensure that it is done rests with the client.

Case study

A diligent client forwarded form F10 to the relevant HSE office 12 weeks before the anticipated start-on-site date.

Approximately 2 weeks before the start-on-site date the principal designer forwarded 'additional information' to the HSE, advising of the confirmed start date and the number of contractors.

The principal designer then received a letter from the HSE alleging late notification of the project and possible prosecution should future projects be so notified.

The principal designer immediately contacted the principal inspector of the HSE office to discuss the matter, and advised that the initial notification was sent 10 weeks previously. The inspector advised that when the receipt of an F10 form and the start-on-site date was less than 15 days apart, his department issued a 'standard' advisory warning letter.

The computer was programmed to automatically generate the letter unless the F10 form clearly identified that it was 'additional' information.

In this instance, the HSE apologised to the principal designer and admitted to a clerical error.

Do exploratory works need to be notified and are they the start of construction works?

Exploratory works will be notifiable if they last longer than 30 days and involve more than 20 workers working simultaneously at any point in the project or involve more than 500 person days.

Exploratory works usually form part of the main project works and, where they do, it is the total length of the project works that is calculated for notification purposes.

Exploratory works are included in the definition of 'construction works' if they involve the following

- excavation
- laying and installing the foundations of the structure.

The 'construction phase' of a project commences when the construction works of a project starts.

An exploration works package can be let as a separate contract on a duration of less than 30 days with less than 20 workers (it will therefore not be notifiable) but the project will come under the requirements of the CDM Regulations, and all but regulation 6 (notifiable projects) will be applicable.

Site survey works are those that involve taking levels, measurements, setting out and any other visual activities that generally do not involve physical activity such as drilling bore holes, taking down fixtures and so on.

What happens if a non-notifiable project becomes notifiable midway through the construction phase?

If a construction project is not notifiable at first but there are subsequent changes to its scope so that it fits the criteria for notification, the client must notify the project to the relevant enforcing authority as soon as possible.

The HSE may investigate and ask why the project was not initially notified, as it is keen to ensure that clients do not avoid their legal duties and, in particular, that clients allow contractors adequate resources in respect of time to complete the works safely.

Does form F10 or another notification document have to be displayed anywhere?

The client must ensure that an up-to-date copy of the notice is displayed in the construction site office so that it is accessible to everyone working on the site and in a form that can be easily understood. The client can either do this themselves or ask the principal contractor, where one is appointed, or the contractor to do it on their behalf.

Appendix to Chapter 6

Notification of a construction project – HSE pro forma (HSE, 2015)

The Data Protection Act 1988 requires the Health and Safety Executive (HSE) to inform you that this form may include information about you (this is called 'personal data' in the act) and that we are a 'data controller' for the purposes of this act. The HSE will process these data for health, safety and environmental purposes. The HSE may disclose these data to any person or organisation for the purposes for which it was collected or where the act allows disclosure. As the data subject, you have the right to ask for a copy of the data and to request any inaccurate data to be corrected.

1. **Is this the initial notification of the project or are you providing additional information not previously available?**

   Initial notification ☐ Additional information ☐

2. **What is the date of forwarding this notification or provision of additional information?**

3. **What is the exact address of the construction site?**
   (*Full address, including postcode*)

4. **What is the name of the local authority where the site is located?**

5. Give a brief description of the project and the construction work it includes.

6. Client contact details

(*Name, full address, postcode, telephone number and any email address*
If more than one client, please attach details on a separate sheet)

Name

Email

Address

Postcode

Telephone No.

7. Principal Designer contact details

(*Name, full address, postcode, telephone number and any email address*)

Name

Email

Address

Postcode

Telephone No.

8. Principal Contractor contact details

(Name, full address, postcode, telephone number and any email address)

Name | Email

Address

Postcode | Telephone No.

9. Please give the name and address of any designer already engaged

(Name, full address, postcode, telephone number and any email address
If more than one designer, please attach details on a separate sheet)

Name | Email

Address

Postcode | Telephone No.

10. What is the time allowed by the client to the principal contractor referred to in regulation 5(1) for the planning and preparation for construction work?

11. Please give your estimates of the following:

Please indicate if these estimates are: ☐ original ☐ revised

(a) The planned date for the start of the construction phase ☐

(b) The planned duration of the construction phase ☐

(c) The maximum number of people at work on the site at any one time ☐

(d) The planned number of contractors on the site ☐

12. Please give the name and address of any contractor already appointed
(Name, full address, postcode, telephone number and any email address If more than one contractor, please attach details on a separate sheet)

Name ☐ Email ☐

Address ☐

Postcode ☐ Telephone No. ☐

13. Declaration of client
I hereby declare that I am aware of my duties under the Construction (Design and Management) Regulations 2015

Signed by or on behalf of the organisation ☐

Print name ☐ Date ☐

Guidance notes for completion of notification of a construction project – HSE form F10

What should this form be used for?

- To notify the enforcing authority for the Construction (Design and Management) Regulations 2015 of any project that is likely to last longer than 30 days and involve more than 20 workers working simultaneously at any one time on the project or involve more than 500 person days of construction work.
- Any day on which construction work is carried out (including holidays and weekends) should be counted, even if the work on that day is of short duration.
- A person day is one individual, including supervisors and specialists, carrying out construction work for one normal working shift.
- Construction work for a domestic client is notifiable if it meets the threshold stated above.

Who should use this form?

- The client for the project.

Where should the completed form be sent?

- The HSE area office covering the site where construction work is to take place. You can get the address by telephoning the HSE Infoline: 0845 345 0055.

When should this form be sent?

- As soon as practicable before for the construction phase commences.

CDM 2015 Questions and Answers: A practical approach
ISBN 978-0-7277-6032-6

http://dx.doi.org/10.1680/cdmqa.60326.107

Chapter 7
Pre-construction information

What is meant by pre-construction information?

Any information that could be important in assisting designers or contractors in carrying out their duties under the CDM Regulations could be classed as pre-construction information.

The information required under the Regulations is really only that which is relevant to health and safety so that hazards and risks can be identified and subsequently addressed.

The client is responsible for deciding what information is available and what can be provided to designers and contractors.

Who has to provide the pre-construction information?

The client has the duty to provide information to every person designing the structure and to every contractor who has been or may be appointed by the client.

When a project has, or is likely to have, one or more contractors working on it, the client could provide the information to the principal designer, and the principal designer has to pass it on to every person designing the structure and to any contractor, including the principal contractor who has been or may be appointed by the client to the project.

Designers have to provide information to the client, other designers, the principal designer and other contractors as is necessary for them to carry out their duties. Some of this information will need to be included in the pre-construction information pack.

Is there a specific format for the pre-construction information pack?

No. The CDM Regulations do not specify the format that the pre-construction information should be in and, indeed, does not really refer to a 'pack' as such.

Information could be available in a wide variety of locations, and there is no requirement to duplicate the information. The Health and Safety Executive (HSE) is keen to see the amount of unnecessary paperwork on projects subject to the Regulations reduce significantly.

Whatever information is available and wherever that information is located, it is important that it is clearly identified and its availability made clear to those who may need it.

An information 'road map' could be created, listing what is available, where and who holds it.

Where apponted, the principal designer could advise the client early in the project on the format and content of the pre-construction information pack.

Information packs should not be generic with lots of 'strike throughs' and irrelevant information.

What type of information specifically is the client required to provide?

The pre-construction information should consist of all the information in the client's possession, or which is reasonably obtainable, including

- any information about or affecting the site or the construction work
- any information concerning the proposed use of the structure as a workplace
- any information about the existing building.

The client must ensure that they have obtained anything about the site, including environmental issues, local traffic, use of the site, previous use of the site, use of adjoining premises/land, overhead power cables, public transport and so on.

Designers and contractors will need to know what the proposed use of the building is, especially if it will be a workplace, because designers, in particular, will need to ensure that they design the structure in accordance with the Workplace (Health, Safety and Welfare) Regulations 1992 and other legislation in order to ensure that the future owners and occupiers of the building do not contravene legal requirements.

Who has to be provided with the pre-construction information?

Very broadly, every person designing the structure and every contractor who has been or may be appointed by the client.

Designers will be

- architects
- quantity surveyors
- structural engineers
- surveyors

- civil engineers
- interior designers
- landscape architects
- specialist contractors (e.g. temporary works)
- design and build contractors.

Contractors will be

- any of those who are tendering
- specialist contractors
- demolition contractors
- mechanical contractors
- electrical contractors
- ground work contractors
- civil and structural contractors.

The list of contractors who *may* be appointed could be considerable, and the requirement for information could generate vast amounts of paperwork.

The client and the principal designer could identify key information that contractors will need to enable them to plan for the job safely.

Once a short list of tenderers has been drawn up, or one contractor chosen, the client could discuss further details of the information that would be specific to the execution of the works.

Does a client have to provide any additional information to a contractor?

Yes. A requirement of the CDM Regulations is that the client must advise the principal contractor – or any contractor – what resources, including time, they will be allowed for the various stages of the project from concept to completion.

Contractors will need to determine the resources they need to carry out the project safely, and they therefore need to have an idea of the time available for further research and mobilisation for the project.

Resources and time are not defined in the Regulations, but they are taken to mean the time period in which the contractor considers exactly how to do the job, exactly what equipment will be needed, what specialist contractors and others need to be procured, what welfare facilities are required and how they will be installed on the site, the number and skill base of operatives and so on. Additionally, the client must ensure that designers have suitable resources, including time, to carry out their design duties.

Unrealistic deadlines and failure to allocate sufficient resources – both financial and human – are two of the main causes of construction site accidents and poor risk management.

Clients should consult with the principal designer where appointed and also with all contractors and discuss the timescales needed, expected and offered.

Clients often believe that contractors are prolonging the period they need for mobilisation and for the construction phase, and will be keen to ensure as a short a timescale as possible.

Contractors often say that clients and their advisers do not understand the full complexities of what they have to do to mobilise and complete a project.

If a contractor says the project can be completed in 24 weeks, the client may counter that it has to be finished in 20, believing that the contractor will be 'dragging their heels for 4 weeks'.

Sensible dialogue is necessary, and a true consensus of opinion and expectation needs to be reached.

Again, the principal designer could assist in these discussions, but if the client is unsure of what timescales would be reasonable, the client could consult the competent person appointed under the Management of Health and Safety at Work Regulations 1999.

How thorough does the client or principal designer have to be in providing information?

The answer from the HSE is likely to be 'very'.

The provision of comprehensive and suitable information about potential and actual hazards and risks relating to health and safety on the project is seen as key in ensuring better standards of health and safety.

The client and principal designer will be expected to do everything that is reasonably practicable to obtain the information that could protect the health and safety of those constructing the building/structure or working in or on it.

It will be unacceptable for clients to leave contractors to find things out for themselves – they must make all reasonable enquiries to obtain the information.

If information is not available, the client must commission surveys to obtain the information.

As an example, many buildings contain asbestos-containing materials, and exposure to asbestos is known to be a serious health hazard. Often, asbestos is hidden in buildings and its presence not easily identifiable. If contractors do not know whether they will be exposed to the substance, they cannot plan the job safely, and may release asbestos fibres – affecting themselves and others.

Clients are thus expected to provide thorough and detailed information on the likelihood of asbestos-containing materials, and should commission surveys if they do not have any information available in an asbestos register.

Surveys need to be thorough, and whenever any demolition works (of any sort) are planned, a refurbishment and demolition survey must be commissioned. If a client commissions only a presumptive visual management survey, this will not be seen as suitable and sufficient information, and the client may well be in breach of the duties imposed on them in regulation 4 of the CDM Regulations.

The client, or the principal designer, should search archives, records, ask maintenance managers, other contractors and so on whether they hold records of, for example, drawings and specifications.

Utilities companies should be contacted for details of underground and overground services.

Contaminated land surveys, radon surveys, atmospheric surveys and so on may be needed.

The client would be wise to communicate with the design team and the principal designer about the type of information they believe would be necessary and to 'brainstorm' what other information might be available. Expected information relating to the project should be clearly listed in the client's brief for the project.

Information will also be required about the minimum standards the client will expect the contractor to meet in relation to managing health and safety. Such standards are often known as employer's requirements, and include standard operating procedures for permit to work systems, accident reporting, fire safety, induction training and so on.

Does the requirement for pre-construction information apply only to notifiable projects?

No. Regulation 4 of the CDM Regulations applies to all projects, and requires that the client provides every person who may design the structure and every contractor who has been or may be appointed to the project to be provided with *all* the information in the client's possession (or which is reasonably obtainable), including

- any information about or affecting the site or the construction work
- any information concerning the proposed use of the structure as a workplace
- the minimum amount of time before the construction phase that will be allowed for preparation and planning
- any information in any existing health and safety file.

The Regulations apply to all construction projects, and so whatever information is relevant to the project must be made available or commissioned.

The HSE expects the information provided to be proportionate to the risks involved in the project, and so will not expect the level of detail in the pre-construction information pack to be the same in all projects. Small refurbishment projects for instance, carried out in buildings that continue to operate as businesses, may have greater hazards and risks than a green field new-build site.

The expectation of the HSE will be that detailed information will be provided to the contractor on the refurbishment project, including on-site health and safety procedures, traffic management, permit to work systems, fire safety, deliveries, access for personnel, site security, residual site hazards and so on.

Case study

A principal designer acted for one client on many similar projects.
A pre-construction information pack format was agreed as a 'core' document for all projects, with one section dedicated to the site-specific information.

The contents of the pre-construction information pack included

- an introduction to and the purpose of the plan
- project details and information
- timescales for preparation and planning
- health and safety objectives for the project
- the responsibilities of all parties
- statutory requirements applicable to the project
- project-specific health and safety information
- site rules and management procedures
- requirements for managing health and safety resources, contractors and so on
- requirements for site welfare facilities
- requirements for co-ordination and co-operation regarding design changes and design risks
- project review procedures
- information for the health and safety file.

What type of information should be available from the client?

The client has a duty to provide information to all duty holders in order to enable them to fulfil their legal duties under the CDM Regulations.

Information will vary according to the complexity of the construction project: for example, a green field site may have little available information whereas a refurbishment of an existing building will require detailed knowledge of the structure, surrounding area, building use and so on.

The client should be able to provide the following information, arranging to have surveys carried out if necessary:

- contaminated land surveys
- existing services locations
- structural/building safety reports
- survey reports for hazardous substances (e.g. asbestos, lead and other toxic substances)
- survey reports for hazardous areas (e.g. confined spaces)
- survey reports for hazardous locations (e.g. fragile roof access)
- local environmental conditions
- local hazardous areas (e.g. schools and major roadways)
- the current health and safety file
- intended occupancy details
- proposed site rules (e.g. existing permit to work systems)
- proximity of watercourses, transport systems and so on.
- any history of previous damage (e.g. from fire or floods).

The client should be encouraged to commission surveys when the information produced would be vital to the planning of the project in respect of health and safety. The HSE will expect clients to have taken responsibility for identifying hazardous conditions or substances (e.g. asbestos), and would consider prosecuting where there has been a blatant disregard to making information available.

Case study

The principal contractor was undertaking refurbishment works, and came across unexpected asbestos lagging to pipe work that had been previously hidden by a partition wall.

The project had to stop while notification for asbestos removal was made to the HSE. The HSE inspector visited the site, and wanted to know why the required 14 day notice had not been submitted. She formed the view that no planning for asbestos had been undertaken and that the principal contractor was trying to 'pull a fast one' to avoid the 14 day delay before works of removal could commence.

The HSE inspector inspected the pre-construction information pack, and noted that an asbestos survey had been carried out but that the asbestos lagging on hidden pipes had not been noted and was unforeseen. The principal contractor had the procedures in place for the removal of any other asbestos found in the building, and the inspector was reasonably satisfied.

However, the investigation turned to the principal designer, who was asked to account for the actions the client took regarding commissioning asbestos surveys and how the client judged the competency of the asbestos-surveying contractor. The inspector formed the view that it was reasonable for the client to have commissioned a more detailed survey that should have included for the removal of wall and ceiling panels to ascertain whether asbestos was in any hidden areas, particularly in view of the building's age.

The inspector issued an advisory letter to the client, stipulating that reasonable enquiries should have been made to obtain relevant information, and that, in her opinion, the client had failed to comply with the duties imposed on him in regulation 4 of CDM 2015. The letter went on to say that any further breaches of the client's duty to provide information relating to a construction project would be dealt with by more formal action. The client incurred a fee for intervention of several hundred pounds.

Does reference have to be made in the pre-construction information pack to 'good practice' and legal requirements?

The pre-construction information pack does not need to repeat legislation applicable to construction safety, but it can include references to good practice where the client and principal designer (if appointed) expect such standards to be adopted in the construction phase health and safety plan.

It may be helpful to include references to legislation where unusual safety hazards exist or where legislation has recently been introduced with which the contractors may not be fully conversant, such as the Work at Height Regulations 2005.

The client or principal designer may have decided to include references to HSE guidance notes on the CDM Regulations or codes of practice within the information pack or to other trade or professional guides. Of particular importance may be standards on fire safety, and reference to the Loss Prevention Council's code of practice on fire safety may stipulate the minimum standard of fire safety on the site, which the contractor should price for.

Rather than include detailed references in the body of the pre-construction information pack, it may be more practical and beneficial to include an appendix to the document, which lists all relevant legislation, codes of practice and guidance available to the project.

Experience has shown that the contractors are sometimes unaware of the extent of health and safety legislation, and appreciate advice and information. The principal designer can create and develop a good working relationship with the principal contractor if they are perceived as helpful and supportive, and likewise the client with designers and contractors where a principal designer appointment is not required.

What does the pre-construction information pack need to include in order to comply with the CDM Regulations?

The pre-construction information pack should include the following:

- a general description of the construction works comprising the project
- details of the time within which it is intended that the project, and any intermediate stages, will be completed
- details of the project team, including the principal designer, designers and other consultants
- details of existing plans, any health and safety file and so on
- details of risks to the health and safety of any person carrying out the construction work so far as such risks are known or are reasonably foreseeable or any such information as has been provided by the designers, including design risk assessments
- details of any client requirements in relation to health and safety (safety goals for the project, site rules, permits to work, emergency procedures, management requirements, etc.)
- such information as the client or principal designer knows or could ascertain by making reasonable enquiries regarding environmental considerations, on-site residual hazards, hazardous buildings, overlap with the client's business operation, site restrictions and so on
- such information as the client or principal designer knows or could ascertain by making reasonable enquiries and which it would be reasonable for any contractor to know in order to understand how they can comply with any requirements placed on them in respect of welfare by or under the relevant statutory provisions (availability of services, number of facilities required to be provided, details of any shared occupancy of the site, etc.)
- the content and format of the expected health and safety file.

The HSE guidance document supporting the CDM Regulations gives information on what to include in the document, and this can be summarised as

- the nature and description of the project
- client considerations and management requirements
- existing environmental and residual on-site risks

- existing drawings
- significant design principles and residual hazards and risks
- significant construction hazards.

The information pack is expected to include only that information that it is reasonable to expect or that could reasonably be ascertained by making enquiries. The Regulations do not expect every aspect of potential hazard and risk to be known at the outset of the project, but they expect reasonable enquiries to be made.

It has become apparent that further investigative works are required on the site to establish whether the land is contaminated. As the client, am I responsible for arranging this?

The question to really ask here is why the likelihood of a contaminated site was *not* discovered prior to the commencement of the construction phase. The client is responsible under the CDM Regulations for providing all persons with information regarding the state or condition of any premises at or on which construction work is to be carried out. Where appointed, the principal designer has a duty to relay this information to the principal contractor in the pre-construction information pack.

Should evidence be discovered that potentially contaminated land exists, the principal contractor should take steps to ensure the health and safety of all operatives and persons resorting to the site.

The client will be responsible for commissioning a suitable survey so that information on hazards and risks can be provided to the principal contractor. The client may prefer to commission the architect, project manager or principal contractor to organise the investigation works, but whether they do so or not, the client is responsible for ensuring that they do not abdicate their responsibilities for health and safety within the project process and must ensure that the principal contractor has adequate resources and the competency to deal with the hazards discovered.

The principal designer should give advice to the client on the steps to take to ensure that any unexpected hazard is properly managed and that information is communicated to the people who need to know.

The discovery of a potentially contaminated site after works have started will be of interest to the HSE. It will, in particular, want to establish what actions the client took prior to the commencement of works. It would also be interested in the role of the design team and the principal designer, and will be looking to see what information was available at design stage and what additional surveys would have been reasonable to have carried out.

Appendix to Chapter 7

Site survey – information that may be appropriate for inclusion in the pre-construction information pack

A site survey is always necessary in order to be familiar with the hazards and risks associated with the building and the site.

A site survey will always be necessary before preparing the following:

- the pre-construction information pack
- the construction phase health and safety plan.

Pre-construction information pack

A pro forma/checklist should be produced that covers the main items of information that will be required for the pre-tender health and safety plan, and guidance on completing such a pro forma follows:

1. Describe the exact location of the premises: for example, if the project is the refurbishment of a catering outlet within a site at a specific hospital, then state the precise location (e.g. 'On the ground floor of the main outpatients admissions building close to the main entrance and reception area').
2. Provide information on the type of building: for example, whether it is a unit within a premises (as in the item 1 above) or a part of a free standing structure, the number of floors and the approximate size of the premises.
3. State what type of project it is: for example, refurbishment of existing premises or fit-out of newly built premises.
4. What type of premises surround the site? For example, a department within a hospital, offices, shops or general public area (e.g. a waiting area or reception in a hospital).
5. What are the main hazards in the area of the site? For example, railway lines, schools (children), overhead restrictions (power cables, etc.), watercourses or members of the public.
6. Is work likely to cause a nuisance to neighbouring areas/premises? Are there any restrictions on working hours, types of work and so on?
7. How will deliveries get to the site? How will changes in level/floors be negotiated? Is the site is located within a larger building? Is there a goods lift available, and is it adequate? Is there a specific area for receiving deliveries (loading bay, etc.)?
8. Can the general public pass around the site? If so, to the front and rear? Could they be put at risk from site activities?
9. Describe the general condition of the building: for example, good (as will probably be the case for most), run down or dilapidated.
10. Is the site below ground level? If so, has it been tanked? Is it on a flood plain or near a river?

11 Is there any equipment/plant that needs to be removed during the project?
12 Is an asbsestos register available?
13 Detail areas where fragile materials may be or are located: for example, roofs, glazed areas or shopfronts.
14 Is there evidence of pests? For example, droppings, feathers or sight of rodents/birds/insects.
15 Are syringes or other equipment associated with drug abuse present?
16 Where will skips be located? If so, how will they be operated (weight and load, permanent skip, etc.)?
17 Is there a likelihood that deliveries will need to be co-ordinated with neighbouring premises. Could deliveries to neighbouring premises (either by nature or size) interfere with operations?
18 Are there any other works being undertaken in the area? If so, who is the principal contractor?
19 Does floor/ground penetration form part of the scope of works?
20 Were photographs taken on site?
21 Will new building services be required? For example, a fan for kitchen extraction.
22 Are there confined spaces, temporary supports, weakened structure, areas where a risk of falling over 2 m exists and so on?
23 Will steel supports be used? Will propping be required?
24 Where are the locations of meters, the main stop valve and so on?

The construction phase health and safety plan

When a principal contractor is appointed to a project, a site survey will be necessary to establish additional information regarding the building and site in order for the construction phase health and safety plan to be developed.

Key information on hazards and risks relating to the project will be contained in the pre-tender health and safety plan. The principal contractor must develop the content of the pre-tender health and safety plan into the construction phase health and safety plan.

A site survey should be used to gather information on

- the use of the building (will the client still have employees, customers and others within the building?)
- safety rules, risk assessments and so on available for the site or activity
- emergency procedures
- restrictions to the site access
- delivery routes
- residual hazards on the site/within the building

- site security procedures to be followed
- the locations of the nearest hospital and other medical services
- the location of underground and above-ground services
- the use of adjacent land/buildings
- the location of site welfare facilities
- access for construction workers
- storage areas
- manual handling hazards (e.g. the need to 'handball' materials)
- the location of any cranes and access equipment
- the risk of unauthorised persons to site
- hazardous substances on the site
- the presence of asbestos
- large delivery items and access routes
- the type of heating required for the site
- the type and location of lighting.

Much of the above information should be contained in the pre-tender health and safety plan, but if it is not, it does not mean that a hazard does not apply to the site. The onus is on the principal contractor to assess site safety hazards themselves and plan to avoid/control them.

Contents page from a pre-construction pack

Contents

Part 1 – Site-specific issues

1. Introduction
2. Brief description of the project
3. Project directory
4. Local information directory
5. Project programme
6. Roles and responsibilities of CDM duty holders
7. Existing environment and site
8. Site survey information
9. Existing information
10. Site-specific hazards
11. Design assumptions and hazards
12. Co-ordination of design during the construction phase
13. Proposed construction works
14. Overlap with the client's undertaking and other users of the building
15. Client direct appointments to the project

Part 2 – Employer's requirements: health and safety

16. Management of health and safety
17. Risk assessments and method statements
18. Commissioning and testing
19. Health and safety file

Part 3 – Construction phase health and safety plan

Appendices

CDM 2015 Questions and Answers: A practical approach
ISBN 978-0-7277-6032-6

http://dx.doi.org/10.1680/cdmqa.60326.121

Chapter 8
The principal contractor

What is, and what are the duties of, the principal contractor under the CDM Regulations?

The principal contractor is a legal appointment that the client has to make under the CDM Regulations.

The principal contractor must be a contractor: that is, someone who either undertakes or manages construction work. A client who normally co-ordinates construction works carried out on their premises, and who is competent and adequately resourced, can be a principal contractor.

The principal contractor would normally be a person carrying out or managing the construction work on the project to which they are appointed: that is, the main or managing contractor. However, where specialist work is involved, it may be appropriate to appoint the specialist contractor as the principal contractor, as they would be more suited to managing the risks of the specialist activity.

The principal contractor has specific duties under the Regulations and, ultimately, carries responsibility for site safety issues.

The main duties can be summed up as follows:

- develop and implement the construction phase health and safety plan
- appoint only competent and properly resourced contractors to the project (e.g. specialist contractors and sub-contractors)
- obtain and check method statements from contractors
- ensure the co-operation and co-ordination of contractors while they are on site (i.e. control multi-occupied site working)
- ensure health and safety training is carried out
- develop appropriate communication arrangements between contractors in respect of site health and safety issues
- make arrangements for discussing health and safety issues relative to the project

- allow only authorised persons onto the site
- display form F10 on site for all operatives to be able to see the details
- monitor health and safety performance on site
- pass information to the planning supervisor for the health and safety file.
- The principal contractor is a critical appointment to the construction project, as the standard of site safety will be determined by the commitment and competency of the appointment.

When is the principal contractor appointed and how is this done?

The client has to appoint a principal contractor on all projects where there is or is likely to be more than one contractor working on a project at any one time and as soon as is practicable after having information about the construction aspects of the project.

If the preferred contractor is known prior to the tendering or negotiating phase of the project, they should be appointed early in the design process.

Usually, the principal contractor is appointed as a result of the tendering process, and is usually the successful tenderer.

The appointment of the principal contractor is often assumed, or is a verbal instruction, but it is best defined in writing from either the client, contract administrator or principal designer (if requested to do so).

The preliminaries in the bill of quantities may stipulate that the successful tenderer will be appointed the principal contractor. Equally, a statement can be made in the pre-construction information pack.

The timing of the appointment should allow the principal contractor the opportunity to develop the construction phase safety plan. The client has to allow the principal contractor sufficient time and other resources before the commencement of construction works, and this must be clearly stated so that adequate time is available for developing the construction phase health and safety plan.

Preparation and planning time will be part of the resources time, and it will vary according to the complexities of the project, but the matter should be discussed with the principal designer and any competent persons appointed by the client. Any concerns about inadequate provision should be raised with the client.

Can anyone be appointed as the principal contractor?

No. The client must appoint only competent persons: that is, those with the requisite skills, knowledge and experience and who normally carry out the business of a contractor.

A contractor is defined in regulation 2 of the CDM regulations as

> Any person (including a non-domestic client) who, in the course or furtherance of a business, carries out, manages or controls construction work.

A principal contractor is defined as the contractor appointed to perform the specified duties in CDM regulations 12 to 14.

A client or any other person could be appointed as the principal contractor if they fall within the above definition.

The traditional 'main' contractor could be appointed as the principal contractor, as could a design and build contractor, management contractor or construction project manager. If a nominated sub-contractor or specialist contractor has more health and safety experience than the main contractor, they could be as the appointed principal contractor.

A principal contractor does *not* have to be the biggest or the main contractor on the site, but whoever is the principal contractor must be able to influence and control site safety throughout the project.

Can a client appoint themselves as the principal contractor?

A principal contractor must be a 'contractor'; that is, someone who carries on a business in which they carry out, manage or control construction work.

If the client has an in-house project management team that will be supervising and managing the construction work, and provided it is competent (i.e. the members have the necessary skills, knowledge and experience for the role, and also the resources), it could appoint itself as the principal contractor.

In-house maintenance and facilities management departments could appoint themselves as the principal contractor if they undertake the works themselves, or if they are managing the construction process.

The complexity of the construction project has to be considered, and appointments made only within core competencies: for example, facilities or maintenance teams may not be competent to manage the health and safety of a new-build office complex or an industrial unit constructed with a portal frame.

A client who appoints the office manager as the principal contractor in respect of a new office extension would be likely to be contravening the CDM Regulations because the office manager would not normally carry out or manage construction works – they would not be a contractor and unlikely to be competent.

The site manager or agent of the principal contractor is key in setting the standards of health and safety for the site, and so the Health and Safety Executive (HSE) will expect them to be able to demonstrate experience.

Do method statements/risk assessments have to be sent to the client or principal designer?

There is unlikely to be any benefit to the safe management of the project from the principal contractor sending method statements or risk assessments to the client or principal designer *unless* the client has instructed the principal designer to assist the principal contractor in such matters by an agreement outside the duties required by the CDM Regulations.

If the method statement contains the sequence of construction or similar information, it could be a useful document for inclusion in the health and safety file in case of future demolition of the structure, and, in this instance, it would be appropriate to send the method statement to the principal designer

The client has a duty to ensure that the arrangements made for managing the project are maintained and reviewed throughout the project. They do not have to personally check each individual aspect of health and safety on the site but must ensure that the management of health and safety is suitably addressed in a proportionate manner by all contractors and designers involved in the project.

The client should attend regular site or progress meetings, and should ensure that health and safety is covered in any agenda and that arrangements are maintained for all workers on the site to be consulted regarding health and safety matters.

Why should method statements/risk assessments be completed and when should they be done and for what type of work?

Method statements are written procedures that outline how a job is to be done so as to ensure the safety of everyone involved with the job, including persons who are in the vicinity.

A method statement equates to a 'safe system of work' that is required under the Health and Safety at Work etc. Act 1974.

Risk assessments are required under the Management of Health and Safety at Work Regulations 1999, and all employers are required to assess the risks to workers and any others who may be affected by their undertaking.

A risk assessment identifies the hazards present, evaluates the risks involved and identifies control measures necessary to eliminate or minimise the risks of injury or ill health.

A method statement can be used as the control measure needed to eliminate or minimise the risks involved in carrying out the job.

The principal contractor should carry out risk assessments for all work activities that they require their *employees* (i.e. their own employed tradesmen) to do.

Also, the principal contractor should carry out risk assessments for all those work activities that involve all operatives on site: that is, communal activities such as accessing routes to places of work, and delivery of materials, plant and equipment and so on.

The principal contractor should receive risk assessments from all the other contractors, sub-contractors and self-employed tradesmen working on the site. These will tell the principal contractor what the hazards associated with their tasks are (e.g. noise from drilling equipment) and will include details of how the risks from the hazards (e.g. noise-induced hearing loss) can be eliminated or reduced.

When the principal contractor has reviewed each of the contractor/sub-contractor risk assessments, they must consider whether they need to do anything else to protect other workers in the area (e.g. forbid certain work activities in certain areas). If so, the principal contractor will need to undertake an additional risk assessment that identifies how they are going to manage and control the combined risks of several contractors.

Risk assessments need only identify *significant* risks involved in carrying out a work activity. Routine risks and everyday risks such as crossing the road to get to the employee car park need not be included.

Where anything unusual or uncommon is to be undertaken on the site, a risk assessment will be essential. Where works involve significant hazards (e.g. working in confined spaces, at heights or with harmful substances), then risk assessments are legally required, and the control measures identified could be incorporated into a method statement that operatives are required to follow.

Case study

The principal contractor was responsible for ensuring the delivery of materials to the site. The delivery area incorporated the rear access road, which was shared by a neighbouring retail premises. There were hazards to both the site operatives and adjoining tenants from the delivery vehicles and the off-loading of materials.

Hazards included moving vehicles, restricted access to the roadway for emergency vehicles, off-loading materials from the lorries, dust, noise and falling objects. The risks from the hazards included being knocked over, being hit by materials, noise-induced hearing loss, and breathing in dust and exhaust fumes.

The principal contractor formulated the risk assessment, identifying the above as the hazards and risks and determining the control measures needed to eliminate or minimise the risks. These included having a banksman to guide in the delivery vehicles, setting specific delivery times, liaising with the adjoining tenants, providing lifting devices, requiring engines to be switched off during delivery and avoiding reversing vehicles wherever possible.

The principal contractor then prepared a short method statement that was given to the site foreman to follow when deliveries occurred.

The preparation of this risk assessment and method statement was the principal contractor's responsibility because they had overall management control of these activities and could co-ordinate everyone else's deliveries to site.

Can generic risk assessments be used as the basis of the health and safety plan?

Yes. However, they may not be sufficient for the principal contractor to demonstrate that they have done everything 'reasonably practicable' to ensure the health and safety of all persons at work carrying out construction works.

Generic risk assessments (i.e. those that cover general work activities such as bricklaying) form the basis of identifying hazards and risks associated with the job, and provided the principal contractor develops the generic risk assessment to include any site-specific hazard (e.g. carrying out brickwork adjacent to a deep watercourse) and the additional control measures they intend to adopt, the principal contractor will have sufficient information to ensure operatives work safely.

When reviewing a construction health and safety plan, an HSE inspector will not be satisfied with a plan that contains only general information, no matter how thick and impressive the plan. Often, a much thinner and more accurate site-specific plan will gain praise from the inspector. They will look for information proportional to the project risks – too much information is confusing and often of little value.

What other requirements in respect of the CDM Regulations does the principal contractor have to comply with?

In addition to developing the construction phase health and safety plan, the principal contractor has specific duties laid down in regulation 13 of the CDM Regulations.

These include

- liaising with all those who may be affected by the project or who are involved with it
- ensuring co-operation between all persons and all contractors on the site or on adjacent sites where there is an overlap (e.g. shared access routes)
- ensuring that all contractors and employees work following the rules contained in the health and safety plan
- ensuring that only authorised persons are allowed onto the premises where construction works are being carried out
- ensuring that the HSE notification of form F10 is displayed, is in a readable condition and is in a position where anyone involved in the construction works can read it
- providing the principal designer with information, particularly if that information would be necessary for inclusion in the health and safety file
- giving reasonable direction to any contractor, so far as is necessary, to enable the principal contractor to comply with their duties
- ensuring that the safety rules contained in the health and safety plan are in writing and brought to the attention of persons who may be affected by them
- ensuring that welfare facilities are provided that comply with the requirements of Schedule 2 of the Regulations
- advising contractors of the time they will be given for preparation and planning.

In order to be able to ensure co-operation between all contractors, it is necessary to have an understanding of regulations 11 and 12 of the Management of Health and Safety at Work Regulations 1999. These regulations require employers and the self-employed to co-ordinate their activities, co-operate with each other and to share information to help each comply with their statutory duties.

For instance, to be effective, risk assessments will need to cover the workplace as a whole, and the duty to co-ordinate these will be the principal contractor's. Information must be provided by all employers/contractors so as to enable the principal contractor to co-ordinate activities.

Another important aspect of co-ordination and co-operation relates to the use of work equipment and tools that are shared by all contractors on the site. The principal contractor may assume responsibility for the provision, maintenance and testing of all common equipment (e.g. lifting devices) or they may pass the responsibility onto another contractor. It does not matter *who* does it as long as someone assumes responsibility and everyone else knows who that someone is.

The principal contractor will need to request the names of all the people (contractors, clients, the design team, etc.) who wish to visit the site as 'authorised persons'. The principal contractor may authorise them to enter all or part of the site. The principal contractor should adopt a formal signing-in procedure that all persons should follow. Unauthorised visitors should be accompanied around the site by a trained operative.

The principal contractor has the responsibility of ensuring that every contractor is provided with comprehensive information on the risks to health and safety to all employees and others on the site.

In addition, the principal contractor must ensure that every employer carries out suitable training for all employees, relative to the works involved, and also that they provide information relating to health and safety issues. The CDM Regulations require all persons appointed to a project to have the necessary skills, knowledge and experience to fulfil the roles they undertake, and it will fall to the principal contractor to check out the competencies of the contractors they appoint.

The most appropriate way to ensure information is available is to include it in the health and safety plan.

Under CDM regulation 14, the principal contractor must ensure that there are procedures in place for any employee or self-employed person to discuss health and safety issues, and that there are arrangements for co-ordinating the views of others in respect of health and safety issues.

Depending on the size of the project, all that may be required is an item for health and safety on a site meeting agenda and a formal process whereby someone can raise health and safety concerns without fear of reprisals.

What other legislation in respect of health and safety does the principal contractor have to comply with?

The CDM Regulations are only one set of legislation governing safety that is applicable to construction sites. Other applicable health and safety legislation includes

- Health and Safety at Work etc. Act 1974
- Health and Safety (First Aid) Regulations 1981
- Control of Asbestos Regulations 2012
- Noise at Work Regulations 2005
- Electricity at Work Regulations 1989
- Management of Health and Safety at Work Regulations 1999
- Provision and Use of Work Equipment Regulations 1998

- Personal Protective Equipment Regulations 1992
- Manual Handling Operations Regulations 1992
- Control of Substances Hazardous to Health Regulations 2002
- Reporting of Injuries, Diseases and Dangerous Occurrences Regulations 2013
- Confined Space Regulations 1997
- The Lifting Operations and Lifting Equipment Regulations 1998
- Control of Lead at Work Regulations 2002
- Regulatory Reform (Fire Safety) Order 2005
- Work at Height Regulations 2005
- Control of Vibration Regulations 2005.

What does the principal contractor have to provide regarding training for operatives on site?

The CDM Regulations do not require the principal contractor to provide training (other than to their own employees) to operatives on site. However, the principal contractor must ensure that every contractor is provided with comprehensible information on the risks to health and safety from work activities on the site: the principal contractor has a duty to provide site inductions to all those working in or visiting the site.

Comprehensive information is information that is understood by everyone. It is meaningless to issue complex site rules and risk assessments if the understanding of written English is poor. Verbal instructions, diagrams and so on may be more comprehensible.

The principal contractor should also ensure that every contractor who is an employer provides all of their employees with information, instruction and training as required by the Management of Health and Safety at Work Regulations 1999, and as required by CDM regulation 15.

Site induction training is considered to be the responsibility of the principal contractor, and the information given should include

- site rules
- emergency procedures
- fire safety
- accident procedures
- permit to work systems
- site security
- welfare facilities on the site
- first aid facilities
- management of health and safety on the site.

All the above information should be included in the construction phase health and safety plan, and a copy of the plan should be given to the site foreman of every contractor on the site.

Whether it is the principal contractor or each individual contractor who carries out the site induction training, it is essential that written records of the training are kept and regularly reviewed and updated.

The principal contractor will require all contractors to provide evidence of competency in the various trades, and they will be entitled to request training certificates for trades such as mobile equipment driving, fork lift truck driving and gas fitting works.

All such documentation should be kept readily available by the principal contractor, and is best kept appended to the construction phase safety plan.

The health and safety plan should set out what level of training site operatives are expected to have, who is to have provided it, how often and to what standard.

The principal contractor's role in respect of training is a co-ordination role unless they are an employer of their own workforce – when the requirements of health and safety training and the provision of information will apply equally to the principal contractor as to others.

Is the Construction Skills Certificate Scheme a requirement of the CDM Regulations?

No, not specifically. However, the Construction Skills Certificate Scheme (CSCS) card does provide the principal contractor, the client or any other person evidence that the holder has met certain criteria with regard to health and safety knowledge.

The CSCS scheme is an independently run and assessed health and safety competency programme that confirms that trades people are competent to undertake their duties and that they have passed a general health and safety test.

The increased requirements for contractor and individual competency (i.e. skills, knowledge and experience under the CDM Regulations) has encouraged both clients and principal contractors to require all contractors and operatives to sign up to the scheme.

Clients may stipulate that the CSCS card scheme should be adopted for the project – and so may principal contractors.

The Regulations do not specifically require membership of any competency and assessment scheme, and individual companies can choose to demonstrate their competency in a number of different ways.

When the HSE inspects a construction site or carries out any investigation into an accident or dangerous occurrence, it will be looking for suitable and sufficient evidence of competency and not necessarily membership of commercial schemes.

Evidence of training, toolbox talks and so on may be just as effective as presenting a CSCS card.

The client has appointed several 'client direct' appointments who are employed by the client and required to access the site to undertake works. Are they exempt from complying with the CDM Regulations?

No. The principal contractor has absolute responsibility for site safety issues, and can specify these in the site rules included in the construction health and safety plan.

Even if the client direct is not a 'contractor' under the CDM Regulations, they will either be employers or employees, and, as such, have legal duties under the Management of Health and Safety at Work Regulations 1999 to comply with the requirements of the principal contractor.

Contractors are defined in CDM regulation 2 as

> Any person (including a non-domestic client) who, in the course or furtherance of a business, carries out, manages or controls construction work.

Construction work includes fitting out, commissioning, alteration, conversion and so on.

The principal contractor has a duty under CDM regulations 13 and 14 to ensure that every contractor at work in connection with the project complies with any rules contained in the health and safety plan. If the health and safety plan contains a rule that states that all client direct appointments and any others conducting a trade, business or undertaking must provide method statements/risk assessments to the principal contractor prior to commencing work, they must legally comply with the rule.

What actions can the principal contractor take where either a contractor or client direct appointment fail to comply with the site rules, the health and safety plan or requirements of the CDM Regulations?

The principal contractor should write to the employer of the contractor or client direct, and advise them that they are in breach of their duties under the CDM Regulations and other health and safety legislation, and that unless they start complying they will be removed from site, incurring any contractual penalties.

Often, non-compliance with a requirement is due to fear or ignorance. Perhaps they do not know how to conduct risk assessments. If this is the case, the principal contractor should provide information and guidance.

The principal contractor should review their procedures for assessing the competency of contractor. Remember that the duty to ensure the competency of contractors rests with *any person*, including principal contractors who let subsidiary work packages to sub-contractors. If the principal contractor knew that the contractor could not provide risk assessments, why did they appoint them?

The principal contractor should consider whether they are asking the contractor/client direct to provide more information than is justifiable. Risk assessments and method statements need to be relevant and cover *significant* risks. Requesting meaningless paperwork from contractors merely compounds their reluctance to produce any.

The contractor or client direct should be issued with a 'yellow' or 'red' card by the principal contractor: that is, a formal way of letting them know that they have breached site rules.

The principal contractor should ask the principal designer to help encourage the contractor or client direct to comply with their statutory duties. The principal designer could give advice to the client that the contractor or client direct is not competent, and recommend that the client moves to formally dismiss the contractor or client direct.

The local HSE inspector should be approached for guidance by the principal contractor. If the site is complying with the CDM Regulations, then there is nothing to fear in seeking advice from the HSE on how to improve health and safety management procedures (e.g. improving contractor compliance).

The principal contractor should de-list the contractor or client direct from their approved list.

Whatever steps the principal contractor takes, they should not allow the breach of safety management procedures go unrecorded. Detailed records of what actions the principal contractor took to ensure compliance with site rules should be kept, together with who they spoke to, when, how often and so on. The principal contractor should make sure that they have provided the contractor or client direct with all the information they could reasonably expect to have regarding the site works (e.g. the health and safety plan).

What happens if the principal contractor does not fulfil their responsibilities under the CDM Regulations?

In this scenario, the principal contractor could be in breach of their duties under regulations 13 and 14 of the CDM Regulations, and could be prosecuted by the HSE.

Also, there could be contraventions of CDM regulations 8, 12 and 15, and criminal charges could apply. The principal contractor could also be in breach of these if they failed to manage the site effectively. Every contractor carrying out construction work must comply with regulations 16 to 35, and failing to do so will be a criminal offence.

The HSE will prosecute principal contractors who fail to discharge their duties under the Regulations effectively.

Prosecutions are taken either in the magistrates' court or in the Crown Court, and fines have the potential to be unlimited.

What notices under the CDM Regulations must the principal contractor display on the site?

The principal contractor must ensure that a copy of the notification of the project (form F10) as sent to the HSE by the client or the principal designer is displayed on the site so that it remains legible and can be read by those working on site.

The notice can be displayed in the site office, but this may restrict the number of operatives who could easily refer to it. A more acceptable place to display form F10 would be in the welfare or messing facilities, where it would be readily available to all operatives. A copy could be kept in the site office or with the construction phase health and safety plan.

The location of the F10 form needs to be brought to the attention of all contractors working on the site, and the easiest way to do this is to inform them during the site induction training.

The principal contractor may be required to display other signs as required by other health and safety legislation (e.g. the 'information for employees' poster).

What type of welfare facilities must be provided by the principal contractor?

Regulation 13 and Schedule 2 of the CDM Regulations set out the requirements for the provision of welfare facilities. The principal contractor must comply with the duty to provide suitable and sufficient welfare facilities, and this will be determined by a risk assessment. The client must be satisfied that adequate provision has been made for welfare facilities on the project.

Schedule 2 of the Regulations sets out the principles to be observed, and these can be summarised as follows:

- Sanitary accommodation:
  - No numbers are specified, but they must be suitable and sufficient for the number of operatives on site.
  - Adequate lighting and ventilation must be provided.
  - Sanitary accommodation must be kept clean.
  - Separate male and female accommodation is not required if each water closet is in a separate room with a door that can be locked from the inside.
- Washing facilities:
  - Washing facilities must include wash hand basins and, if necessary due to the type of work, showers.
  - Facilities must be in the immediate vicinity of the sanitary accommodation and reasonably accessible throughout the site.
  - Clean hot and cold, or warm, water must be provided, preferably as running water.
  - Soap or other hand-cleaning chemicals must be provided, together with hand-drying facilities.
  - The facilities must be well lit and ventilated, and kept clean.
  - Unless washing facilities are provided in individual cubicles, there must be separate male and female facilities unless they are used only for washing the hands, face and forearms.
- Drinking water:
  - A clean, wholesome supply of water for drinking must be provided in readily accessible locations.
  - Drinking water supplies must be clearly labelled or signed as suitable for consumption.
  - Drinking vessels must be provided if the water is not from a drinking fountain.
- Changing rooms and lockers:
  - Changing facilities must be provided if specialised clothing is required to be worn, with separate male and female rooms.
  - Seating must be installed, and drying facilities must be provided for clothing.
  - Facilities must be made available to lock away everyday clothes not worn on the site, personal effects and special clothing worn at work but not taken home.
- Rest facilities:
  - Suitable and sufficient rest facilities with tables and seating and maintained at an appropriate temperature must be provided
  - The facilities must include areas for eating and preparing meals and a means of boiling water.
  - Where necessary, facilities for pregnant or nursing mothers must be provided.

What training must be provided by the principal contractor?

The CDM Regulations do not specifically require the principal contractor to carry out training for persons other than their own employees.

The principal contractor does have to carry out site inductions for workers, although this may be information giving and not necessarily training.

The principal contractor is responsible for ensuring that all contractors working on the site are given comprehensible information about the risks they are likely to face during their work on site.

In addition, the principal contractor must ensure that contractors provide their employees with relevant training that covers new or increased risks from working on the project.

Induction training arranged by the principal contractor may be the simplest way for the principal contractor to discharge their duties in respect of making information available to all contractors. An organised, planned induction programme would give the principal contractor a mechanism to record attendees and to keep training records. In the event of any accident or incident, the principal contractor would have attendance records to show that they had discharged their duties regarding the sharing of information relevant to health and safety risks comprehensively.

What subjects would be appropriate to include in induction training organised by the principal contractor?

The duty of the principal contractor is to convey information on risks relating to the carrying out of construction works within the designated site. One of the most effective ways to communicate information is by way of a structured training programme. Subjects to cover in induction training would be the following:

- Outline of the project – who's who (the client, the design team, etc.).
- Statement on health and safety and commitment of senior management to high standards of health and safety.
- Site-specific risks, such as
  - access routes
  - contaminated land
  - overhead power cables
  - underground services
  - the proximity of water
  - unstable buildings
  - hazardous materials (e.g. asbestos).
- Site-specific control measures for identified risks.

- Site rules.
- Welfare facilities, location maintenance and cleaning provision and so on.
- First aid facilities.
- Accident and near-miss reporting procedures.
- Emergency procedures, such as
  - fire evacuation
  - raising the fire alarm
  - names and addresses of emergency services
  - assembly points
  - fire marshals/wardens
  - for building collapse
  - for flood, chemical escape and gas escape
  - for release of hazardous substances (e.g. asbestos).
- Responsible persons.
- Requirements for protective equipment and clothing, such as
  - use of hard hats
  - use of safety footwear
  - use of ear defenders.
- General site safety controls, such as
  - permit to work
  - permit to enter
  - use of banksmen.
- Arrangements for communicating with all the workforce in respect of health and safety, such as
  - weekly site safety meetings
  - notice boards
  - display of safety notices
  - use of other aids (e.g. visual aids to assist those with language difficulties).
- Names of safety representatives, competent persons and so on.
- Site security and access procedures.

All of the above information should be available in the construction phase health and safety plan, and key details should be displayed on an information board at the entrance to the site or in the messing facilities.

Case study

A new shopping centre was being constructed for a consortium of developers. The developers nominated the main developer as the client under the CDM Regulations. As the client, the developer appointed a management contractor to oversee the entire construction works. The management contractor was designated the principal contractor, even though they were not undertaking any actual construction works, but because they managed the construction process they met the qualification requirement under the Regulations for a principal contractor.

The client required the principal contractor to have overall responsibility for the site, and although each tenant shop fit-out had a main contractor, only the management contractor was designated as the principal contractor.

The client imposed a duty on the principal contractor to carry out induction training for all persons entering the site.

The principal contractor set up a separate training room within his site office compound and by way of a strict security control point. All persons entering the site for the first time had to report to the training office for induction.

Each induction programme ran for 30 min, and three sessions were undertaken each day – two in the morning and one in the early afternoon. If persons wanted to gain access to the site at other times, they had to wait until the next induction session. Every person going through the induction process was registered, and when they had completed the course, they signed a declaration to that effect. These records were kept centrally by the management contractor.

Each attendee received a photo ID card, which indicated that they had been inducted and when. On the reverse of the ID card were the basic emergency rules of the site.

Each individual main contractor undertaking their own client's shop fit-out was required to provide additional training to operatives about the specific hazards and risks found on the site.

Each main contractor had to regularly provide information to the principal contractor on any health and safety issue on their site that could affect the safety of the whole site (e.g. liquid petroleum gas storage). The principal contractor then ensured that this information was added to his induction training programme.

In addition, the principal contractor issued a site directive that every contractor must carry out a toolbox talk every month, and issued every contractor with a topic timetable and a 5 min presentation pack to assist them in delivering a consistent message across the site.

CDM 2015 Questions and Answers: A practical approach
ISBN 978-0-7277-6032-6

http://dx.doi.org/10.1680/cdmqa.60326.139

Chapter 9
The construction phase health and safety plan

What is the construction phase health and safety plan?

The construction phase health and safety plan is the document produced by the principal contractor where appointed or, where not, by the only contractor appointed who develops any information contained in the pre-construction information pack and sets out the arrangements that the principal contractor will take to ensure management of health and safety on the site.

The document is required by regulation 12 of the CDM Regulations.

The health and safety plan is the foundation upon which the health and safety management of the construction phase needs to be based. A written plan clarifies who does what, who is responsible for what, what hazards and risks have been identified, how works will be controlled and so on.

The contractor appointed as the principal contractor must develop the construction phase plan *before* construction works start, so that it outlines the health and safety procedures that will be adopted during the construction phase.

Where a principal contractor has not been appointed to a project, the sole contractor appointed must prepare and develop the construction phase health and safety plan, as required by CDM regulation 15.

The construction phase health and safety plan must be site specific and reflect the site hazards and risks. It needs to be focused and suitable for the works to be undertaken. Generic, weighty documents written by people who have never been to the site are not recommended, and the Health and Safety Executive (HSE) is increasingly identifying these plans as being unsuitable, leading them to serve improvement notices under the Health and Safety at Work etc. Act 1974.

The CDM Regulations do not specifically list the information that should be included in the construction phase health and safety plan, but it must address the requirements of regulations 16–35, and therefore the information that could be included will be

- A description of the project
- Details of the project team
- Details of any existing information and the location of records and so on.
- Management of the work, including the roles and responsibilities of key people.
- Goals and objectives regarding health and safety for the project.
- Health and safety arrangements for the project once on site:
  - liaison and co-operation between all parties
  - consultation procedures with the workforce
  - exchange of design information
  - handling design changes during the progress of work
  - selection and control of contractors
  - exchange of health and safety information
  - site security, including preventing unauthorised access
  - site induction
  - site training
  - welfare facilities
  - first aid
  - accident reporting
  - procedures for preparing, issuing, reviewing risk assessments and method statements
  - site rules
  - emergency procedures
  - fire safety arrangements.
- Arrangements for controlling significant site risks:
  - delivery procedures
  - removing waste
  - the location and management of services
  - the stability of structures
  - contaminated land issues
  - environmental conditions
  - overhead and underground power supplies
  - temporary structures
  - work at height
  - work on or near fragile materials
  - hazardous substances
  - control of lifting operations
  - maintenance of equipment

  - excavations, earthworks, tunnels and so on
  - demolitions
  - work near water courses
  - use of compressed air
  - working with explosives
  - traffic routes and vehicle management
  - storage of materials.
- Arrangements for controlling health hazards on site:
  - asbestos
  - other hazardous materials
  - manual handling
  - noise
  - stress
  - exposure to radiation.
- Details to be included in the health and safety file, including the layout, the format of information, timescales to provide information and so on.

What format does the construction phase health and safety plan require?

The construction phase health and safety plan should be in a format that is

- easy to use and to refer to
- understandable to those who need to use it
- easy to update
- easy to duplicate
- clear, concise and logical.

The construction phase health and safety plan is needed *on site*, and should therefore be physically robust. An A4 ring binder is a popular choice for keeping the information in order, as pages can be easily removed and photocopied for other contractors as necessary, and, importantly, it can be readily updated.

The construction phase health and safety plan would not be particularly useful as a computer file on the site because access to its information may be restricted.

The construction phase health and safety plan need not be all written words – often the use of diagrams, pictograms and cartoons are very effective at explaining health and safety messages.

The plan should not contain every conceivable health and safety procedure, but only those *applicable* to the site. This should prevent it from becoming unwieldy, and will not discourage people from reading it. It is perfectly acceptable to refer to other manuals, such as the site Risk Register or risk assessment manual or to the COSHH (Control of Substances Hazardous to Health Regulations 2002) Manual. If site safety procedures are reliant on safety procedures specified in other documents, then copies of these must be available on site.

The construction phase health and safety plan should contain all the site-specific information, including risk assessments and method statements, not quantities of generic forms.

The construction phase health and safety plan should be a 'road map' to point all operatives on site to the key information on how health and safety will be managed on the site.

Does a copy of the health and safety plan have to be given to every person working on the site – the client, the principal designer and so on?

The client must be given a copy of the plan so that they can be satisfied that it has been prepared and complies with the CDM Regulations.

There is no duty to give a copy to the principal designer unless they are acting on behalf of the client in assessing its adequacy before construction works can start. In this case, there would be no need to send one to the client, unless specifically required to do so.

The Regulations require the principal contractor to provide information to all persons working on or resorting to the site in respect of health and safety issues.

As the health and safety plan contains valuable information on site health and safety matters, it makes sense to issue the document to as many people as practicable. However, that may become expensive, and some operatives may only be on site for a few days. Information could be issued to the site foreman. Relevant information could be displayed around the site in poster format.

Site safety rules should be issued to all individual operatives, and should be issued to all employers, contractors and employees during site induction training.

Specific risk assessments should be issued to the ganger or foreman, with instructions that they are responsible for ensuring that their gang/team is made aware of the hazards and risks and the protective measures needed to control the risks.

Copies of relevant information could be displayed in the mess room, site office and site canteen (e.g. the location of first aid kits and the names of companies with trained first aiders).

Key aspects of the health and safety plan can be issued to safety representatives, site foremen and so on, with guidance on how and where they can access the full health and safety plan and supporting information, documentation (the company safety policy, HSE codes of practice, etc.).

A practical way of disseminating site health and safety information is to convene a weekly site safety forum or committee, requiring a foreman or representative from every contractor or self-employed person on site to attend, using the meeting to review site health and safety issues and to discuss, for example, forthcoming works on the programme and new site safety rules.

It is important to remember that the CDM Regulations place a duty on all persons to ensure co-operation, co-ordination and communication between all those engaged in the construction project.

There will be no one set way to disseminate the information in the construction phase health and safety plan, and it will be for the client, the principal designer and the principal contractor to discuss the plan and the methods of sharing the information during any preparation and planning time.

The construction phase plan must be site specific: that is, it must cover issues that apply to the works to be carried out, include actual site personnel, site-specific emergency procedures, and so on.

The HSE is increasingly identifying construction phase plans that are generic, and serving improvement notices on the basis that the plans are not suitable and sufficient.

The information that should be included in the construction phase plan is as follows:

- A description of the project.
- Arrangements for the project (including, where necessary, for the management of construction work and monitoring compliance with the relevant statutory provisions) that will ensure, so far as is reasonably practicable, the health and safety of all persons at work carrying out the construction work and all persons who may be affected by the work of such persons at work, taking account of
  - the risks involved in the construction work
  - any activity of persons at work that is carried out, or will be carried out, on or in premises where construction work is undertaken
  - any activity that may affect the health and safety of persons at work or other persons in the vicinity.

- Sufficient information about arrangements for the welfare of persons at work by virtue of the project to enable any contractor to understand how they can comply with any requirements placed upon them in respect of welfare by or under the relevant statutory provision.

A construction phase health and safety plan outline for major projects is provided in the appendix to this chapter.

Case study

A major shop fitting contractor acted as the principal contractor and developed a comprehensive health and safety plan. A summary version was produced that covered site rules, emergency procedures, fire safety, personal protective equipment, signing in procedures and so on, and this document was given to all operatives and visitors to the site who underwent the site safety induction training. Additional notices were displayed in the canteen, mess room, site office and at the entrance to the site.

Each major contractor on the site was given a full copy of the health and safety plan, and was required to sign a record to that effect. They were then required to ensure that all relevant information regarding health and safety had been provided to their site operatives.

The principal contractor introduced an auditing system that regularly checked how, to whom and when the contractor issued the information.

The principal contractor provided a number of site posters in a variety of languages and with cartoons and pictograms, and encouraged all contractors to take note of them. This helped to disseminate key health and safety information to all operatives, and met the principal contractor's duty to ensure that information was suitable for those who had difficulty understanding English.

As the principal contractor I intend to revise the original construction phase health and safety plan. Do I need to tell the principal designer or the client?

Only if your revisions are due to design changes. The principal designer has to ensure that designers prepare design risk assessments, and need to be sure that these have been done. They rely on the principal contractor to inform them. It would be beneficial to have a discussion with the principal designer regarding the proposed changes.

There is no duty to advise the client of any changes to the construction phase health and safety plan under the CDM Regulations, and there is no duty on the client to check that you, as the principal contractor, have made changes or updates to the plan, although they must be satisfied that you have procedures in place to revise, amend and monitor the plan, and they must be satisfied that arrangements are maintained and reviewed (CDM regulation 4).

A good way of helping the client to discharge their duties will be to send them updates of your construction phase health and safety plan or ensure that you cover the matter and discuss the revisions at the project meetings. If the client does not attend, then copies of the minutes could be sent.

The client and the design team or other professional advisers will need to be briefed by you on any changes to the construction phase health and safety plan that could affect their safety (changes to site rules, restricted areas of the site, etc.).

What is the best way to update the construction phase health and safety plan without it becoming complicated or confusing?

The construction phase health and safety plan is the document that sets out the health and safety management of the construction phase of the project. It must be a document that outlines what special health and safety precautions are to be taken to ensure the safety of everyone on the construction project, and must be updated to reflect any changes to the working procedures, management systems, welfare facilities and so on that happen on the site.

If the majority of the construction health and safety plan has been agreed before the commencement of the construction works, there will be little need to change substantial parts of it. Details on site management, emergency procedures and welfare facilities may not change during the construction phase if they have been well thought through at the beginning.

If the construction health and safety plan needs to be updated, it should be done by adding information clearly and removing old information so as to avoid confusion. For instance, if the names of the trained first aiders change, the new ones should be added to the plan and the old ones removed. If the location of the first aid kit has changed, this should be included too.

The most important thing about the construction health and safety plan is that the information contained in it is made available to all operatives on the site – the simpler the updates the easier things will be understood.

Risk assessments and method statements could be included as a separate document to the plan, making it easy to add new information to risk assessments without changing the overall plan.

If the principal contractor decides to implement a new permit to work procedure for a specific activity that has only recently come to light, then this permit to work system must be clearly explained in the construction health and safety plan.

An aspect of the construction health and safety plan that will need to be constantly kept under review, and updated when necessary, will be the fire safety plan. As construction work progresses, site exit routes may become altered (e.g. by permanent partitioning). Alterations must be clearly depicted on the fire safety plan.

The principal contractor should ensure that, perhaps once a week, time is set aside to review the construction health and safety plan, and any relevant changes that are made must be *communicated* to site operatives by way of the arrangements made for ensuring health and safety issues are considered (e.g. at the weekly site contractors' meeting).

Feedback from site operatives on health and safety matters should be considered, and the construction health and safety plan amended or updated to take into account operatives' concerns, ideas and suggestions as to how the site could be improved from a health and safety point of view.

Generic risk assessments will need to be reviewed and updated to incorporate site-specific issues. These should then be kept in a separate document to the construction health and safety plan, together with any associated method statements. Individual risk assessments can be issued to specific operatives as necessary, or, importantly, to the contractor foreman so that they can assess what safety precautions need to be followed by their team.

Is the fire safety plan a separate document?

No. The fire safety plan can be an integral part of the construction phase health and safety plan, as fire safety matters are itemised in regulations 16–35 of the CDM Regulations.

The fire safety plan should identify fire risks throughout the site, such as

- combustible materials
- the use of hot flame equipment
- the use of liquid petroleum gas
- the use of combustible substances
- storage and use of any explosive materials and substances
- sources of ignition (e.g. smoking)
- the use of heaters.

Once the potential fire risks are identified (i.e. where, when, why and how a fire *could* start on site – or in the surrounding area, yards and outbuildings), the fire safety plan should include precautions and procedures to be adopted to *reduce* the risks of fire. These could include

- operating a hot works permit system
- banning smoking on site
- controlling and authorising the use of combustible materials and substances
- providing non-combustible storage boxes for chemicals
- minimising the use of liquid petroleum gas and designating external storage areas
- controlling the siting and use of heaters and drying equipment
- operating a permit to work system for gas and electrical works.

Having identified the potential risks and the ways to minimise them, there will always be some residual risk of fire. The fire safety plan should then contain the emergency procedures for dealing with an outbreak of fire, namely

- the types and location of fire notices
- the location, number and type of fire extinguishers provided throughout the site
- the means of raising the alarm
- the identification of fire exit routes from the site and surrounding areas
- access routes for emergency services
- the procedure for raising the alarm
- the assembly point/muster point.

The fire safety plan should also contain the procedures to be taken on site to protect against arson, such as

- the erection of high fencing/hoarding to prevent unauthorised entry
- fenced or caged storage areas for all materials, particularly those that are combustible
- site lighting (e.g. infra-red)
- the use of CCTV
- continuous fire checks of the site, particularly at night if site security is used.

Procedures for the storage and disposal of waste need to be included, as waste is one of the main sources of fire on construction sites.

Materials used for the construction of temporary buildings should be fire protected or non-combustible whenever possible (e.g. 30 min fire protection). The siting of temporary buildings must be considered early in the site planning stage, as it is best to locate them at least 10 m away from the building being constructed or renovated.

Having completed the fire safety plan, a sketch plan of the building indicating fire points, the assembly point, fire exit routes, the emergency services access route to site and so on should be completed and attached to the plan. The sketch plan (which could be an architect's outline existing drawing) should be displayed at all fire points and main fire exit routes, and must be included in any site rules/information handed out at induction training.

What criticisms does the HSE have of construction phase health and safety plans?

The HSE has raised many concerns about the quality of construction phase health and safety plans, in particular regarding the general content, which is often not relevant to the project in hand. They would prefer thinner but more site-specific documents.

Some of the common deficiencies are itemised as follows:

- Activities not assessed: that is, those activities with health and safety risks that affect the whole site or specific trades (the storage and distribution of materials, the movement of vehicles, pedestrian access ways, the removal of waste, the provision and use of common mechanical plant, the provision and use of temporary services, commissioning and testing procedures, etc.).
- Management arrangements do not focus sufficiently on the role of risk assessments.
- Site supervisors and managers do not have reasonable knowledge of safety, health and welfare requirements and standards.
- Site supervisors and managers are not familiar with the contents of the construction phase health and safety plan.
- Monitoring arrangements are overlooked or the 'competent' person performing this role is not suitably qualified.
- Details of welfare provision are limited to just a few lines of the plan. Welfare provisions should cover in explicit detail the requirements and implementation of Schedule 2 of the CDM Regulations.
- Fire precautions, including arrangements for the fire alarm system (if required) and emergency lighting, are often overlooked.
- The implication for health and safety of tight timescales for the project are not fully addressed in the plan. The plan often fails to recognise that shortening a construction programme increases the amount of material stored on site and increases the number of operatives on site, both of which lead to restricted work space, inadequate supervision, poor co-ordination and control and so on. This issue should now be clearly set out in the section of the plan that covers 'preparation and planning time'.

The HSE requires that all of the above must be considered before works commence on site, and that if a construction phase health and safety plan does not adequately address them, a client should consider delaying the start of the project until any shortcomings in the plan are rectified.

What other health and safety legislation must the construction phase health and safety plan address?

The construction health and safety plan is a specific requirement of the CDM Regulations, but the Regulations are not the only health and safety legislation applicable to construction projects.

The construction phase health and safety plan must make reference to the monitoring of compliance with the relevant statutory provisions that are applicable to the construction site, namely, but not exhaustively, the following:

- Health and Safety at Work etc. Act 1974
- Health and Safety (First Aid) Regulations 1981
- Control of Asbestos Regulations 2012
- Noise at Work Regulations 2005
- Electricity at Work Regulations 1989
- Manual Handling Operations Regulations 1992
- Personal Protective Equipment Regulations 1992
- Gas Safety (Installation and Use) Regulations 1998
- Reporting of Injuries, Diseases and Dangerous Occurrences Regulations 2013
- Health and Safety (Consultation with Employees) Regulations 1996
- Health and Safety (Safety Signs – Signals) Regulations 1996
- Control of Lead at Work Regulations 2002
- Confined Spaces Regulations 1997
- Provision and Use of Work Equipment Regulations 1998
- The Lifting Operations and Lifting Equipment Regulations 1998
- Management of Health and Safety at Work Regulations 1999
- Control of Substances Hazardous to Health Regulations 2002
- Work at Height Regulations 2005
- Regulatory Reform (Fire Safety) Order 2005.

Not all of the above legislation will apply to a construction site – it depends on the complexity of the project and the type of work to be carried out. The principal contractor must be aware of which legislation is applicable to the works, and must ensure that both their own employees and other contractors are complying with the requirements where necessary.

If the principal contractor is not familiar with the requirements of the legislation, they should seek expert guidance from in-house safety officers or external consultants. The principal designer should be able to give good practical guidance if asked to do so.

The duty to comply with some of the legislation may fall to the principal contractor where the activity will affect all operatives and visitors to site (e.g. conducting a COSHH assessment on dust that affects the whole of the site.

Case study

The principal contractor identified that several trades would be using power tools at the same time, and although individual contractors had provided ear defenders to their own employees, there were other operatives in the vicinity who would be subjected to high noise levels over a prolonged period. The principal contractor carried out a noise assessment of all the tools in operation at any one time, and concluded that the noise level in the site was over the legal limit, and that she must therefore take action to reduce the noise levels. She re-organised the work programme so that only half the number of tools were in use at any one time, thereby reducing the overall noise level to below the legal limit. This action was preferable to issuing all workers on the site with ear defenders: that is, she had controlled the noise at source – step 1 of the hierarchy of risk control.

The principal contractor is responsible for applying the 'principles of prevention'. How should these be addressed and recorded in the construction phase health and safety plan?

The 'principles of prevention' are set out in the Management of Health and Safety at Work Regulations 1999 in regulation 4 and Schedule 1 as follows:

- avoid risks
- evaluate unavoidable risks
- combat risks at source
- adapt work to the individual, especially the design of places of work
- adapt the place of work to technical progress
- replace dangerous articles, substances or systems of work by non-dangerous or less-dangerous articles, substances or systems
- use collective protective measures over individual measures
- develop an adequate prevention policy
- give appropriate training and instruction to employees.

They are also set out in Appendix 1 of the HSE guidance document on the CDM Regulations.

The construction phase health and safety plan should address all the site-specific hazards and risks identified in the pre-construction information and describe how they will be eliminated, reduced or otherwise managed.

Consider the major hazards on a construction site such as working at height, managing asbestos materials, manual handling, slips, trips and falls. How can these hazards be eliminated from the site? If they cannot, how can the hazards be significantly reduced to an acceptable risk factor?

The construction phase plan should contain these procedures and itemise the steps needed or should refer individuals to other documents that contain the information.

A construction phase health and safety plan is needed for smaller domestic projects and for projects where there is only one contractor. Does it have to be in a particular format?

No. The plan can be in any format as long as it is 'drawn up' as far as is practicable before works start on site.

The plan should be simple and appropriate for the complexity of the site, and can be prepared using text, pictures, photographs or diagrams. The purpose of it is to clearly share health and safety information among the workers on the site, and it should reflect the best way to get that information across.

Consider those workers whose first language is not English, and make sure that whatever information is included can be understood. Consider translating key elements of the plan.

The plan needs to be in a format that can be easily updated, which means an electronic version could be sensible – as long as copies are made available to those who will need to have the information.

Devise a template plan and use it for all projects. This will help you, as the principal contractor, to become familiar with the matters to consider on every project and will enable those workers who work regularly for you to recognise key information relating to their safety.

A template for small projects is included in the appendix to this chapter.

The principal contractor, or where one is not appointed, the contractor, must take steps to prevent unauthorised access to the site. What information needs to be included in the construction phase health and safety plan?

Preventing unauthorised access to the construction site is a key duty for the persons charged with being responsible for managing the site. Many accidents occur to people who enter a site and who do not recognise the hazards present.

The plan should set out how the site will be protected by fencing or hoarding, what entry recording system will be in place (e.g. signing in to the site) and what security will be present.

Entry doors/routes must be locked once works have ceased for the day, and the plan should clearly describe what needs to be done and who will do it.

The plan should describe any monitoring checks that should be carried out to ensure that the fencing/hoarding is secure and safe.

A major responsibility of the principal contractor is to communicate with all workers on the project and to ensure that they all co-operate with one another. What should be included in the construction phase health and safety plan?

Successful health and safety management is based on worker engagement and communication, and the HSE has produced tools and guidance on the steps that clients, designers and contractors can take to improve the standard of health and safety and occupational health on sites.

The Leadership and Worker Involvement Toolkit contains seven steps that form the basis of the approach, and the principal contractor could include these in the construction phase plan.

The seven steps are

1. Assess how you're doing.
2. Find the root of the issues.
3. Make it fit with what you do.
4. Lead this in your company.
5. What's in it for your team.
6. How your team can carry it out.
7. Make it last.

The plan should contain information about how workers will be included in site-wide initiatives. The site safety meeting is only one way to engage with workers, and the principal contractor could outline other methods that will adopted on the site.

Examples from the seven steps that can be included in the construction plan are

- *Don't walk by initiative.* It is everyone's responsibility on site to prevent any unsafe acts and conditions that they witness from turning into accidents as soon as they see them. Talk to the person(s) involved and draw their attention to the risks. Encourage everyone to adopt the same approach.
- *Stop.* All workers should be encouraged to stop working whenever they feel unsafe, no matter if their reasons for doing so turn out to be unfounded. Better to stop than to have an accident.
- *A safe working environment drives safe behaviour.* If you expect your workers to work in a safe way, you need to make sure that you do all you can to make the environment they work in as safe as possible. Identify a time when all work can stop for a period of time while the site is cleared and good housekeeping attended to.
- *Don't blame the worker until you have accounted for all causes.* The causes of unsafe ways of working, accidents, incidents and ill health do not always stop with the worker. The problem can often be traced back to less obvious causes such as decisions made by management and the wider organisation. Avoid blaming the worker without having considered the full range of possible causes. Instigate a full 'root cause analysis' review of accidents, incidents and poor practices.
- *Use your workforce for ideas.* Your workers can have a more accurate idea of which efforts to improve health and safety may or may not work than you, your management or other experts. They have to deal with the issues every day. Use and include them. Set up suggestion schemes and mechanisms for workers to raise concerns. Give good feedback on any suggestions.
- *Knowledge is not enough.* Simply telling workers that something is wrong, or is a risk, is not enough. They also need to know why, and how to avoid harm if they are to act on the information that you provide. Provide a resource of information for workers so that they can learn more about safe ways of working. Include information in the plan. Put up a 'signpost' to where workers can obtain information. Share the principal contractor's health and safety officer with sub-contractors so that they can benefit from the wisdom and experience.
- *Encourage co-operation.* Treat your sub-contractors in the same way as direct employees. Encourage different sub-contractors and trades to proactively communicate with each other. Getting consistency in standards will then be that much easier. Include sub-contractors in toolbox talks and other training initiatives that are carried out on site.

(Source: HSE, Leadership and Worker Involvement, Key Principles)

Appendix to Chapter 9

Construction phase health and safety plan – major projects

Project details
Address of premises:
Client:
Architect:
Quantity Surveyor:
Structural Engineer:
Building Services Consultants:
Principal Designer:
Principal Contractor:
Project timescales Start date: Completion: Partial handover dates:
Description of the project
Time allowed for preparation and planning

Maximum number of site operatives envisaged on site
Males: Females:
Health and safety objectives for the project
Organisation and management for health and safety on site
Site Agent:
Contracts Manager:
Health and Safety Manager:
Safety Director:
Health and Safety Consultants:
Attach specific job responsibilities for health and safety
Site-specific conditions that may affect health and safety
Emergency procedures
Evacuation procedure:
Emergency rescue procedures:

Explosion, gas release:
Building/structure collapse:
Major chemical release:
Discovery of asbestos or other prohibited substance:
Scaffold collapse:
Flood:
Fire Plan *A fire plan showing means of escape, fire points and assembly point must be appended*
Method of raising the alarm (audible throughout the site):
Names and number of fire wardens:
Number and location of fire extinguishers:
Means of escape:
Emergency lighting:
Fire signage:
Assembly point:

Fire induction training:
Specific procedures for the storage of highly flammable material:
Smoking policy:
Provision of temporary accommodation:
Fire risk assessments:
Will LPG be used on site? YES/NO How and where will it be stored? Specific control measures:
Site management/co-ordination
How will contractors and sub-contractors be selected and how will their competency and resources be assessed?
Who will undertake risk assessments on site?
Detail procedures for approving sub-contractors' risk assessments and method statements:
How will information in relation to risk assessments and general health and safety issues be communicated to site operatives, sub-contractors and consultants?
What site-specific hazards and risks have been identified and how will they be controlled? *Include risk assessments and method statements as appropriate*
Where will the project notification (form F10) be displayed?

Site-specific control measures for hazards and risks
Are any of the following activities envisaged? If so, permit to work systems must be implemented ■ Confined space working* ■ Hot works* ■ Excavations, cofferdams, etc.* ■ Earth moving* ■ Working at height* ■ Work in risers, lift shafts* ■ Working over water* ■ Working with electricity, gas services* ■ Using mobile elevating platforms* ■ Scaffolding* ■ Working with hazardous substances* ■ Vehicle movements* **Delete as applicable*
Site set up
Welfare facilities: Number and location of WCs: Number and location of urinals: Number and location of wash hand basins? How will washing facilities be provided? How will drinking water be provided? Location of drying room facilities: Location of mess room facilities: Location of canteen facilities (if applicable):
First aid
Names of trained first aiders:

Location of first aid kits:
Location of first aid room (if applicable):
Accident/incident reporting
How will accidents/incidents be notified and by whom?
What records/systems will be kept on site?
Who will investigate site accidents/incidents?
How will site personnel be made aware of site safety issues raised by accident reporting?
Facilities for clients and site visitors
Provision of toilet facilities – what and where (including female facilities)?
Meeting room:
Signing in procedure:
Site safety meetings
How often will safety co-ordination meetings be held and who will attend?
Co-ordination of a multi-occupied site
Will the site be multi-occupied? If so, how will the sites be co-ordinated?

Site access, storage and waste disposal arrangements
Pedestrian operatives/visitors access:
Vehicle access:
Delivery access:
Location of and route to the site office:
Details and location of site storage facilities:
Site waste disposal facilities:
Training
What induction training will be given?
Who will conduct induction training?
Where will it take place?
Where will the training records be kept and by whom?
How often will induction and other training take place?
How will contractors' and sub-contractors' safety training be checked?
What training and communication in respect of site risk assessments will be undertaken and by whom?

Statutory inspections of equipment
What equipment will be inspected, how often and by whom?
Where will records be kept?
Asbestos materials and the Control of Substances Hazardous to Health (COSHH) Regulations 2002
What asbestos materials and hazardous substances are likely to be encountered on site?
Who will carry out asbestos and COSHH assessments?
How will asbestos and COSHH assessments be checked?
How will any residual asbestos materials be managed on the site?
How will use of substances, generation of dust, etc., be controlled on site?
What personal protective equipment (PPE) will be provided for dealing with harmful substances?
Control of noise
When will noise assessments be carried out and by whom?
How will noise be controlled (e.g. at source, PPE)?

Monitoring and auditing site safety
Who will review health and safety?
What procedures will be adopted?
Who will audit records?
Site rules
What site rules are envisaged to deal with the following (the list is not exhaustive)? Identify what activities are relevant to this site. ■ Working at heights ■ Excavations ■ Demolitions ■ Erecting scaffolding, etc. ■ Using hoists, lifts, cranes ■ Maintaining 110 V supply ■ Using portable electrical equipment ■ Manual handling ■ Using welding equipment ■ Hot works permits ■ Permit to work systems ■ Working with polluted water ■ Rodent/insect/pigeon infestations ■ Temporary support works formwork ■ Roof works ■ Removal/installation of glazing ■ Provision and use of work equipment ■ Removal of waste ■ Use of hazardous substances ■ Controlling dust emissions ■ Controlling noise emissions ■ Testing and commissioning equipment *List any specific activities as detailed in the Pre-construction Health and Safety Information Pack/documents*

Health and Safety File
Who will be responsible for collating information for the Health and Safety File and forwarding it to the Principal Designer?

Pre-construction safety information and construction phase health and safety plan – small projects (the 'combi' plan)

Site address:

Description of project:

This Plan is concerned with only the most significant or any unusual hazards that will be encountered during the construction phase of this project.

Approved Contractors are expected to manage commonplace hazards encountered frequently on construction sites by standard methods of control familiar to a competent contractor. The Health and Safety Specification provides details of the minimum standards expected of Approved Contractors as well as general rules and considerations required from all Contractors.

This Plan must be read together with the Health and Safety Specification.

Contact the following person with any queries relating to this document		
Principal Designer:	Email address:	Mobile No.:

The Principal Contractor is to complete the 'PC response' boxes and return the finalised document to the Principal Designer/Client/Client Advisors for review

Principal Contractor	
Person completing the PC response	

Section 1
Summary of significant hazards

Listed below is a summary of the most significant hazards and the most significant issues that apply to this project. These must be addressed by the Principal Contractor.

Hazard	Description	Photo number
[*Hazard or most significant issues*]	[*Principal Designer/Client/Client Advisor comments*	
	PC response [details of how the hazard will be managed]	
[*Hazard or most significant issues*]	[*Principal Designer/Client/Client Advisor comments*]	
	PC response	
[*Hazard or most significant issues*]	[*Principal Designer/Client/Client Advisor comments*]	
	PC response	
[*Hazard or most significant issues*]	[*Principal Designer/Client/Client Advisor comments*]	
	PC response	

<table>
<tr><th colspan="3">Section 2
The project</th></tr>
<tr><th colspan="3">Description of the works</th></tr>
<tr><td>Project description</td><td colspan="2">[Principal Designer/Client/Client Advisor comments. For example, summary of the work proposed, where will works be carried out?]</td></tr>
<tr><td>Trades</td><td colspan="2">PC to confirm trades</td></tr>
<tr><td rowspan="2">Programme for the construction phase</td><td>Start date:</td><td></td></tr>
<tr><td>Completion date:</td><td></td></tr>
<tr><td rowspan="3">Phasing of the works</td><td colspan="2">There will be no phasing of the works</td></tr>
<tr><td colspan="2">The works will be phased as follows [delete if N/A]

[Principal Designer/Client/Client Advisor comments]</td></tr>
<tr><td colspan="2">PC response</td></tr>
<tr><th colspan="3">Existing information and records</th></tr>
<tr><td>Existing Health and Safety File</td><td colspan="2">(a) Is there an existing Health and Safety File?
(b) An existing Health and Safety File is available on the premises</td></tr>
</table>

Asbestos Report	(a) The Asbestos Register is appended [delete if N/A] (b) A refurbishment and demolition survey has been requested (c) The Asbestos Register has been checked, and other information sources (a) There is no asbestos identified on the premises (b) There is asbestos present on the premises as follows: How it will be managed to avoid exposure? How will details be passed on to operatives? What emergency procedures will you have in place? [*Principal Designer/Client/Client Advisor comments*]	
	PC response (management procedure details to consider)	

Section 3 **Client's consideration and management requirements**	
Emergency procedures	The Principal Contractor to develop emergency procedures compatible with those operated by the Client or building operator and to brief the premises manager accordingly
Security arrangements	(a) The Manager will have overall responsibility for security and will secure the site at the end of the working day. The Principal Contractor must, however, prevent unauthorised access into the site (b) The Principal Contractor is responsible for security of the site
	(a) The premises intruder alarm shall be operable outside working/accessible areas (zones) at all times. (b) The premises intruder alarm shall be operable in working/accessible areas (zones) at all times outside working hours
Client's use of the premises during the project	
Trading status during the construction phase	(a) The premises will not trade during the construction phase
	(b) The premises will continue to trade during the construction phase, as follows: [*Principal Designer/Client/Client Advisor Comments*]
	PC response

Occupation by house management/staff	(a) Management/staff will not be in residence during the construction phase
	(b) Management/staff will remain in residence during the construction phase [*Principal Designer/Client/Client Advisor Comments*]
	PC response
Compliance with disability discrimination legislation	Has compliance been considered during construction? Access and egress for disabled persons as well as to suitable toileting facilities need consideration; access to be provided to all facilities that are available to able-bodied customers where possible
	PC response

Section 4 **Information on the existing environment**	
Historical information	
General description of the property, premises or site	[*Principal Designer/Client/Client Advisor comments*]
Local environment	
Type of neighbourhood (e.g. residential) and notable features	[*Principal Designer/Client/Client Advisor comments to all sections*]
Safety-critical facilities such as schools or hospitals	
Local roads and highways	
General parking	
Where are the main controls for services? ■ Electricity ■ Gas ■ Water ■ Telecom	
Other (including residual hazards from previous projects)	[*Principal Designer comments (low ceiling heights, openable cellar hatches, stock areas, etc.)*] PC response

Site restrictions and access	
Access/shared access	The site is accessed as follows:
	(a) There is no shared access
	(b) Access is shared as follows:
	PC response
Manager's access	(a) The Manager has separate access into the accommodation that does not require entry into the trading areas
	(b) The Manager's access is through the trading area or work area
	PC response
Plant rooms	(a) No work is anticipated in the plant rooms
	(b) Work is required in the plant rooms. There are access difficulties as follows: [*Principal Designer/Client/Client Advisor Comments*]
	PC response

Cellar access (for information only unless work to be undertaken in cellar)	Beer deliveries take place on:
	(a) No work is anticipated in the cellar
	(b) Work is required in the cellar. There are access difficulties as follows:
	PC response
Roof access	(a) No work is anticipated on the roof
	(b) Work is required on the roof. There are access difficulties as follows: [*Principal Designer comments – are any additional controls required for any work at height over and above those that a competent contractor should be aware of (e.g. shutting off access/egress route due to siting of tower scaffold, ladder, etc., use of banksman)*]
	PC response
Fire precautions	
Fire exits	(a) Fire exits will not be affected by these works
	(b) Fire exits will be affected by these works as follows:

Fire exits (*cont.*)	PC response
	A fire plan detailing fire access/emergency exits/routes/ fire points and assembly points shall be displayed on site and details shall be included in all site inductions The emergency muster point is:
Fire Risk Assessment	Confirm that the Fire Risk Assessment has been reviewed to take into account the reinvestment works. If yes, attach updated copy. If no, confirm who will complete and when [*Principal Designer comments*]
	PC response
Means of escape during construction	The fire alarm system for the premises shall be kept operational for the duration of the project. An alternative system shall be provided, where necessary, for the construction work area. Whatever system is established, it must be capable of alerting those in other parts of the building of the danger of fire. Similarly, the Contractor must be satisfied that the system for alerting those in occupied areas of the site, also alerts those working on site

Safe storage of plant and materials	[*Principal Designer comments (e.g. where will materials be stored for the duration of the project?)*] All flammable materials shall be kept secure in fireproof containers, away from fences and boundaries and areas within the premises, away from flammable materials/construction
	PC response
Access for emergency vehicles	Access to the site and surrounding roads, properties and businesses shall be kept free and open at all times, to allow free access for emergency vehicles
Health hazards	
Pest infestation	(a) No evidence of pest infestation
	(b) There is pest infestation in this building [*Principal Designer Client/Client Advisor comments*]
	PC response
Contaminated soil	(a) Not relevant in view of the extent of works planned for this project (b) Details of contaminants present on site are provided in (*give details*): [*Principal Designer/Client/Client Advisor comments*]
	PC response

Other	For example, accumulated waste, needlestick injuries, drain seepage [*Principal Designer/Client/Client Advisor comments*]

Section 5 **General construction considerations**	
Welfare facilities available during the project	(a) There are no facilities available on site. The Contractor shall provide adequate toilets, mess rooms and washing facilities for the workforce and visitors to the site, including the provision of hot or warm water, means of heating food, etc. Details shall be identified fully in the Construction Stage Health and Safety Plan (b) Existing facilities on site can be used subject to the agreement of the Premises Manager. It will be the Contractor's responsibility to ensure that the facilities comply with all relevant health and safety regulations and guidelines. All welfare facilities are to be kept in good order at all times
	PC response
Storage of construction materials and plant	Areas for storage of materials and plant shall be agreed with the Premises Manager
	PC response
Waste management	(a) Space is available for skips to be placed in the car park (b) Space is available for skips to be placed in the public highway, providing approval is obtained from the local authority (c) Location of skips will be a problem. Waste may need to be bagged and removed by tipper truck. Rubbish must not be allowed to build up to the extent that it poses a health or safety risk
	PC response
Temporary site accommodation	PC response
Delivery and unloading areas	PC response

Section 6 **Information on significant risks identified during design**	
Protection of the general public/obstructions to public highway	(a) There should be no significant hazards to the general public (b) Work will be required to the external elevation, which will require scaffolding the building or using mobile towers. This area is used by the general public. Adequate control measures must be put in place. Describe the measures below (c) Include appropriate Risk Assessments and Method Statements as well as scaffold design and any other relevant documentation
	PC response
Signage	Describe any difficulties in removing/erecting signage e.g. depressed gables
	PC response
Installation of kitchen equipment	Kitchen equipment needs to be installed in the kitchen, which is located (*ground*): This will involve transferring the equipment through *single/double* width doors The route is *flat/involves steps*
	PC response

Materials requiring particular precautions	
Hazardous materials	The Principal Contractor to carry out a Control of Substances Hazardous to Health (COSHH) Regulations 2002 assessment on all hazardous materials and select the most reasonably practicable, least hazardous material for use. All such materials to be subject to the controls identified by the COSHH assessment
	PC response

Project directory		
Client		
Name	Contact: Mobile No.: Email address:	
Project Development Manager		
Name	Contact: Mobile No.: Email address:	
Project Administrator/Architect		
Name	Contact: Mobile No.: Email address:	
Principal Designer		
Name	Contact: Mobile No.: Email address:	
Quantity Surveyor		
Name	Contact: Mobile No.: Email address:	

Structural Engineer		
Name	Contact: Mobile No.: Email address:	
Building Services Engineer		
Name	Contact: Mobile No.: Email address:	
Principal Contractor		
Name	Contract Manager: Mobile No.: Email address:	
Name	Site Manager: Mobile No.: Email address:	
Name	First-aider: Mobile No.: Email address:	
Designer		
Name	Contact: Mobile No.: Email address:	

CDM 2015 Questions and Answers: A practical approach
ISBN 978-0-7277-6032-6

http://dx.doi.org/10.1680/cdmqa.60326.181

Chapter 10
Contractors

What are the main duties of contractors under the CDM Regulations?

Contractors are given specific duties under regulation 15 of the CDM Regulations as follows:

> (1) A contractor must not carry out construction work in relation to a project unless satisfied that the client is aware of the duties owed by the client under these Regulations.
>
> (2) A contractor must plan, manage and monitor construction work carried out either by the contractor or by workers under the contractor's control, to ensure that, so far as is reasonably practicable, it is carried out without risks to health and safety.
>
> (3) Where there is more than one contractor working on a project, a contractor must comply with—
>
> (a) any directions given by the principal designer or the principal contractor; and
> (b) the parts of the construction phase plan that are relevant to that contractor's work on the project.
>
> (4) If there is only one contractor working on the project, the contractor must take account of the general principles of prevention when—
>
> (a) design, technical and organisational aspects are being decided in order to plan the various items or stages of work which are to take place simultaneously or in succession; and
> (b) estimating the period of time required to complete the work or work stages.
>
> (5) If there is only one contractor working on the project, the contractor must draw up a construction phase plan, or make arrangements for a construction phase plan to be drawn up, as soon as is practicable prior to setting up a construction site.

(6) The construction phase plan must fulfil the requirements of regulation 12(2).

(7) A contractor must not employ or appoint a person to work on a construction site unless that person has, or is in the process of obtaining, the necessary skills, knowledge, training and experience to carry out the tasks allocated to that person in a manner that secures the health and safety of any person working on the construction site.

(8) A contractor must provide each worker under their control with appropriate supervision, instructions and information so that construction work can be carried out, so far as is reasonably practicable, without risks to health and safety.

(9) The information provided must include—

(a) a suitable site induction, where not already provided by the principal contractor;
(b) the procedures to be followed in the event of serious and imminent danger to health and safety;
(c) information on risks to health and safety—
  (i) identified by the risk assessment under regulation 3 of the Management Regulations, or
  (ii) arising out of the conduct of another contractor's undertaking and of which the contractor in control of the worker ought reasonably to be aware; and
(d) any other information necessary to enable the worker to comply with the relevant statutory provisions.

(10) A contractor must not begin work on a construction site unless reasonable steps have been taken to prevent access by unauthorised persons to that site.

(11) A contractor must ensure, so far as is reasonably practicable, that the requirements of Schedule 2 are complied with so far as they affect the contractor or any worker under that contractor's control.

All contractors must have relevant parts of the construction phase health and safety plan that relate to the work they will carry out on the site.

Contractors are 'persons' under the Regulations, and therefore have a duty to ensure that they provide information to others, only appoint persons with the necessary information, instruction and training to carry out tasks, and co-operate and co-ordinate works with all others to whom it may be relevant.

In certain circumstances, contractors may also be designers, and will need to comply with the specific duties placed on designers.

What responsibilities do contractors have for on-site training of their operatives when the project is non-notifiable?

The CDM Regulations make no differentiation between notifiable projects and non-notifiable projects in respect of specific duties regarding training.

Where contractors are employers, they have general duties under health and safety legislation to ensure that their employees receive suitable and sufficient training, or information, instruction and training.

A general duty for training employees is contained in regulation 13 of the Management of Health and Safety at Work Regulations 1999, especially where they are likely to face new or increased risks within the work environment.

CDM regulation 15 contains specific duties for contractors in relation to providing information and training to every worker carrying out construction work under their control, so that the worker can carry out the particular work safely.

Contractors will need to provide any worker on any project with

- induction training
- information on the risks to their health and safety
- information on their risk assessments provided under regulation 3 of the Management of Health and Safety at Work Regulations 1999
- information relating to the activities of any other contractor working on the site that may be relevant to ensuring their health and safety
- any site rules
- emergency procedures, including fire
- names of the people who are responsible for managing the site or carrying out certain tasks.

The above will probably best be covered by a pre-start meeting and formal training session. Simple site rules can be issued and various posters displayed in the mess room, office and so on to advise workers of the safety controls in place on site.

As the contractor has a duty to ensure that every worker is informed of relevant site safety matters, it would be sensible to ensure that a record of attendance is kept for every induction or meeting. Workers should preferably sign their attendance.

The principal contractor has not organised induction training. What should contractors do?

Good health and safety starts with communication and co-operation, so the first step for the contractors to take is to discuss the matter with the principal contractor.

The contractors should review the construction phase health and safety plan to see how the process of induction training was to be implemented, and inform the principal contractor if there seems to be a discrepancy.

Under regulation 14 of the CDM Regulations, the principal contractor must ensure that there is a mechanism for employees and the self-employed at work on the construction site to discuss and offer advice in respect of health and safety.

Failure to have an appropriate induction training programme will affect the health and safety of the site and its operatives. The principal contractor has to *ensure* that contractors have provided their employees with information about risks, control measures and emergency procedures.

If the principal contractor still fails to fulfil their legal duties, the next step is for the contractors to seek advice from the client or principal designer regarding the performance of the principal contractor.

Contractors are entitled to receive parts of the construction phase health and safety plan that directly affect them, and this would include information on welfare facilities, fire precautions, emergency planning, accident management, site security and so on. If this information is made available to the contractors, they should be able to organise their own induction training rather than leave their operatives without it.

In some circumstances, the principal contractor may provide the relevant information to each contractor's site foreman for them to give suitable 'toolbox talks' on site induction information.

Contractors must remember that the duty to ensure that their employees are adequately trained rests with them as employers, and that this duty cannot be delegated to someone else.

The CDM Regulations increase the expectation on all persons, and especially duty holders, to co-operate, co-ordinate and communicate with one another.

The principal designer's role is specifically to help co-ordinate the flow of information across the project, including any information required for the construction phase.

Also, the client, and the principal designer, need to be satisfied that management arrangements are in place for the duration of the project, and it should become very evident to them early on in the project if the principal contractor is failing in their duties.

If an assessment of the construction phase health and safety plan is carried out by the client, the principal designer or health and safety advisors, a key review area should be to ensure that the principal contractor has included details on how they will conduct induction training for all workers on the site.

Do contractors have to provide risk assessments for the principal contractor?

The principal contractor has a duty to co-ordinate health and safety across the construction site and, in particular, must address hazards and risks that affect all operatives on the site, no matter who their employer is.

Under regulation 15 of the CDM Regulations, a contractor must provide relevant information to the principal contractor on any activity that might affect the safety of operatives or others working on or resorting to the site. Relevant information includes any part of a risk assessment made under the general provisions of the Management of Health and Safety at Work Regulations 1999.

Principal contractors should not require risk assessments for tasks that only affect the contractors' operatives, but often, in order to ensure that a culture of health and safety pervades the site, the principal contractor may include a site rule that states that risk assessments for all activities must be provided to them. Contractors have a duty to comply with site rules.

As the only contractor appointed to a project, do we have any specific responsibilities under the CDM Regulations?

Where there is only one contractor appointed to a project, that contractor takes overall responsibility for the health and safety on the project and effectively acts as the principal contractor.

It will be important to establish whether there will be any other contractors appointed, such as self-employed tradesmen, as this could trigger the *more than one contractor* threshold. The definition of a contractor under CDM regulations 2 is

> any person (including a non-domestic client) who, in the course or furtherance of a business, carries out, manages or controls construction work.

A building contractor bringing in sub-contractors to carry out specialist trades would not be classed as a single contractor. There would be multiple contractors, and one will need to be appointed as the principal contractor. This appointment would be made by the client on commercial projects. On domestic projects the appointment of the principal contractor, unless made by the domestic client, would fall by default to the first contractor appointed.

If one contractor is appointed to a project and all the operatives are employees, then that would be classed as one contractor appointment, and health and safety and co-ordination responsibilities would fall to that contractor.

All construction projects will have to have a written construction phase health and safety plan, and all sites must meet the requirements for welfare facilities as laid out in Schedule 2 of the Regulations.

The construction plan must be regularly reviewed and updated during the progress of the works.

The contractor must ensure that the requirements of Part 4 of the Regulations, covering general requirements for all construction sites, are met and that the principles of prevention are also met.

Part 4 covers the following site safety topics:

- safe places of construction work
- good order and site security
- stability of structures
- demolition or dismantling
- explosives
- excavations
- cofferdams and caissons
- reports of inspections
- energy distribution installations
- prevention of drowning
- traffic routes
- vehicles
- prevention of risk from fire, flooding or asphyxiation
- emergency procedures
- emergency routes and exits
- fire detection and fire-fighting
- fresh air
- temperature and weather protection
- lighting.

Not all topics will be relevant. Those that are must be covered in the construction phase plan.

The main duty of the contractor is to plan for health and safety on the project and to put in place procedures to manage any identified risks.

All persons involved with a project have a duty to report matters relating to health and safety. What does this mean?

All persons working on a project under the control of another must report to that person anything that they are aware of in relation to the project that is likely to endanger their own health or safety or that of others.

The requirement is specific that where anyone is seen to be working unsafely the person observing the unsafe work must report it to the person in charge. The person in control must obviously do something to improve the safety of the site or take steps to safeguard the health of workers on the site (e.g. reducing noise levels or controlling dust). The person in control must encourage workers to stop work and report dangerous conditions when they see them.

Turning a blind eye or connivance is an offence under the Health and Safety at Work etc. Act 1974, and the above-described duty under the CDM Regulations strengthens the requirement to take action to improve health and safety matters when they are noted.

All persons are covered by the duty, including any member of a professional teams, other contractors, client representatives and, of course, the duty holders under the Regulations.

Unsafe working practices could include poor working at height, unsafe edge protection, poor housekeeping, unguarded openings, dangerous excavations and so on.

The principal contractor must ensure that a procedure has been put in place to enable full consultation with all workers on the site regarding matters of health and safety, and they must ensure that the process set up on the site has been properly communicated to all workers by way of the construction phase health and safety plan.

Individuals can be prosecuted under both the CDM Regulations and the Health and Safety at Work etc. Act 1974 for failing to raise health and safety issues, and fines could be imposed in the courts.

CDM 2015 Questions and Answers: A practical approach
ISBN 978-0-7277-6032-6

http://dx.doi.org/10.1680/cdmqa.60326.189

ice
Institution of Civil Engineers
publishing

Chapter 11
The health and safety file

What is the health and safety file as required by the CDM Regulations?

A health and safety file is required by the CDM Regulations for all projects for which a principal designer has been appointed: that is, all projects where there is more than one contractor appointed to the project.

The health and safety file provides information that will be needed by anyone who is preparing for construction work or cleaning work on an existing structure, including maintenance, repair, renovation, modification or demolition.

If prepared well, the health and safety file should be an invaluable document for all building owners and/or occupiers. It should contain information about the building that is relevant to health and safety.

There is no specific format to the health and safety file laid down by the Regulations, and it can therefore be in various forms (e.g. text or drawings, on paper or as a computer file).

The Regulations do not specify the contents of the health and safety file except to require that it contains (regulation 12(5))

> information relating to the project which is likely to be needed during any subsequent construction work to ensure the health and safety of any person.

The information to be provided for the health and safety file could include all or some of the following:

- a record of or 'as-built' drawings and plans used and produced throughout the construction process along with design criteria
- general details of the construction methods and materials used
- details of the structure's equipment and maintenance facilities
- maintenance procedures and requirements for the structure

- manuals produced by specialist contractors and suppliers that outline operating and maintenance procedures and schedules for plant and equipment installed as part of the structure
- details and the location of utilities and services, including emergency and fire-fighting systems
- residual hazards and risks within the structure, such as the location of hazardous substances and materials (e.g. asbestos-containing materials).

The information to be contained in the health and safety file should be agreed between the client and the principal designer. The document should, after all, provide information for the building owner, and they will have a valuable input into describing what information they believe will be relevant. Not all clients will know what to include in the health and safety file, and in these instances the principal designer should give advice.

Who prepares the health and safety file and where does the information come from?

The principal designer has to prepare the health and safety file where one does not already exist, or must review and update any existing file. Regulation 12 of the CDM Regulations requires the principal designer to prepare the health and safety file at the pre-construction information stage of the project.

Information for the health and safety file comes from a variety of sources, but key contributors are

- designers – the design risk assessments are important, particularly where they indicate a 'residual risk' associated with the design
- the principal designer – who will have collected important information regarding previous uses of the site and so on (whether contaminated land is present, environmental hazards, etc.)
- the principal contractor – who will have prepared construction sequences for the works and have details of the materials and substances required
- structural engineers – who will have details of load-bearing structures, imposed loadings on floors, guardrails and, and details and the location of foundations
- building services engineers – who will have details of plant and equipment, operating and maintenance manuals, the location of services and so on
- specialist contractors (e.g. architectural glaziers) – who will have details of types of glazing, fitting details, weight loadings, cleaning methods and so on.

It is the responsibility of the principal designer to co-ordinate the information to ensure that it is included in the health and safety file. They do not have to write the information although it would be sensible to compile an introduction and index to the information and detail of where such information is located if not all in one volume.

In many cases it can be extremely practical to have the principal contractor compile a draft health and safety file for the project, as they see most of the information at first hand: for example, amended drawings that show what was actually constructed and where, and operations and maintenance manuals from specialist contractors.

If the principal contractor compiles a draft health and safety file, it must be handed over to the principal designer for checking, revisions and final drafting.

The agreement as to who is responsible for providing information for the health and safety file must be reached at the beginning of the project. Reference should be made in the pre-construction information pack as to who are the responsible persons, what the health and safety file is to contain and how many copies of it there will be, and, importantly, how soon after construction will it be compiled and completed.

What procedures should be followed to put the health and safety file together?

If put together properly, the health and safety file should be one of the most beneficial requirements of the CDM Regulations.

How often have you taken ownership of a building (even a new house) and not had any idea where the main services are located, whether certain materials have been used or how and when to maintain equipment and service plant? A good, detailed health and safety file should include all the information necessary to understand how, what, when, where and why a building is to be used and maintained safely.

A written procedure and checklist is a good starting point for the principal designer to begin compiling the health and safety file. The following steps could be considered:

- Define who will contribute information to the health and safety file.
- Agree with the client what information they want in the file, how they want it compiled and how many copies are to be produced.
- Discuss with the designers key information that will need to be included:
  - design risk assessments
  - details of residual risk
  - specific construction methods
  - structural details (e.g. floor loadings).
- Agree the procedure for advising contractors to compile the file (e.g. the principal designer).
- Discuss with building services consultants/contractors key information in respect of services that will need to be provided (operations and maintenance manuals, how many manuals and what format, drawings, etc.).

- Advise the designers and the principal contractor of additional information that the client has requested.
- Advise the designers and principal contractor of additional information that the client has requested.
- Agree the procedure for site visits to obtain relevant information for the file during the construction phase.
- Obtain a list of all 'client direct' appointments and establish what information they may have that will be relevant (e.g. specialist installers).
- Write to all parties outlining the information required and the timescales and deadlines it must be provided by.
- Issue reminders through the principal contractor and site meeting minutes.
- Discuss with the client formal procedures for withholding payment of accounts if information is not received.
- Visit the site towards the end of the construction phase and conduct a hazard and risk assessment of the premises to identify information that should be included in the file.
- Agree with the client a handover procedure for the file, including a timescale as to when *realistically* it will be available after the construction works are completed.

Who keeps the health and safety file?

The client must retain a copy of the health and safety file. It must be delivered to them by the principal designer at the end of the construction works for each structure.

The client has a duty under regulation 4 of the CDM Regulations to ensure that any information contained in the health and safety file is kept available for inspection by any person who may need information in the health and safety file for the purposes of complying with any statutory provisions.

The client could delegate the responsibility for storing the health and safety file to another person (e.g. the principal designer or project architect), but the statutory duty to ensure that it is available will rest with the client.

Where should the health and safety file be kept and how many copies of it should there be?

The client has to retain a copy of the health and safety file in a format that can be easily used by other persons who may need the information.

The information in the health and safety file relates to health and safety issues, and it is therefore essential that it is readily available, in both location and format, for people using, maintaining or cleaning the building or structure.

A copy of the health and safety file should be kept on the premises to which it relates so that it can be easily referenced by staff and maintenance personnel. However, site copies of documents have a tendency to be mislaid, and if the health and safety file goes missing the client will not be able to fulfil their duty under CDM regulation 4.

A practical solution is to prepare two health and safety files – one a detailed master copy held at the client's head office (e.g. in the property or legal department) and a second summary version within the actual premises that contains essential health and safety information for day-to-day use and maintenance of the building.

The information in the health and safety file must be accessible to anyone who needs it. This will include maintenance contractors. They should be required to consult the health and safety file for methods of access to plant and equipment, maintenance procedures, potential hazards and so on. If a copy is not available for them on site and they undertake a task without being made aware of any residual hazards or procedures to be taken, and subsequently have an accident, the resulting accident investigation could conclude that the accident would not have occurred had they had access to the information, and the client could be charged with contravening the CDM Regulations. The maintenance company could then sue for negligence in the civil courts.

The location, format and numbers of the health and safety file should be discussed and agreed between the principal designer and the client at the outset of the project. The resulting decisions should be recorded in the pre-construction information pack.

With whom does liability for the health and safety file rest?

The client is legally responsible for ensuring that the health and safety file is available to any person who may need the information contained within it in order to comply with relevant statutory provisions.

The client is legally responsible for ensuring that the file is transferred to the structure's new owners on disposal of their interests in the 'property of the structure'.

The client is also legally responsible for ensuring that leaseholders of any parts of the building have access to the information contained within the health and safety file. This may include issuing all leaseholders with a copy, but this is not mandatory.

The principal designer is legally responsible for preparing the health and safety file.

The principal designer is, in addition, legally responsible for ensuring that the health and safety file is reviewed, updated and revised in order to ensure that the information referred to in the CDM Regulations is contained within the file.

The principal designer is also legally responsible for ensuring that, on completion of the construction work on each structure comprised in the project, the health and safety file is delivered to the client.

The duty placed on the principal designer will be 'as far as is reasonably practicable'. It would be unreasonable and impractical to expect a principal designer to be fully conversant with the detailed knowledge necessary to approve specialist contractor information. What would be reasonable is to expect the principal designer to ask the specialist critical questions in respect of health and safety, such as

- Has safe access been designed?
- How is access to be gained?
- Have any fragile materials been used, and if so where?
- What hazardous materials/substances have been used, and where?
- What residual risks remain?
- Have safety notices been displayed?
- Have any hazardous areas been created (e.g. confined spaces)?
- What health and safety management systems are recommended to mitigate residual risk?
- What personal protective equipment is expected to be used?
- How and when is maintenance to be carried out?
- How and when is cleaning to be carried out?

The principal designer should be skilled at picking out unusual and key site-specific health and safety issues and *must* ensure that relevant information is included in the file. A competent health and safety professional does not have to have detailed knowledge of every construction task to be able to identify common hazards and risks and to apply the principles of the hierarchy of risk control to the process.

Once the health and safety file has been completed and handed over to the client, who is responsible for keeping it up to date?

The client retains responsibility for the health and safety file once it is handed over to them.

The file will need to be handed over to future principal designers and designers and any other contractors whenever the client commissions works that fall within the jurisdiction of the CDM Regulations.

It will be the duty of the principal designer, where one is appointed, to keep the file amended, reviewed and added to in respect of all new works. When these works are complete, the file will be handed back to the client in its amended form.

Clients need to be mindful of the fact that health and safety files may not be updated when minor works are carried out, and that they could feasibly give contractors and designers inadequate information: for example, an original health and safety file could include references to encapsulated asbestos material that subsequent works removed. If a contractor were to be provided with this old health and safety file, they may make unnecessary plans to deal with and manage the risk of asbestos on site.

Whenever a health and safety file is provided for a building, the client or subsequent building owner/occupier should develop a robust procedure for ensuring it is kept up to date. This will help reduce site safety hazards caused by lack of knowledge and so on.

There is a growing tendency as buildings change hands for the legal profession to require existing owners to confirm in writing that the health and safety file is available and up to date. If a subsequent owner found it to be deficient, they could instigate civil proceedings. Equally, both freeholders and mortgage lenders (e.g. commercial banks) are making the health and safety file a legal item in conveyancing and lease drafting work.

Having gone to the not inconsiderable expense of having a health and safety file prepared, it seems foolish not to ensure that it is kept up to date when improvement and repair works are undertaken within the building.

What happens if certain information that should be included in the health and safety file is not available?

The information required for the health and safety file should be made clear at the beginning of the project, including who is responsible for providing what.

If all the information necessary to complete the file is not available at the end of the construction works, the principal designer may have to provide an incomplete health and safety file to the client, with details of the outstanding information.

The information required for the file is that which is, 'so far as is reasonably practicable, foreseeable' to ensure the health and safety of any person at work (CDM regulation 11(3)). Certain information may *not* have been reasonably foreseeable during the project and may only be collected retrospectively. This information could reasonably be expected to be excluded from the health and safety file when first compiled, but the principal designer would be expected to 'add to' the file by providing the information as soon as it is available.

Contractors and 'client direct' appointments who fail to provide relevant information as requested should be held in breach of contract, and financial penalties should be imposed (e.g. retention fees increased to 20% of the contract value).

The principal designer should do all that is reasonably practicable to obtain the information, such as contacting contractors directly, and keep records to that effect.

Those who fail to comply with the requirements to provide information should be judged as lacking either competency or the resources to fulfil their obligations under the CDM Regulations. The client should be informed of such instances, and advice given to 'de-list' the contractor. The client could write to the contractor/company advising them that they will be de-listed.

Contractors could be in breach of CDM regulations 8 and 15 if they fail to provide information promptly to the principal contractor. The principal contractor will be in breach of regulation 12 if they fail to provide information promptly to the principal designer.

Case study

The electrical contractor on a project went into liquidation shortly after the completion of the construction works on a new restaurant project, and before the electrical operations and maintenance manual had been completed and handed over to the principal designer.

A major, though not fatal, electrical accident occurred to a staff member, and, following the reporting of the accident, the local environmental health officer conducted an investigation. He asked for electrical test certificates to verify the correct installation of the works under the Electricity at Work Regulations 1989. Those certificates could not be produced. The environmental health officer shut down the entire kitchen operation until such time as he could be satisfied that the electrics were safe. The client commissioned another electrical contractor to undertake a full test and survey of all of the electrics and to produce the appropriate test certificates, together with a suitable operations and maintenance manual. When the subsequent test showed all the electrical wiring installation to be satisfactory, the environmental health officer lifted the prohibition notice, and the kitchen re-opened. The suspect piece of kitchen equipment was seized for examination.

Had the original test certificates been readily available, the client would have been able to demonstrate immediately that the electrical installation met all the safety criteria. As the documents were not available, valuable trade was lost for 3 days – the time taken to organise re-testing and commissioning of all electrical equipment and installations within the premises.

The client and the principal designer reviewed their procedures for the handing over of key information at the completion of works, and agreed that, on future projects, unless a copy of the electrical installation test certificate was provided at the handover meeting or was faxed to the principal designer prior to the handover meeting, project completion would not be achieved and liquidated damages and other contractual financial penalties would be imposed. The principal contractor would be held responsible for co-ordinating the information from the electrical contractor.

Does the construction phase health and safety plan have to be included in the client's health and safety file?

No. The construction phase health and safety plan details what safety precautions are to be adopted during the course of construction. Once the construction work has finished, the plan is, to all intents and purposes, redundant, and provides little useful information to the future use and occupancy of the building.

The construction phase health and safety plan should contain a section on how information will be collected for the health and safety file and who will be responsible for relaying it to the principal designer.

If the health and safety plan identifies, for instance, that an area within the site will be a confined space and includes a system of work (method statement) of how to enter and work in the confined space safely, this information could be passed on to the principal designer for inclusion in the health and safety file because it would have relevance to the safe use of the building/land in the future.

The principal designer should issue guidance to the design team and all contractors, early on in the project, on the type and detail of the information required for the health and safety file.

Are there any specific duties to include information on asbestos in the health and safety file?

The CDM Regulations do not specifically contain duties to include asbestos material information in the health and safety file because other legislation deals with the matter and either the client or the principal designer is expected to know that such duties exist. Nevertheless, information about any residual asbestos materials or any other hazardous materials should be included in the health and safety file.

The information required by regulation 4 in the Control of Asbestos Regulations 2012 relates to the duty to manage asbestos-containing materials. It requires the person who has this duty (i.e. the 'duty holder') to

- take reasonable steps to find out if there are materials containing asbestos in non-domestic premises, and, if so, its amount, where it is and what condition it is in
- presume materials contain asbestos unless there is strong evidence that they do not
- make, and keep up to date, a record of the location and condition of the asbestos-containing materials (or materials that are presumed to contain asbestos)
- assess the risk of anyone being exposed to fibres from the materials identified
- prepare a plan that sets out in detail how the risks from these materials will be managed

- take the necessary steps to put the plan into action
- periodically review and monitor the plan and the arrangements to act on it so that the plan remains relevant and up to date
- provide information on the location and condition of the materials to anyone who is liable to work on or disturb them.

There is also a requirement on others to co-operate as far as is necessary to allow the duty holder to comply with the above requirements.

Principal designers must therefore ensure that they have a good knowledge of asbestos legislation.

The principal designer has finished involvement in the project but construction works have not been completed and the principal contractor is still on site. What happens about the completion of the health and safety file?

If the principal designer's appointment concludes before the end of the project, the principal designer must pass the health and safety file to the principal contractor, and, where this is done, the principal contractor must ensure that the health and safety file is appropriately reviewed, updated and revised from time to time to take account of the work and any changes that have occurred.

Once all the relevant information about residual health and safety risks, safe operating procedures, construction details and so on has been collected and collated, the principal contractor must hand the file over to the client.

The client would be wise to include details of how they expect the health and safety file to be progressed during the project and what steps need to be taken whenever there are duty holder changes, clearly outlining roles and responsibilities, in the client brief or pre-construction information document.

Clients may not be aware of these duties, and all designers and contractors should be able to advise them of the steps they need to take to comply with the law. In particular, the principal designer must advise the client about their duties under the CDM Regulations because they have a specific legal duty to do so.

Appendix to Chapter 11

Contents page and covering note from a health and safety file

Contents

Section A	Introduction Brief description of the project, including Consultants, Contractors, etc.
Section B	Preliminary survey information relevant for future occupancy of the building
Section C	Asbestos materials and any hazardous substances
Section D	Design Risk Assessments/Design Risk Registers
Section E	Construction work Structural work Mechanical and electrical services
Section F	Cleaning work
Section G	Maintenance work
Section H	Hazardous areas/equipment within the building Safety precautions to be followed
Section I	As-built drawings (construction) Schedule of materials used Schedule of operating and maintenance manuals Schedule of equipment with health and safety implications As-installed mechanical and electrical drawings Structural specification Miscellaneous information

Note

The information contained in the Health and Safety File that is kept on the premises must be read in conjunction with the operating and maintenance manuals for mechanical and electrical that are also kept on the premises.

This Health and Safety File is a legal document, and is required by law to be made available to any person who requires information relating to the construction of the building.

It must not be altered or amended without authorisation.

CDM 2015 Questions and Answers: A practical approach
ISBN 978-0-7277-6032-6

http://dx.doi.org/10.1680/cdmqa.60326.201

Chapter 12
Competency and training in health and safety

The 2007 CDM Regulations required competent persons to be appointed to projects. What do the 2015 Regulations require?

The 2015 CDM Regulations require that any person appointing a designer or contractor to work on a project must take reasonable steps to satisfy themselves that those who will carry out the work have the necessary skills, knowledge, experience and, where they are an organisation, the organisational capability to carry out the work in a way that secures health and safety. 'Reasonable steps' will depend on the complexity of the project and the range and nature of the risks involved. This means that a designer or contractor should be capable of understanding how to

- identify the significant risks likely to arise during either the design process or the construction work
- prevent those risks or manage or control them to acceptable levels.

Organisational capability is effectively the policies and systems that an organisation has in place to set acceptable health and safety standards that comply with the law, and the resources and people to ensure that the standards are delivered.

'Skills', 'knowledge' and 'experience' are the three important words that a client, designer or contractors needs to remember when considering an appointment under the 2015 CDM Regulations.

Persons – clients in particular – making any appointments must satisfy themselves that those they intend to appoint have a good understanding of health and safety and that they understand the complexities of the project. People must have experience of similar projects, so, for instance, if the project involves lots of temporary works and demolitions, it is essential that anyone appointed can demonstrate experience of managing such projects: for example, they operate as demolition experts and have qualifications in structural engineering.

What approach should I take when looking to appoint persons to work on my project?

When looking to appoint a designer or a contractor, sensible enquiries should be made about their organisational capability to carry out the work involved. Those making appointments should carry out pre-qualification checks. A good example of what to cover is included in Publicly Available Specification (PAS) 91:2013, 'Construction pre-qualification questionnaires'. PAS 91 sets out standardised pre-qualification questions that include health and safety questions along with other questions on matters such as financial information. This is one way of assessing organisational capability.

In addition to carrying out pre-qualification checks on organisations, those responsible for making appointments should also check that the designer or contractor as an individual has sufficient experience and a good track record in managing the health and safety risks involved in the project. These checks are ideally carried out at the final stage after the pre-qualification checks have been completed and before appointments are made.

When appointing designers, architects and other construction professionals, due consideration should be given to membership of an established professional institution or body: for example, professional bodies that have arrangements in place that provide reassurance that health and safety form part of the route to their membership. However, questions should be asked of individuals to ensure that they have sufficient skills, knowledge and experience to carry out the work involved, and how they keep those capabilities up to date.

What are the requirements regarding competency for those who are seeking appointments?

Designers and contractors (including individuals and sole traders) must be able to demonstrate they have the necessary health and safety skills, knowledge and experience to carry out the work. This is the case for individuals working either for larger organisations or for themselves – in particular, self-employed designers.

Designers or contractors can use the services of an independent (third-party) assessor to assess their organisational capability. If they do so, there are companies that provide pre-qualification assessment services, including those who are members of the Safety Schemes in Procurement (SSIP) Forum. The SSIP Forum is an umbrella body with binding agreements in place to ensure member schemes recognise each other's pre-qualification assessments. SSIP assessment is one way in which a designer or contractor can demonstrate organisational capability at the prequalification stage of the appointment process, but not the only way.

The CDM Regulations do not require any individual or business to use the services of a third party to help them in bidding for work. Rather than use the services of a third party, they have the option to assess their own capability and supply relevant documentation to a client in support of a bid for work.

Any person or organisation will need to keep all their health and safety policies and documents up to date and relevant for the jobs undertaken. They should ensure that training is up to date and that refresher courses are undertaken on a regular basis.

We already have membership of several of the third-party assessment schemes. Do we need to keep our memberships current?

The requirement for demonstrating competency under the 2015 CDM Regulations is not as onerous as that contained in the 2007 Regulations, and the universal acceptance that membership of these third-party schemes is the only way to demonstrate competency no longer applies. Some schemes are expensive to join and require annual re-assessments yet they provide little real evidence that organisations and the individuals within those organisations have the skills and experience necessary to satisfy the Regulations.

Some schemes are better than others and, if real benefits can be demonstrated, then continued membership would be worthwhile, and clients or others would accept membership as evidence of suitable skills and knowledge. However, experience may need to be demonstrated by way of references of previous work and, perhaps, visits to the site to check the standards adopted on the site regarding health and safety.

Some clients may still demand that membership of certain schemes is essential, and apart from having a sensible discussion with them about what is actually needed by the Regulations, you may have no alternative but to join or continue in the scheme.

Joining all the third-party schemes available should not be necessary, and it would be sensible to choose one scheme that is a member of the SSIP and forgo all others.

How can I ensure the competency or experience and knowledge of the site-based workforce?

For basic construction trades, such as bricklayers, carpenters, and painters and decorators, the achievement of a National Vocational Qualification (NVQ) (or SVQ in Scotland) level 2 qualification or higher will ensure a level of competence and skills in respect of health and safety in line with the requirements of the CDM Regulations.

The Construction Industry Training Board (CITB) Health and Safety Test, a computer-based multiple choice test for a basic level of health and safety knowledge, provides a good way of making sure that new entrants have a threshold knowledge of health and safety. This provides a Construction Skills Certificate Scheme (CSCS) card at various levels, from operative through to health and safety manager.

Possession of a CSCS card does at least signify that the holder has passed the CITB Health and Safety Test, or an equivalent basic test of health and safety awareness. However, employers need to be careful that the trade or work occupation on the card matches the work activity to be carried out, and that the actual level of qualification, experience and training of the individual is sufficient. Also, cards can remain valid for 5 years but knowledge can deteriorate over that period of time, so those appointing contractors and workers, and even professionals, need to check that they have had up-to-date refresher training.

Where less experienced staff are engaged to carry out construction work, additional supervision must be provided to make sure that adequate risk control is achieved.

What is an employer responsible for in respect of training employees while they are at work?

The Health and Safety at Work etc. Act 1974 (Section 2) sets out the duties of employers as

> the provision of such information, instruction, training and supervision as is necessary to ensure, so far as is reasonably practicable, the health and safety at work of his employees.

Training should be considered as a risk control measure for hazards identified within the workplace.

The employer can provide their employees with any combination of instruction, information, training and supervision as is appropriate.

The Management of Health and Safety at Work Regulations 1999 (regulation 13) states that an employer must

- provide training upon recruitment and induction
- provide training whenever an employee is exposed to new or altered risks in respect of people, machinery, processes, materials and so on
- provide continuous, repeated training so that employees are given information on current best practice
- provide training in methods that are flexible and adaptable and which meets the needs of special groups of workers (e.g. those with disabilities, literacy problems or language problems)
- provide training in working hours as a business necessity and without charge to their employees.

What are the 'five steps' to information, instruction and training?

The Health and Safety Executive (HSE) publishes many guidance documents on health and safety matters, and in its useful 'five steps' series it covers training as

- Step 1: Determine who needs training.
- Step 2: Decide what training is needed and what the objectives are.
- Step 3: Decide how the training should be given, carried out.
- Step 4: Decide when training should be carried out.
- Step 5: Check that the training has worked.

Generally, all employers need to have a policy on health and safety training: that is,

- what you are going to do
- when you are going to do it
- what subjects will be covered
- who will do it
- how often
- what assessment tests will be made.

Increasingly, training records are vitally important to prove that, as an employer, you have discharged your duties. If an accident occurs, the first documents that the investigating officer will want to see will probably be

- risk assessments
- training records.

Employees who have been able to demonstrate that they have *not* been given adequate training are more likely to be successful in civil claims than those who have been trained.

Case study

A contractor's employee had his hand crushed between the lid of a container and its frame when the lid crashed down while he was using it. The HSE inspector was planning to prosecute for the accident (i.e. an unsafe system of work because the container lid had not been secured in position), but the contractor was able to demonstrate that the employee had been given training and did know how to secure the lid into the upright position. The HSE inspector did not prosecute because the employer had suitable and sufficient risk assessments and had training records that proved that he had implemented a robust toolbox talk training programme for all of his workers.

What are the legal requirements for instruction and training under the various health and safety regulations?

- *Management of Health and Safety at Work Regulations 1999*
  Health and safety training:
  - on recruitment
  - on being exposed to new or increased risks
  - repeated as appropriate.
- *Health and Safety (First Aid) Regulations 1981*
  First aiders provided under the regulations must have received training approved by the HSE.
- *Health and Safety (Safety Signs and Signals) Regulations 1996*
  Each employee must be given instruction and training on
  - the meaning of safety signs
  - measures to be taken in connection with safety signs.
- *Health and Safety (Consultation with Employees) Regulations 1996*
  Training for employee representatives in their functions as representatives (as far as is reasonable). The employer is required to meet the costs of this training, including travel and subsistence and giving time off with pay for training.
- *Safety Representatives and Safety Committees Regulations 1977*
  Sufficient time off with pay for safety representatives to receive adequate training in their functions as a safety representative.
- *Control of Substances Hazardous to Health Regulations 2002*
  Instruction and training in
  - risks created by exposure to substances hazardous to health (e.g. high-hazard biological agents), and precautions
  - the results of any required exposure monitoring
  - the collective results of any required health surveillance.
- *Health and Safety (Display Screen Equipment) Regulations 1992*
  Adequate health and safety training in the use of any workstation to be used.
- *Noise at Work Regulations 2005*
  Instruction and training for employees likely to be exposed to daily personal noise levels at 85 dB(A) or above in
  - noise exposure: the level, the risk of damage to hearing and the action employees can take to minimise that risk
  - personal ear protectors (to be provided by employers): how to get them, where and when they should be worn, how to look after them and how to report defective ear protectors/noise control equipment
  - when to seek medical advice on loss of hearing
  - employees' duties under the regulations.
- *Control of Asbestos Regulations 2012*
  Instruction and training about the risks and precautions for

- employees liable to be exposed to asbestos
- employees who carry out any work connected with the employer's duties under these regulations.

- *Control of Lead at Work Regulations 2002*
  Instruction and training about the risks and precautions for
  - employees liable to be exposed to lead
  - employees who carry out any work connected with the employer's duties under these regulations.
- *Ionising Radiations Regulations 1999*
  Instruction and training to enable employees working with ionising radiation to meet the requirements of the regulations (e.g. in radiation protection for particular groups of employees classified in the regulations).
- *Provision and Use of Work Equipment Regulations 1998*
  Employees who use work equipment (including hand tools) and those who manage or supervise the use of work equipment need health and safety training in
  - methods which must be used
  - any risks from use and precautions.
- *Personal Protective Equipment at Work Regulations 1992*
  Employees who must be provided with personal protective equipment (PPE) need instruction and training in
  - risk(s) that the PPE will avoid or limit
  - the purpose of the PPE and the way it must be used
  - how to keep the PPE in working order and good repair.

What instruction and training needs to be given on fire safety?

Fire training, to the extent that employees should know what action to take when fire alarms sound, should be given to all employees, and should be included in the induction training. Knowledge of particular emergency plans and how to tackle fires with equipment available may be given in specific training at the workplace. At whatever point training is given, the following key points should be covered:

- the evacuation plan for the building in case of fire, including assembly points(s)
- how to use the fire-fighting appliances available
- how to use other protective equipment, including sprinkler and other protection systems, and the need for fire doors to be unobstructed
- how to raise the alarm and operate the alarm system from call points
- workplace smoking rules
- housekeeping practices that could permit a fire to start and spread if not carried out (waste disposal, use of ash bins, handling of flammable liquids, etc.).

Fire training should be accompanied by practices, including regular fire drills and evacuation procedures. No exceptions should be permitted at these.

What specifically must employers provide in the way of information on health and safety issues to employees?

The requirements for information are set out in legislation as follows.

- *Management of Health and Safety at Work Regulations 1999*
  Information on
  - risks to health and safety
  - preventative and protective measures
  - emergency procedures, including evacuation
  - specific health and safety risks for temporary employees
  - requirements for any health surveillance
  - competent persons
  - risks created by other employers.
- *Control of Substances Hazardous to Health Regulations 2002*
  Information on
  - risks to health created by exposure to substances hazardous to health (including, for example, high-hazard biological agents)
  - precautions
  - the results of any required exposure monitoring
  - the collective results of any required health surveillance
  - safety data sheets.
- *Chemicals (Hazards Information and Packaging) Regulations 2002*
  Provision of safety data sheets or the information they contain to be made available to employees (or to their appointed representatives).
- *Manual Handling Operations Regulations 1992*
  Information on
  - the weight of loads for employees undertaking manual handling
  - the heaviest side of any load whose centre of gravity is not positioned centrally.
- *Health and Safety (Display Screen Equipment) Regulations 1992*
  Health and safety information about display screen work for both operators and users (the regulations define who is an operator and who is a user).
- *Health and Safety (First Aid) Regulations 1981*
  First aid arrangements, including facilities, responsible personnel and where first aid equipment is kept.
- *Health and Safety (Safety Signs and Signals) Regulations 1996*
  Information on the measures to be taken in connection with safety signs.
- *Health and Safety Information for Employees Regulations 1989*
  Information about employees' health, safety and welfare in the form of
  - an approved poster to be displayed where it can be easily read as soon as is reasonably practicable after any employees are taken on or

  - an approved leaflet to be given to employees as soon as practicable after they start.
- *Health and Safety (Consultation with Employees) Regulations 1996*
  Information to enable employees to fully take part in consultation and to understand
  - what the likely risks and hazards arising from their work, or changes to their work, might be
  - the measures in place, or to be introduced, to eliminate or reduce them
  - what employees ought to do when encountering risks and hazards.
- *Safety Representatives and Safety Committees Regulations 1977*
  Necessary information to assist the work of safety representatives nominated in writing by a recognised trade union.
- *Ionising Radiations Regulations 1999*
  Information
  - to enable employees working with ionising radiations to meet the requirements of the regulations
  - on health hazards for particular employees classified in the regulations, the precautions to be taken and the importance of complying with medical and technical requirements
  - for female employees on the possible hazard to the unborn child and the importance of telling the employer as soon as they find out they are pregnant.
- *Control of Pesticides Regulations 1986*
  Information on the risks to health from exposure to pesticides and precautions.
- *Provision and Use of Work Equipment Regulations 1998*
  Information on
  - conditions and methods of use of work equipment (including hand tools)
  - foreseeable abnormal situations (what to do and lessons learned from previous experience).
- *Personal Protective Equipment at Work Regulations 1992*
  Information on
  - risk(s) that the PPE will avoid or limit
  - the purpose of the PPE and the way it must be used
  - what the employee needs to do to keep the PPE in working order and good repair.
- *Control of Asbestos Regulations 2012*
  Information about the risks and precautions for
  - employees liable to be exposed to asbestos
  - employees who carry out any work connected with the employer's duties under these regulations.
- *Control of Lead at Work Regulations 2002*
  Information about the risks and precautions for

  - employees liable to be exposed to lead; and
  - employees who carry out any work connected with the employer's duties under these regulations.
- *Noise at Work Regulations 2005*
  Information on
  - the risk of damage to hearing
  - what steps are to be taken to minimise risk
  - steps the employee must take to obtain personal ear protection
  - employees' obligations.

Are toolbox talks a suitable way of training operatives on a construction site?

Yes. Toolbox talks can be very effective in getting key messages across to a range of contractors.

Toolbox talks are short-duration – approximately 5–10 min – presentations that focus on just one topic and update/remind or inform operatives about key health and safety messages that should be practised on a construction site.

Good training is provided in a medium that is understood by the recipients – verbal instructions, posters, quiz games, videos and so on may be much better presentation methods than a long lecture.

Training should be practical and commensurate with the needs of the job: for example, there is no need to train an electrician about the hazards of steel erection if they will never undertake this role, but an electrician may need to know that there might be a hazard to their safety while steel girders are moved around or delivered to the site.

Toolbox talks can be supported by pictorial aids, and this will help overcome any language barriers.

The principal contractor or the contractor should keep a record of all those who attend toolbox talk sessions, the topic and the date of the training.

Records are important as they demonstrate the steps taken by the employer, or other person as appropriate, to comply with their legal duties of 'information, instruction and training'.

ice
Institution of Civil Engineers
publishing

Part 2

Construction safety

CDM 2015 Questions and Answers: A practical approach
ISBN 978-0-7277-6032-6

http://dx.doi.org/10.1680/cdmqa.60326.213

Chapter 13
Health and safety legislation

What legislation applies to the construction industry and its projects?

All legislation that places duties on employers and others to ensure the safety of their workers and those affected by their undertaking is relevant to the construction industry.

The following legislation can be applied to construction projects:

- Health and Safety at Work etc. Act 1974
- Health and Safety (First Aid) Regulations 1981
- Electricity at Work Regulations 1989
- Noise at Work Regulations 2005
- Personal Protective Equipment Regulations 1992
- Manual Handling Operations Regulations 1992
- Reporting of Injuries, Diseases and Dangerous Occurrences Regulations 2013
- Control of Vibration at Work Regulations 2005
- Health and Safety (Safety Signs and Signals) Regulations 1996
- Confined Spaces Regulations 1997
- Provision and Use of Work Equipment Regulations 1998
- Lifting Operations and Lifting Equipment Regulations 1998
- Management of Health and Safety at Work Regulations 1999
- Regulatory Reform (Fire Safety) Order 2005
- Control of Asbestos at Work Regulations 2012
- Control of Lead at Work Regulations 2002
- Control of Substances Hazardous to Health Regulations 2002
- Work at Height Regulations 2005
- Corporate Manslaughter and Corporate Homicide Act 2007.

In addition to the above list, one of the key pieces of legislation for construction safety is the following:

- Construction (Design and Management) Regulations 2015.

What are the main duties covered by the Health and Safety at Work etc. Act 1974?

The Health and Safety at Work etc. Act (HSWA) 1974 is the 'underpinning' legislation that governs virtually all other health and safety law.

The act sets out the general parameters of what is expected of employers and other persons in respect of ensuring their health, safety and welfare.

Regulations are subsidiary legislation made under the enabling powers of the HSWA. Contravening regulations is an offence, and prosecutions can be brought regarding each breach. In addition, there may be a breach of the more general principles of health and safety enshrined in the HSWA, and additional charges could be brought under various sections.

The main sections of the HSWA are summarised as follows.

Section 2: General duty of employers to their employees

It is the duty of every employer to ensure, so far as is reasonably practicable, the health, safety and welfare at work of all their employees. The matters to which that duty extends include

- the provision and maintenance of plant and equipment, and systems of work that are safe and without risks to health
- arrangements for ensuring safety and the absence of risks to health in connection with the use, handling, storage and transport of articles and substances
- the provision of such information, instruction and training as is necessary to ensure the health and safety at work of employees
- the maintenance of any place of work under the employer's control in a condition that is safe and without risk to health, and the provision of means of access and egress that are safe
- the provision and maintenance of a working environment that is safe, without risks to health and adequate with regards to facilities and arrangements for the welfare of employees at work
- the provision of a written statement of the employer's policy in respect of the health and safety of their employees.

Section 3: General duties of employers and the self-employed to persons other than their employees

It is the duty of every employer to conduct their undertaking in such a way as to ensure, so far as is reasonably practicable, that persons not in their employment are not exposed to risks to their health and safety.

Section 4: General duties of persons concerned with premises to persons other than their employees

It is the duty of each person who has control of premises to take such measures to ensure that premises, means of access to and egress from available for use by persons using the premises are safe and without risks to health.

Section 7: General duties of employees at work

It is the duty of every employee while at work

- to take reasonable care for the health and safety of themselves and of other persons who may be affected by their acts or omissions at work
- to co-operate with their employer so far as is necessary so as to enable the employer to undertake statutory duties.

Offences, penalties and prosecutions

Offences

These include the following:

- failing to comply with the general duties on employers, employees, the self-employed, persons in control of premises, manufacturers and so on
- failing to comply with any requirement imposed by regulations made under the HSWA
- obstructing or failing to comply with any requirements imposed by inspectors in the exercise of their powers
- failing to comply with an improvement or prohibition notice
- failing to supply information as required by a notice served by the Health and Safety Executive (HSE) (e.g. investigations into major accidents)
- failing to comply with a court order to remedy the cause of an offence.

Penalties

The Health and Safety (Offences) Act 2008 increased the penalties under various health and safety legislation and in March 2015 section 85 of the Legal Aid, Sentencing and Punishment of Offenders Act 2012 came into force, which increased fines.

Most offences under health and safety law are 'triable either way' (i.e. summarily or on indictment, i.e. either in the magistrates court or the Crown Court).

Breaches of employers' duties under section 2 of the HSWA carry an unlimited fine *per offence* if tried summarily.

Breaches of employers' duties under the numerous regulations enacted under the HSWA carry fines of up to £20 000 per offence if tried summarily. In some circumstances, custodial sentences can be imposed of up to 12 months if the case is heard in the magistrates' court. If the case is heard in the Crown Court, fines can be unlimited and custodial sentences can be up to 2 years.

Breaches of improvement or prohibition notices can now carry an unlimited fine and 6 months' imprisonment if heard in the magistrates' court, or unlimited fines and up to 2 years' imprisonment if heard in the Crown Court.

Prosecutions

OFFENCES BY COMPANIES, CORPORATE BODIES AND DIRECTORS (HSWA, SECTION 37)

The health and safety statutes place duties upon limited companies and/or functional directors.

Where an offence is committed by a body corporate, senior persons in the hierarchy of the company may be *individually liable*.

If the offence was committed with the consent or connivance of or was attributable to any neglect on the part of any of the following persons, that person is guilty of an offence and liable to be prosecuted:

- a functional director
- a manager
- a secretary (company)
- other similar officer of the company
- anyone purporting to act as the above.

The conditions for liability under section 37 are

- Did the person act as the company?
- If they acted in that capacity, did they act with neglect?

Directors, managers and company secretaries are personally liable for ensuring that corporate safety duties are performed throughout the company. They may be able to delegate the specific responsibilities, but that does not absolve them of liability.

OFFENCES DUE TO THE ACT OF 'ANOTHER PERSON'

Section 36 of the HSWA states that where an offence is due to the act or default of another person, then 'that person shall be guilty of the offence'.

A second person (e.g. body corporate) can be charged and convicted whether or not proceedings are taken against the first-mentioned person.

Case studies

Imprisonment for a developer

A developer was sent to prison for 30 months after repeatedly breaching prohibition notices that were put in place to ensure the safety of workers during the development of residential accommodation.

Workers had no protection from falling from height, and debris was falling from the site: members of the public, as well as the operatives, were at risk of injury.

The developer ignored the prohibition notices served by the HSE on more than one occasion, and continued work on site in an unsafe manor.

The case was heard in the Crown Court, and the judge imposed a 30 month custodial sentence and a costs order for over £5000.

Overturning telehandler causes death

A construction worker was using a telehandler to lift a pallet of tiles on to the fourth floor of a building when the telehandler started to topple over due to instability. The operative could not jump clear, and the machine fell on top of him.

The investigation found that the safety of the machine was compromised due to lack of space and the inability of the spreaders to be fully extended. The boom was fully raised but not extended, making the machine unstable, and there was an inadequate turning circle. It was deemed impossible for the machine to be used safely under these conditions: there was no safe system of work for the activity, the vehicle was not suited to the activity and the space available was inadequate.

The construction company was ordered to pay over £615 000 in fines and costs for contravening both section 2 of the HSWA and the Management of Health and Safety at Work Regulations 1999.

What are the main duties contained in the Management of Health and Safety at Work Regulations 1999?

Employers must make suitable and sufficient assessment of the risks to health and safety of their employees and to non-employees affected by their work. Each risk assessment must identify the measures necessary to comply with relevant statutory provisions.

Risk assessments must be in writing where there are more than five employees, and they must be reviewed regularly.

Employers must introduce appropriate arrangements for effective planning, organisation, control, monitoring and review of the preventative and protective measures. These arrangements must be in writing where there are five or more employees.

Where appropriate, employees must have health surveillance if they are exposed to hazards that could affect their health and safety and well-being.

Employers must establish and effect appropriate procedures to deal with emergencies (evacuation, major chemical spillage, explosion, etc.). Employees must stop work immediately and proceed to a place of safety, if exposed to serious imminent and unavoidable danger.

Employers must provide comprehensive and understandable information to all employees on risks identified in the risk assessments, emergency procedures, preventative and protective measures, and competent personnel.

Employees must receive appropriate health and safety training on recruitment and throughout their employment.

Employers must appoint a 'competent person' to assist in undertaking the measures necessary to comply with statutory provisions. Competent persons can be employees or external advisors/consultants (if there are no suitable employees).

Employers must consider the health and safety of young persons at work, pregnant women and nursing mothers.

Contact must be made by employers with external services (e.g. fire-fighting, police and other emergency services) so that any necessary measure can be taken in the event of an emergency or rescue.

Multi-occupied sites must have a plan for co-operation and co-ordination, with one employer taking the lead role in respect of health and safety.

Employers must take into account the capabilities and training of all employees before assigning them tasks and so on.

Employees are under a duty to use equipment, materials and so on provided to them by their employer in accordance with safe systems of work, any training and so on.

Employees must inform their employer of any matter relating to their own or others' health and safety.

Temporary workers are to be afforded health and safety protection, information on hazards risks, health surveillance as appropriate and so on.

What are the key things I need to know about the Health and Safety (First Aid) Regulations 1981?

Every employer must provide equipment and facilities that are adequate and appropriate in the circumstances for administering first aid to employees.

Employers must make an assessment to determine the needs of their workplace. First aid precautions will depend on the type of work, and therefore the risk, being carried out.

Employers should consider the need for first aid rooms, employees working away from the premises, employees of more than one employer working together, and non-employees.

Once an assessment is made, the employer can work out the number of first aid kits necessary by referring to the HSE's guidance document L74, *First Aid at Work: The Health and Safety (First Aid) Regulations 1981* (HSE, 2013).

Employers must ensure that adequate numbers of 'suitable persons' are provided to administer first aid. A 'suitable person' is someone trained in first aid to an appropriate standard.

In appropriate circumstances the employer can appoint an 'appointed person' instead of a first aider. This person will take charge of any situation (e.g. call an ambulance) and should be able to administer emergency first aid.

Employers must inform all employees of their first aid arrangements and identify trained personnel.

What are the main duties in the Noise at Work Regulations 2005?

Employers must carry out noise assessments where employees are likely to be exposed to

- noise above 80 dB – known as the first action level
- noise above 85 dB – known as the second action level

Competent persons must carry out the assessments and identify employees exposed to the noise.

Noise assessments must be reviewed on a regular basis and whenever circumstances change.

Records must be kept of all noise assessments.

Employers have a duty to reduce the noise to which employees are exposed, if it measures 85 dB(A) or above.

Ear protection zones must be designated where appropriate.

Information, instruction and training must be given to all employees.

What are the main provisions of the Electricity at Work Regulations 1989?

The main provisions of the regulation are as follows:

- All systems, plant and equipment to be designed to ensure maximum practical level of safety.
- Installation and maintenance to reflect specific safety requirements.
- Access, light and working space to be adequate.
- Means of cutting off power and isolating equipment to be available.
- Precautions to be taken to prevent charging.
- No live working unless absolutely essential.
- Specific precautions to be taken where live working is essential.
- All persons to be effectively trained and supervised.
- Responsibility for observing safety policy to be clearly defined.
- All equipment and tools to be appropriate for safe working.
- All personnel working on electrical systems to be technically competent and have sufficient experience.
- Work activity must be carried out so as not to give rise to danger.
- Electrical systems must be constructed and maintained to prevent danger. 'Danger' means the risk of injury.

What are the key things I need to know about the Manual Handling Operations Regulations 1992?

Employers must, as far as is reasonably practicable, avoid the need for employees to undertake any manual handling operations at work that involve a risk of injury.

Employers must carry out risk assessments of all manual handling operations that cannot be avoided. The assessments must be in writing.

Employers must take appropriate steps, following assessments, to reduce the risk of injury to the lowest level reasonably practicable.

Employers must take appropriate steps to provide employees undertaking manual handling operations with general indications and/or precise information on the weight and so on of each load and handling activity undertaken.

Employers must review assessments regularly, especially if there is any cause to believe that there has been any significant change in the handling operations.

Employees must make full and proper use of any system of work provided by the employer concerning steps to reduce risks.

Consideration must be given to the health hazards from the repetitive action of handling small loads.

What are the main duties I need to know about in the Personal Protective Equipment Regulations 1992?

Employers must ensure that suitable personal protective equipment (PPE) is provided to employees who may be exposed to risks to their health and safety, except where the risk has been controlled adequately by other means.

PPE must be suitable and appropriate to the risks and workplace conditions. It must suit the worker due to wear it, and afford adequate protection.

If more than one sort of PPE must be worn, they must be compatible.

PPE must be assessed to ensure that it is suitable for the tasks. Any assessments must be reviewed regularly.

PPE must be maintained in an efficient state, in efficient working order and in good repair.

Where PPE is needed, suitable accommodation must be provided for employees to change into any clothing or put on equipment.

Adequate and appropriate instruction and training must be given to employees in the use of PPE.

Employees must use any PPE provided, return it to its accommodation after use, and report loss or defects.

Employers must consider PPE as a 'last resort' and must address all other risk reduction methods.

What are the main duties in the Reporting of Injuries, Diseases and Dangerous Occurrences Regulations 2013?

Where any person dies or suffers any of the injuries or conditions specified in regulations 4, 5, 6 or 7 of the Reporting of Injuries, Diseases and Dangerous Occurrences

Regulations 2013, or where there is a 'dangerous occurrence' as specified in Schedule 2, as a result of work activities, the 'responsible person' must notify the relevant enforcing authority.

Notification must be online or by telephone, and confirmed in writing within 15 days.

Where any person suffers an injury not classed as a specified injury but which results in an absence from work of more than 7 calendar days (i.e. a minor injury), the 'responsible person' must notify the enforcing authority in writing.

The 'responsible person' may be the employer, a self-employed person, someone in control of the premises where work is carried out or someone who provided training for employment.

Where death of any person results within 1 year of any notifiable work accident, the employer must inform the relevant enforcing authority.

When reporting injuries, diseases or dangerous occurrences the approved forms must be used: either F2508 or F2508A.

Records of all injuries, diseases and dangerous occurrences that require reporting must be kept for at least 3 years from the date they were made.

Accidents to members of the public that result in them being taken to hospital as a result of the work activity must be reported.

Incidents of violence to employees that result in injury or absence from work must be reported.

What are the key things I need to know about the Lifting Operations and Lifting Equipment Regulations 1998?

Lifting equipment and operations are covered by the Lifting Operations and Lifting Equipment Regulations 1998, including lifts, hoists, eye bolts, chains and slings.

Lifting equipment must be adequate in strength and stability for each load.

All equipment used for lifting persons must be safe so that they cannot be crushed, trapped, or struck by or fall from the lifting carrier.

Where safety ropes and chains are used, they must have a safety coefficient of at least twice that required for general lifting operations.

Lifting equipment must be installed or positioned in such a way that it reduces the risk of the equipment striking a person or of the load drifting, falling freely or being released unintentionally.

Suitable devices must be available to prevent persons from falling down a shaft or hoist way.

Equipment used for lifting must be marked with safe working loads, including that used for lifting people.

If equipment is not suitable for lifting persons, it must be marked accordingly.

Lifting operations must be properly planned, appropriately supervised and carried out in a safe manner.

Lifting equipment must be regularly inspected and tested, and a report of its condition produced

- before it is used for the first time
- after its assembly and being put into service for the first time at a new site.

Lifting equipment used for lifting *people* must be examined and tested every 6 months.

All other lifting equipment (e.g. goods hoists) must be examined and tested every 12 months, unless a competent person deems them to need inspecting more frequently.

Records must be kept. These can be electronic.

Persons carrying out examination and testing must be competent.

What are the key provisions of the Provision and Use of Work Equipment Regulations 1998?

Employers must ensure that work equipment is suitable by design, construction or adaptation for the work for which it is provided.

Employers must take into account any risks in the location where the equipment will be used (e.g. wet areas and electrical equipment).

Work equipment must be maintained in a suitable condition and in good working repair. Where there is a maintenance log it must be kept up to date.

Equipment will have to be guarded where necessary, be able to be isolated from power sources, be used in the right environment, have display warnings and so on.

Employees must be given adequate health and safety information on the use of equipment, the dangers and so on. They must also be given suitable training.

Persons must be nominated to carry out maintenance and repairs on equipment. They must be suitably trained and competent.

Work equipment must be capable of isolation from any power source.

Maintenance operations are to be carried out when equipment is stopped, unless work can be done without exposure to health and safety risks.

Warning notices must be displayed as necessary, adjacent to equipment.

Work equipment includes installations.

Mobile work equipment is included in the regulations, as are woodworking machines and power presses.

What are the key things I need to know about the Control of Substances Hazardous to Health Regulations 2002?

Employers must carry out assessments of all work activities involving the use of substances hazardous to health that might pose health risks to employees or others.

All assessments must be suitable and sufficient, and must be in writing.

Exposure of employees to hazardous substances must be prevented or otherwise controlled – the hierarchy of risk control must be followed.

PPE is only permitted when other control measures are not practicable.

Control measures must be maintained in efficient working order and good repair. Ventilation equipment must be checked annually.

Monitoring of exposure of employees to substances must be carried out where appropriate.

Employees must be given health surveillance where appropriate, and records kept for 40 years.

All employees exposed to hazardous substances must receive adequate information, instruction and training on the health risks created by their exposure to substances.

Information must be made available from manufacturers. Hazardous substances with an occupational exposure limit or a maximum exposure limit must be used in strict compliance with codes of practice.

What are the main provisions of the Regulatory Reform (Fire Safety) Order 2005?

The Regulatory Reform (Fire Safety) Order 2005 requires all responsible persons to assess fire hazards within their workplace (i.e. produce risk assessments).

The order applies to all employers and persons in control of premises, including principal contractors, clients, landlords and others.

Appropriate controls must be implemented to either eliminate or control the risks.

Suitable provision must be made for fire-fighting equipment.

All fire safety equipment must be maintained and tested regularly.

Emergency plans must be in place for raising the alarm in the event of a fire.

Staff must receive adequate training in fire safety matters.

There must be a fire detection system available that is capable of warning all persons of the risk of fire.

Suitable means of escape in case of fire must be provided from all areas of the building.

The fire risk assessment, procedures and so on must be regularly monitored and updated when circumstances require.

What are the main provisions of the Control of Vibration at Work Regulations 2005?

The Control of Vibration at Work Regulations 2005 require employers to

- assess the vibration risk to employees
- decide if employees are likely to be exposed above the daily exposure value (EAV), and, if they are,
  - to introduce a programme of controls to as low a level as is reasonably practicable

  - to provide health surveillance (regular health checks) to those employees who continue to be regularly exposed above the action value or otherwise continue to be at risk
- decide if employees are likely to be exposed above the daily exposure limit (ELV), and, if they are, to take immediate action to reduce their exposure below the limit value
- provide information and training to employees on health risks and the actions the employer is taking to control those risks
- consult their trade union safety representative or employee representative on the employer's proposals to control risk and to provide health surveillance
- keep a record of their risk assessment and control actions
- keep health records for employees under health surveillance
- review and update their risk assessment regularly.

What are the main provisions of the Confined Spaces Regulations 1997?

The definition of confined space covers, among other things, the following:

- trenches
- vats
- silos
- pits
- chambers
- sewers
- vaults
- wells
- internal rooms.

Employers have a duty to ensure that their employees comply with the Confined Spaces Regulations 1997, and that they, as employers, ensure that employees are not exposed to risks to their health, safety and welfare.

Any environment that could give rise to 'specified risks' can be covered by the regulations. Specified risk includes

- injury from fire or explosion
- loss of consciousness through a rise in body temperature or asphyxiation
- drowning
- injury from free-flowing solids causing asphyxiation
- anything that prevents an escape from space.

Risk assessments are required for work in confined spaces.

Wherever practicable, work in any confined space must be avoided.

Entry into a confined space is prohibited unless suitable rescue arrangements have been put in place.

Workers and rescuers must be trained in the hazards and risks associated with confined spaces.

What are the key things I need to know about the Control of Lead at Work Regulations 2002?

The Control of Lead at Work Regulations 2002 apply to any type of work activity that is liable to expose employees or any other persons to lead.

Lead is any form of lead – including lead alkyls, lead alloys, any compounds of lead and lead as a constituent of any substance – that is liable to be inhaled, ingested or otherwise absorbed by persons. Lead given off in exhaust fumes from road traffic vehicles is excluded.

Employers must

- assess the risk to health from lead
- carry out risk assessments
- identify and implement measures or prevent or adequately control exposure to lead
- record the significant findings of the assessments
- protect employees where exposure to lead is deemed to be significant
- issue employees liable to be exposed to lead with protective clothing
- monitor lead in air concentrations
- place employees liable to be exposed to lead under medical surveillance
- ensure high standards of personal hygiene for employees liable to be exposed to lead
- provide employees liable to be exposed to lead with information, instruction and training
- identify the contents of containers and pipes
- prepare procedures for emergencies, accidents and incidents in connection with the presence of lead.

What are the key things I need to know about the Control of Asbestos Regulations 2012?

Employers are responsible for the health and safety of all persons, whether at work or not.

Employers are not responsible for providing information, instruction and training to persons who are *not* their employees unless those persons are on the premises where the work is to be carried out.

A duty holder must manage any risks from asbestos and must arrange for assessments to be carried out to determine whether asbestos is, or is likely to be, present on the premises.

A duty holder is anyone who has a responsibility or contract for maintenance or repair of the premises, or who is or has control of any part of the premises.

More than one duty holder can exist for a premises, and the level of responsibility will be determined by how much they have to contribute to repairs and maintenance.

Premises have to be inspected to determine whether asbestos is present.

Risk assessments must be made.

A premises register must be developed and maintained as to the whereabouts of asbestos, how it is to be managed and so on. The register must be kept up to date and available to anyone who may need to see it.

No employer is to carry out work on asbestos until its type has been identified.

No employer is to carry out work on asbestos unless a risk assessment has been carried out.

Findings must be recorded.

Asbestos types must be identified.

Control measures need to be assessed.

A suitable written plan must be produced for any work with asbestos. The plan must be kept on the premises for as long as the work continues.

Notification must be made at least 14 days before works are proposed to commence.

Every employer has to ensure that adequate information, instruction and training is provided for employees.

Information on the risk assessment must be given, including risks to health, precautions to be taken, control and action levels and so on.

Information, instruction and training must be given at regular intervals.

Employers must ensure that exposure to asbestos is prevented and, where this is not possible, reduced to the lowest practical level. The number of employees so exposed must be as low as possible.

Substances may be used as a substitution for asbestos if they pose lesser risks.

Any control measures must be used correctly. All control measures must be properly maintained, repaired and so on.

Ventilation equipment must be thoroughly tested and examined.

Suitable records of tests, maintenance and so on must be kept for at least 5 years.

Adequate and suitable protective clothing must be provided for employees working with asbestos. Clothing so exposed to asbestos must be disposed of or cleaned. Clothing must be removed from the site in suitable containers.

If any personal clothing is exposed to asbestos, it must be treated as protective clothing or either disposed of or cleaned.

Emergency procedures must be developed and tested (e.g. safety drills). Information on emergency arrangements must be made available. Suitable warning and communication systems must also be available.

Any unplanned release of asbestos must be dealt with immediately so as to mitigate the effects, and restore the situation to normal. Persons who may be affected must be informed.

Every employer must prevent or reduce to the lowest level reasonably practicable the spread of asbestos from any place where work is carried out under their control.

Premises and plant must be kept clean. Premises, or parts, must be cleaned after works are completed.

All employers must ensure that any area where asbestos work is being undertaken is designated an 'asbestos area', or a 'respirator zone' where the level of asbestos exceeds the control level. Notices designating these areas must be displayed. All employees, other than those needing to work in these areas, must be excluded from the areas. No eating, drinking or smoking must take place in a designated area, and other places for these activities must be available.

Asbestos fibres in the air must be monitored by employers at regular intervals and when changes occur that may affect exposure. Air monitoring is not required if exposure will be below the action level. Records of air monitoring must be kept for 5 years. However, health records must be retained for 40 years for employees exposed to asbestos, unless exposure is less than the action level. Records must be available to those who wish to see them.

A medical examination must be given every 2 years to employees exposed to asbestos.

Adequate washing and changing facilities must be provided to those employees exposed to asbestos.

All asbestos must be stored or transported in sealed containers, properly labelled.

No person must supply products containing asbestos.

An employer may exercise the defence of 'due diligence' or that they took all reasonable precautions to avoid the commission of the offence.

What do I need to know about the Health and Safety (Safety Signs and Signals) Regulations 1996?

The Health and Safety (Safety Signs and Signals) Regulations 1996 require an employer to use safety signs where there is a significant risk to health and safety that has not been avoided or controlled by methods required under other relevant laws, provided the use of a sign can help reduce the risk.

Safety signs are not a substitute for controlling hazards by elimination or substitution, and must be considered as a last resort.

The Health and Safety (Safety Signs and Signals) Regulations 1996 apply to all workplaces and to all activities carried out by employees.

A risk assessment must be undertaken to determine the type and location of any safety signage.

Safety signs must be of the correct type (i.e. prohibition, hazard or mandatory) or indicate a safe place.

Employees and others must have information, instruction and training about safety signs and signals and understand their significance in controlling risks.

Safety signals can be audible alarms, hand signals or voice commands, and are used when signs are not appropriate but hazards need to be conveyed to people in the vicinity.

Employers must consider any language barriers when using signs, and pictogram signage has common understanding across the globe.

What are the key provisions of the Work at Height Regulations 2005?

The principal provision of the Work at Height Regulations 2005 is that work at height must be avoided wherever possible.

The Work at Height Regulations 2005 not only apply to employers but also to 'duty holders'. Duty holders will often be employers, but may also be any person who controls the way work at height is undertaken (e.g. clients commissioning construction work, managing agents and building owners).

Duty holders are required to ensure that

- all work at height is properly planned and organised
- those involved in working at height are competent
- the risks from work at height are assessed, and appropriate work equipment is selected and used
- the risks from fragile materials/surfaces are properly controlled
- equipment for work at height is properly inspected and maintained.

What duties do employers have for the health and safety of non-employees?

Under health and safety legislation, employers are responsible for the health and safety of persons who are not their employees or not in their employ.

The courts consider the employment status of self-employed and casual/contract labour differently to that of the Inland Revenue. A person can be self-employed for tax reasons but employed for health and safety reasons, especially where the employer instructs such persons as part of their business.

Three key legal cases confirm employers' responsibilities for others:

- *R* v. *Associated Octel* (criminal court)
- *Lane* v. *Shire Roofing Company (Oxford) Ltd* (civil court)
- *Nelhams* v. *Sandells Maintenance Limited and Another* (civil court).

Case studies

Associated Octel

Associated Octel operated a chemical plant and instructed a specialist contractor to maintain and repair the plant during its annual shut down. During a repair operation, a light bulb operated by one of the contractor's employees burst, lighting the cleaning fluid he was using and badly burning him.

Both the contractor and Associated Octel (the client) were prosecuted, Associated Octel for failing to ensure the safety of persons other than its employees. Found guilty, Associated Octel appealed to the Court of Appeal. The Court of Appeal agreed with the initial Crown Court decision.

An employer's 'undertaking' can include the activities of a third party over which the employer had no control. The employer must show that it has done enough to protect the third party's safety. The court decided that an employer, by instructing a contractor to perform, would be 'conducting his undertaking' for the purposes of section 3 of the Health and Safety at Work etc. Act 1974, and, as such, has duties to ensure the safety of that contractor's employees.

Shire Roofing Company (Oxford) Ltd

Lane was a self-employed (for tax purposes) roofing contractor who was hired by Shire Roofing to re-roof a porch at a private house. Lane told Shire Roofing that he could do the job for £200 'all in'. To make the job profitable, he needed to use a ladder and not a scaffold or tower scaffold, which was offered by the roofing company. Shire Roofing did not get involved with Lane or the job (did not supervise him, etc.).

During the course of his work, Lane slipped and fell, and suffered severe head injuries.

Lane brought a civil action for damages against Shire Roofing, claiming it had breached a common law duty of care as an employer to ensure his health and safety as an employee.

The court found that Lane was an independent contractor, and was thus not entitled to any common law duty by another.

Lane appealed to the Court of Appeal (civil division). The Court of Appeal reversed the decision, and stated that Lane had been hired as a labourer rather than as an independent contractor, and that his tax status was irrelevant for health and safety purposes.

The Court of Appeal reviewed the 'control' test which the earlier court had applied. The control test obliges the court to ask questions about the relationship between the parties, such as who decides what activities are to be performed and how they are to be carried out, who allocates the resources for the job, who sets the timetable and so on.

The Court of Appeal decided that the first 'control test' was inconclusive, as it had failed to consider the professional skills and discretion that Lane had in making decisions about the work. When determining 'employment', the correct question was not 'who was in control?' but rather 'whose business was it?'. Shire Roofing's supervisor felt that it was his responsibility to provide safety aids, materials and plant and so on to short-term labourers, such as Lane.

The Court of Appeal decided that Shire Roofing had a duty of care to Lane, and awarded him £102 000 damages, plus costs.

Sandells Maintenance Ltd

Nelhams was a permanent employee of Sandells Maintenance. Sandells Maintenance 'loaned' Nelhams to another company – Gillespie – telling him he was under the 'complete control' of Gillespie.

Gillespie asked Nelhams to do some painting. He was told he would have to do the job from a ladder, as scaffolding could not be used. Nelhams asked for the ladder to be footed, but was told that no-one was available. While climbing the ladder, it slipped and Nelhams fell.

Civil claims were brought against Sandells Maintenance and Gillespie by Nelhams. At trial, only Gillespie was found guilty, as Sandells Maintenance had handed over control of Nelhams.

Gillespie appealed to the Court of Appeal. The Court of Appeal found that both employers were equally responsible for Nelhams' safety.

Even though there was an arrangement and 'loan' of an employee from one employer to another, the primary employer remains liable for the safety of their employees, despite the fact that they might loan them to another. As stated by the Court of Appeal, 'The general employer cannot escape liability if the duty has been delegated and then not properly performed.'

Mr Nelhams was awarded damages, and both employers were found responsible. However, the Court of Appeal went on to rule that only Gillespie was wholly responsible for the accident, as it failed to find labour to foot the ladder. Gillespie should therefore pay all the damages and should indemnify Sandells Maintenance completely thereof.

What legal precedence has been set by the Associated Octel, Shire Roofing and Sandells Maintenance cases?
An employer can now be prosecuted for failing to secure the safety of an independent contractor's employees, whether or not they have been involved with that contractor's activities.

An employer can be held liable for the safety of a person holding themselves out to be independent and self-employed if they instruct such persons as part of their business.

A temporary employer can be held financially liable for the injuries suffered by the employee of another.

What do I need to know about the Corporate Manslaughter and Corporate Homicide Act 2007?

Corporate manslaughter is a crime that can be committed by a company in relation to work-related deaths.

The Corporate Manslaughter and Corporate Homicide Act 2007 enables a company to be prosecuted for negligent conduct that leads to a person's death.

Every offence has a legal test

- the accused had a legal duty to care for the deceased
- the accused breached the duty of care
- the breach of duty caused the death and was so severe a breach of duty to be a crime (i.e. gross negligence).

If the breach of duty falls considerably below what would be expected of the organisation, it will be classed as gross negligence.

Consideration is given to

- how serious the failure was
- how real the risk of death was
- the attitudes, policies and accepted practices of the organisation that lead to the failure
- any health and safety guidance relating to the breach.

In broad terms, the prosecution has to be able to prove that there were management failings in respect of health and safety standards.

Penalties are unlimited and must reflect the severity of the case and the affordability of the accused. In addition, publicity orders can be imposed, requiring the company to publicise the case and the actions it has taken to remedy its gross negligence.

What is the HSE's fee for intervention scheme?

The Health and Safety Executive (HSE) introduced the fees for intervention (FFI) scheme on 1 October 2012.

Under the FFI, employers contravening health and safety law must pay the HSE's enforcement costs at the rate of £124 per hour until the breach has been rectified. Invoices are rendered every 2 months.

This scheme only applies to organisations where health and safety regulation is enforced by the HSE, rather than by local authorities.

In October 2010, the Department of Work and Pensions announced that following the comprehensive spending review it was cutting the HSE's grant by 35% over 4 years from April 2011 – roughly equivalent to £80 million. To recoup some of this loss, the government introduced the FFI to recover the costs from businesses failing to comply with health and safety regulation.

The scheme applies to all businesses and organisations inspected by the HSE. Inspections by other regulators, such as local authority environmental health officers, are not affected.

The FFI does not apply to businesses already paying fees to the HSE for their work through other arrangements, such as charges under the Control of Major Accident Hazards Regulations 1999.

The intention is that the HSE recovers the cost of its regulatory work from duty holders found to be in 'material breach' of health and safety law. This will be either from a proactive or a reactive visit where an HSE inspector judges that there has been a material breach serious enough for them to notify the duty holder in writing of that contravention.

What is a 'material breach' of health and safety legislation?

A material breach is defined in HSE guidance as where an HSE inspector is 'of the opinion that there is or has been a contravention of health and safety law that requires him or her to issue notice in writing to that effect'; therefore, it includes a letter, an improvement or prohibition notice or a prosecution. Before deciding to notify the duty holder, an inspector must apply the principles of the HSE's Enforcement Policy

Statement and its Enforcement Management Model and have regard to guidance issued (all of which can be accessed via the HSE website) to ensure the decision on the level of any enforcement action is proportionate in all the circumstances.

Any notification must make clear which contraventions are considered material breaches.

HSE inspectors have invoiced for the time they have spent identifying the breach, advising on putting it right, investigating and taking enforcement action. Included is all the time spent on carrying out visits, all the time on the site during which the material breach was identified, the writing of letters, notices, reports, taking statements and getting specialist reports for complex issues.

Chargeable time runs from the start of the visit when the material breach is identified until it is corrected. Payment is due within 30 days. Non-payment is a civil debt, not a criminal penalty.

There is a dispute procedure in place. If a dispute is raised, the enforcement action taken will be reviewed by a senior manager in the HSE. If the person who raised the dispute is not satisfied with the response, then a further appeal lies with a panel comprising two HSE staff and an independent representative.

If this appeal is unsuccessful, the time spent by the HSE in dealing with the dispute will be charged for at the FFI hourly rate. If the dispute is upheld, the HSE will refund any invoices or part invoices that have been paid.

CDM 2015 Questions and Answers: A practical approach
ISBN 978-0-7277-6032-6

http://dx.doi.org/10.1680/cdmqa.60326.237

Chapter 14
Risk assessments

What is hazard and risk in relation to health and safety?

Hazard means the potential to cause harm.

Risk means the likelihood that harm will occur.

A typical *hazard* on a construction site is falling from heights. The *risk* of falling is dependent on what control measures have been implemented: for example,

- the risk of falling is less if guardrails are provided
- the risk of falling is less if safety harnesses are used
- the risk of falling is less if safe means of access is provided (stairs and not ladders, tower scaffolds and not ladders, etc.)
- the risk of falling is less if materials do not need to be carried
- the risk of falling is less if wind and weather conditions are considered
- the risk of falling is less if adequate working space is provided (e.g. not perched on a ledge).

The *hazard* of falling from heights will be high but the *risk* of falling will be low if guardrails are used.

What are method statements?

Method statements are written procedures that outline how a job is to be done so as to ensure the safety of everyone involved with the job, including persons who are in the vicinity.

A method statement equates to a 'safe system of work', which is required under the Health and Safety at Work etc. Act 1974.

A method statement can be used as the control measure needed to eliminate or minimise the risks involved in carrying out the job.

Who must provide risk assessments?

A principal contractor, where one is appointed under the CDM Regulations, should receive risk assessments from all the other contractors, sub-contractors and self-employed tradesmen working on the site. These will detail what the hazards associated with their tasks are (e.g. noise from drilling equipment) and will include details of how the risks from the hazards (e.g. noise-induced hearing loss) can be eliminated or reduced.

When each of the contractor/sub-contractor risk assessments have been reviewed, the principal contractor must consider whether they need to do anything else to protect other workers in the area (forbid certain work activities in certain areas, etc.). If so, they will need to do an additional risk assessment that identifies how, as the principal contractor, they are going to manage and control the combined risks of several contractors.

Risk assessments need only identify *significant* risks involved in carrying out a work activity. Routine risks and everyday risks such as crossing the road to get to the employee car park do not need to be included.

Where anything unusual or uncommon is to be undertaken on the site, a risk assessment will be essential. Where works involve significant hazards (e.g. working in confined spaces, at heights or with harmful substances), then risk assessments are legally required, and the control measures identified could be incorporated into a method statement, which operatives are required to follow.

Where only one contractor is working on a construction project, there is still a requirement to produce risk assessments and to manage health and safety on the construction site. Any self-employed contractors must manage risks and general health and safety both for themselves and for others who may resort to the site or work area. Paperwork need only be appropriate to the risks identified, and provide added benefit to the management of safety.

What responsibilities do employers have in respect of completing risk assessments?

The Management of Health and Safety at Work Regulations 1999 require employers to complete risk assessments for all activities undertaken at work for which there is a risk of injury or ill health.

The Management of Health and Safety at Work Regulations 1999 place an absolute duty on employers to undertaken suitable and sufficient assessments of

- the risks to health and safety of employees to which they are exposed while at work

- the risks to health and safety of persons not in their employment arising out of or in connection with the conduct by them of their undertaking.

The key details are for employers

- to complete risk assessments
- to ensure that risk assessments are suitable and sufficient.

Any significant findings must be put in writing.

Employees must be given information, instruction and training on the risk assessments so that they understand the hazards and risks of the tasks to be undertaken. Employees must also know what control measures the employer has determined as being appropriate to minimise the risks of injury from the hazards identified.

The law refers to 'suitable and sufficient' risk assessments? What does this mean?

A risk assessment should

- identify the significant risks arising out of the work
- enable the employer to identify and prioritise the measures needed to be taken to ensure that all relevant statutory provisions are complied with
- be appropriate to the nature of the work
- remain in force for the duration of the work
- be regularly reviewed.

A risk assessment does not need to be perfect, but it must be 'the best that can be produced given the knowledge available at the time'. It must be 'proportionate' to the risks associated with the tasks. It must therefore be suitable and sufficient for the hazards and risks that need to be addressed.

Who is best placed to carry out a risk assessment?

There are no legally required qualifications for carrying out risk assessments. The Management of Health and Safety at Work Regulations 1999 requires that whoever carries out a risk assessment must be 'competent' to do so.

Competency is not defined in any of the regulations relating to construction work, but guidance that accompanies various regulations refers to 'appropriate knowledge, training and experience'.

The person carrying out the risk assessment must be able to

- identify hazards
- judge the consequences of the hazards (e.g. likely injury)
- determine how likely it is that the harm will be realised
- identify what control measures are needed to reduce the risks
- determine whether there is imminent risk of serious personal injury and prohibit the continuation of the task.

It would be sensible for the person carrying out the risk assessment to have had some training in the techniques and processes involved in conducting the assessment.

Site agents should be trained in the methodology of risk assessments. Site agents should also enquire about the competency of any sub-contractors who purport to be experienced in carrying out risk assessments.

Under the CDM Regulations, any person who appoints a contractor must ensure that they have suitable information, instruction and training in respect of health and safety and that they have suitable experience and expertise to carry out the task safely.

What are the five steps to risk assessment?

One of the best and most simple of guides to understanding risk assessment is the Health and Safety Executive's (HSE) free leaflet *Five Steps to Risk Assessment* (INDG 163).

This very useful guide lists the five steps as

- Step 1: look for hazards.
- Step 2: decide who might be harmed and how.
- Step 3: evaluate the risks and decide whether existing precautions are adequate or more should be done.
- Step 4: record the findings.
- Step 5: review and revise the assessment if necessary.

Step 1: look for hazards

Walk around the site, and look for anything that could cause harm. Ignore the trivial things, and concentrate on those hazards that could cause serious harm.

Site hazards will include

- trip hazards
- working at heights

- using electrical equipment
- manual handling
- noise
- harmful substances
- unguarded drops and voids
- easy access to site for unauthorised persons
- dust
- low temperatures
- vehicles and mobile plant
- insufficiently trained operatives
- using lifting equipment
- falling objects
- unstable structures
- tower scaffolds
- general scaffolding
- inadequate working platforms
- exposure to live services
- use of welding equipment
- the likelihood of fire.

Step 2: decide who might be harmed and how

Consider anyone who could be exposed to the hazard and who could be harmed as a result. Injury can be minor or major – it really shouldn't matter, as good health and safety is about preventing injury and ill health – no matter how insignificant.

Consider, in particular, people unfamiliar with the site, namely

- visitors
- delivery drivers
- clients
- short-duration visiting contractors.

Also consider

- young workers
- trainees and apprentices
- trespassers on the site
- members of the public
- anyone working on site, or visiting, with a disability
- women who may be pregnant (not an unusual event, as more women are training in the building trades).

Step 3: evaluate the risks and decide whether existing precautions are adequate

Having identified the hazards on site and who might be harmed, review what steps are currently in place to reduce the risk of injury. Usually, something is in place either consciously or not (e.g. 110 V tools are used to reduce the risk of electric shock/electrocution from using 230 V tools).

Are the steps that are in place enough to prevent someone being injured at all?

These steps are called *control measures*.

If more could be done to reduce risk, then the risk assessment process requires additional controls to be implemented when possible. For example, a ladder has been identified as having the potential to cause harm – people could fall off. Their injuries could be severe. The current 'control' is to tie the ladder at the top to stop it moving: the risk of people falling off the ladder due to the ladder moving unexpectedly has been reduced.

However, more could be done, to reduce the risk of falling from ladders altogether by requiring all operatives to use a mobile tower scaffold. This is a safer means of access than a ladder. But mobile tower scaffolds create their own hazards, and so these will need to be controlled too.

Has everything been done, and is being done, that is 'reasonably practicable' to keep the construction site safe? An action list of the controls that could be put in place should be drawn up. Consider the following:

- eliminating the hazard completely
- trying a less risky option
- preventing access to the hazard (e.g. guarding)
- removing people from the hazard
- issuing personal protective equipment
- providing methods with which to deal with the hazard immediately (e.g. washing facilities to reduce skin contamination from substances).

Step 4: record the findings

It is always sensible to keep a record of what steps have been taken to address hazards and risks. If employers have five employees or more, risk assessments must be in writing, although only 'significant' risks have to be recorded.

This means writing down the significant hazards and conclusions for who might be harmed, how and how often, and what control measures are in place, or *are* to be put in place, to control the risks.

These written records are often referred to as 'risk assessments'. They do *not* need to be rocket science, and they do *not* need to be perfect. Risk assessments need be only 'suitable and sufficient'.

You need to be able to show that

- hazards have been reviewed
- what could happen, why, how and when has been thought about
- who could be harmed and why has been considered
- all *known* steps that were practicable, which could be introduced or followed to reduce the hazards, have been considered
- the residual or remaining risk is lower than before the assessment
- people have been informed of the hazards and risks and what safety steps they should follow to reduce the hazards to acceptable levels
- it is recognised that the hazard and risk situation needs to be reviewed from time to time.

If one of the statutory enforcement agencies visits the site and note hazards, they will want to see, almost without fail, the risk assessments.

Risk assessments should be kept until at least the end of the project. If anyone has had an accident of any significance, keep the appropriate risk assessment for at least 6 years – remember that they can launch a civil claim in the courts many years after they had the accident.

It is acceptable to refer to any other health and safety records, procedures, manuals and so on in the risk assessment.

Risk assessments do not have to be in any specific format, but must cover the information required.

Step 5: review and revise the assessment

Site conditions change frequently, and the controls that were in place last week may no longer be appropriate to new working areas, procedures and so on.

Review the risk assessment

- Has anything changed on site?
- Has the weather affected anything?
- Have new operatives started work?
- Have new openings been made or new work areas created?

- Has new plant been brought to site?
- Has the site layout changed?

Check that the precautions introduced to reduce risk are working effectively. If accidents are occurring on the site, it could indicate that the controls are not working. Review and investigate all accidents, and then review the risk assessments.

Case study

A principal contractor was responsible for ensuring the delivery of materials to a site. The delivery area incorporated the rear access road, which was shared by a neighbouring retail premises. There were hazards to both the site operatives and adjoining tenants from the delivery vehicles and the off-loading of materials. Hazards included moving vehicles, restricted access to the roadway for emergency vehicles, off-loading materials from the lorries, dust, noise and falling objects. The risks from the hazards included being knocked over, being hit by materials, noise-induced hearing loss, breathing in dust and exhaust fumes, and so on.

The principal contractor formulated the risk assessment, identifying the above as the hazards and risks, and determining the control measures needed to eliminate or minimise the risks. These included having a banksman to guide in the delivery vehicles, setting specific delivery times, liaising with the adjoining tenants, providing lifting devices, requiring engines to be switched off during delivery, and avoiding reversing vehicles wherever possible.

The principal contractor then prepared a short method statement, which was given to the site foreman to follow when deliveries occurred.

The preparation of this risk assessment and method statement was the principal contractor's responsibility because he had overall management control of these activities and could co-ordinate everyone else's deliveries to the site.

What are 'site-specific' risk assessments?

HSE inspectors prefer to see site-specific risk assessments on every construction site.

There are risk assessments that relate to the actual site conditions and the actual type of construction project and are not general or 'generic' risk assessments that address all sorts of hazards not especially relevant to the site in question.

The value of risk assessments is that they consider the hazards and risks of an activity that employees, or others, are undertaking, and they refer to actual site or workplace conditions that actually *exist*.

Many construction contractors have manuals of 'generic' risk assessments, and often these are referred to as the site risk assessments. They are not.

The 'generic' risk assessments can be used as the basis of the risk assessment, but a competent person must still review the actual site conditions to see if the hazards identified on a risk assessment are less or more likely to cause harm due to the local circumstances. For example, the hazards of moving vehicles on a construction site are quite common and can be generically assessed *but*, on your site, the risks of injury may be increased because of 'blind corners'. The generic risk assessment may not address the hazard of 'blind corners', but the HSE inspector will expect the 'site-specific' risk assessment to do so.

Does one risk assessment satisfy all the regulations?

Risk assessments for various work activities are required under the following regulations:

- Management of Health and Safety at Work Regulations 1999
- Manual Handling Operations Regulations 1992
- Personal Protective Equipment at Work Regulations 1992
- Health and Safety (Display Screen Equipment) Regulations 1992
- Noise at Work Regulations 2005
- Control of Substances Hazardous to Health Regulations 2002
- Control of Lead at Work Regulations 2002
- Control of Asbestos Regulations 2012
- Dangerous Substances and Explosive Atmospheres Regulations 2002.

The requirements of the Management of Health and Safety at Work Regulations 1999 are the overriding superior regulations, and they are superimposed over all the other regulations.

A thorough risk assessment process under the Management of Health and Safety at Work Regulations 1999 will probably satisfy the requirements for all the other regulations, but other regulations may contain specific control measures that will need to be considered: for example, exposure to hazardous substances under the Control of Substances Hazardous to Health Regulations 2002 and whether health surveillance is needed.

When considering the hazards of working in confined spaces, the requirements of the Confined Space Regulations 1997 need to be considered, even though the regulations themselves do not require specific risk assessments.

CDM 2015 Questions and Answers: A practical approach
ISBN 978-0-7277-6032-6

http://dx.doi.org/10.1680/cdmqa.60326.247

Chapter 15
Management of health and safety on construction projects

What aspects of the Management of Health and Safety at Work Regulations 1999 apply to construction sites?

The Management of Health and Safety at Work Regulations (MHSWR) 1999 set out general duties for employers and employees in all non-domestic work activities, and aim to improve health and safety management by developing the general principles set out in the Health and Safety at Work etc. Act 1974.

MHSWR duties overlap with duties contained in several other pieces of health and safety legislation, including the CDM Regulations, by developing the general principles set out in the Health and Safety at Work etc. Act 1974.

Compliance with other legislation normally implies compliance with the MHSWR, but sometimes the duties in the MHSWR go beyond those of other regulations. In these instances, the duties imposed by the MHSWR take precedence over others.

The MHSWR place duties on employers (and the self-employed), including clients, designers, planning supervisors, principal contractors and other contractors.

Under the MHSWR, employers must

- Assess the risks to the health and safety of their employees and others who may be affected by the work activity (regulation 3).
- Identify what actions are necessary to eliminate or reduce the risks to health and safety of their employees and others.
- Apply the principles of prevention and protection.
- Carry out and record in writing, if they have five employees or more, a risk assessment.
- Make appropriate arrangements for managing health and safety, including planning, organisation, control, monitoring and review of preventative and

protective measures. Arrangements must be recorded if employers have five or more employees.

- Provide appropriate health surveillance for employees whenever the risk assessment shows it necessary (e.g. to check for skin dermatitis).
- Appoint competent persons to assist with the measures needed to comply with health and safety laws. Competent persons should ideally be from within the employer's own organisation. Where more than one competent person is appointed, the employer must ensure that adequate co-operation exists between them.
- Set up procedures to deal with emergencies and liaise, if necessary, with medical and rescue services.
- Provide employees with relevant information on health and safety in an understandable form.
- Co-operate with other employees sharing a common workplace, and co-ordinate preventative and protective measures for the benefit of all employees and others.
- Make sure that employees are not given tasks beyond their capabilities and competence.
- Ensure that employees are given suitable training.
- Ensure that any temporary workers are provided with relevant health and safety information in order to carry out their work safely.

Employers have duties under the MHSWR to

- use equipment in accordance with training and instruction
- report dangerous situations
- report any shortcomings in health and safety arrangements
- take reasonable care of their own and other's health and safety.

The principal contractor will carry the bulk of the responsibility for the MHSWR on a construction site, and, as the site will be 'multi-occupied', the principal contractor must ensure co-operation and co-ordination between employers. This will be laid out in the construction phase health and safety plan. Contractors must carry out their own risk assessments, but the principal contractor must complete these where hazards and risks affect the whole workforce (e.g. site access routes and communal lifting operations). As a multi-occupied site, the principal contractor will assume overall responsibility for the management of health and safety, and will co-ordinate and arrange emergency procedures and so on. Information on such procedures must be given to all persons using the site by the principal contractor. Information must be comprehensible and understandable, so may need to be in picture form, posters, foreign languages and so on.

The CDM Regulations strengthen the statutory duties on all persons regarding information, instruction and training, and there is an expectation of co-operation and co-ordination among duty holders. These are the key principles underpinning good health and safety management.

Where no principal contractor is appointed under the CDM Regulations, the contractor appointed will assume broadly the same responsibilities as a principal contractor for managing health and safety on the site. One of the new duties the under the 2015 CDM Regulations is that contractors must produce a construction phase health and safety plan for all projects, irrespective of whether they are notifiable or not. The plan should include the procedures to be followed for monitoring safety on the site.

What are the principles of successful health and safety management?

The key principals to successful health and safety management are exactly the same steps required for the successful management of any project, namely

- Step 1: plan.
- Step 2: do.
- Step 3: do.
- Step 4: check.
- Step 5: act and review.

Step 1: set your own policy/plan for implementation (plan)

All employers who have employees must have a health and safety policy, and those who have five or more employees must have it written down and available for their employees to consult.

All main contractors will have a health and safety policy, and this sets out the framework for the management of health and safety on any site on which their employees are to work.

The site agent, manager or contracts manager should review the company policy, and see whether amendments or additions are necessary for the actual site in question. Does the main policy reflect how you will manage health and safety on site? Have you identified your own safety objectives?

Step 2: organise for health and safety/implement plan (do)

Having identified the overall health and safety policy and objectives for the site, the site and its operatives need to be organised to deliver the policy and objectives. You will need to create a positive health and safety culture for the site – setting standards, enforcing standards, taking a strong lead and so on.

There are four components to a health and safety culture:

1. competence – recruitment, training and advisory support
2. control – allocating responsibilities, securing commitment, instruction and supervision
3. co-operation – between individuals, other contractors and employers
4. communication – spoken, written and visible.

In organising the site health and safety policy and procedures, have you

- Allocated responsibility for health and safety to specific people? Are they clear on what they have to do and are they held accountable?
- Consulted and involved all operatives, contractors, sub-contractors, the self-employed and other employers, trades union representatives and so on ?
- Given everyone sufficient information on the health and safety standards, objectives, hazards and risks of the site?
- The right level of expertise on site to manage all tasks safely and effectively?
- A properly trained workforce and fully inducted operatives?
- Specialist advice available to assist you in managing health and safety?

All of the above should be considered for each construction site before the work actually commences.

Step 3: plan and set standards (do)

When the CDM Regulations apply to a construction project, there must be a construction phase health and safety plan before works start on site. Even if the project is not notifiable under the CDM Regulations, it is still a requirement to formulate a construction phase health and safety plan that is proportionate to the risks associated with the project.

- What objectives do you want for the project?
- Has the client set these? Under the CDM Regulations the client has a greater role in setting health and safety standards for projects. Some such standards could be
  - no major injury accidents
  - lost working days to be no more than 1% of construction days
  - risk assessments to be submitted 24 h before works commence
  - all operatives to be inducted to the site safety standard within the first hour of commencement on day 1
  - no visitors to be let into the construction area unaccompanied.
  - all persons to wear as a minimum
    - a safety helmet
    - safety shoes
    - a high-visibility vest.

Set some achievable objectives. Decide how you are going to monitor that you have achieved them. For example, will you require all accidents to be recorded – or will you only require reportable accidents to be notified?

What mechanism will you introduce so that health and safety can be considered before any major work activity or site alteration.

What plans have you put in place for emergency procedures?

Step 4: measure performance (check)

Consider how you will monitor the performance of health and safety controls and standards

- Who will carry out safety checks – how often?
- Will there be a sub-contractors safety meeting?
- Who will review contractors' risk assessments and method statements?
- How will accident statistics be reviewed?
- Who will worry about near misses?
- Will the number of HSE inspections be recorded? Is there a system?
- What is the lost time rate caused by accidents and ill health?
- Can any costs be quantified for lost time due to work stopping to investigate an accident?
- What are the standards required by the client, and how well do you meet them (e.g. permit to work procedures)?

Step 5: audit and review (act and review)

- How well does the health and safety plan work? Where can it be improved?
- What can be learnt from mistakes and successes?
- If on-site safety checks are carried out, who implements action plans?
- Can performance on this project be reviewed against other projects?
- Can the health and safety compliance be benchmarked against competitors?

Health and safety management is a journey of continuous improvement: never be satisfied with 'making do' – the better the health and safety management, the safer the site and the more efficient and profitable the job. And the greater the contractor's reputation.

Does the principal contractor have to carry out site safety audits?

The principal contractor has a duty to plan, manage and monitor the construction phase in a way that ensures that, so far as is reasonably practicable, it is carried out without risk to the health and safety of persons working on the site or who may be affected by the works.

Also, the principal contractor has a duty to ensure that the principles of prevention are followed.

In order to ensure that the construction phase is monitored, the principal contractor would have to carry out some kind of site safety audit or check, or would need to ensure that others complete the checks.

Many principal contractors will monitor standards themselves by carrying out site safety checks as part of their own management standards, or they will employ third-party safety inspectors/auditors to visit the site and carry out independent checks.

There is no specific legal requirement to carry out checks, but a principal contractor may have difficulty demonstrating 'due diligence' if things go wrong on the site: they will have no records to prove that they have a mechanism in place for monitoring site safety.

An alternative way for the principal contractor to monitor site safety standards is to require all sub-contractors to carry out their own site safety audits and forward checklists to the principal contractor for review, action and filing.

Site safety audit results could be discussed at the site meetings held with all contractors on site, and could be part of the communication process and worker involvement in health and safety.

Does the client have to carry out site safety audits?

No, not specifically. The client has to ensure that management arrangements are in place for health and safety during the course of the project, and must ensure that there is a process for reviewing those arrangements during the project.

Some clients will require the principal designer to monitor and review site safety conditions and issue a report to all relevant parties. In this way, the client will be reassured that the principal contractor is doing what they should do.

Any procedure or record-keeping system that shows that a client is aware of their legal responsibilities and has a system in place to ensure that they are fulfilled will be invaluable should any situation lead to a court appearance or involve enforcement activity.

If the client didn't want to carry out site safety audits themselves or commission others to do so on their behalf, they could ensure that a health and safety item was included on the site minutes and that a regular update of key issues relating to health and safety was issued.

Where only one contractor is appointed to a project, the client will still need to ensure that health and safety is being properly managed and that risks are being addressed. It would be sensible to review health and safety at each site visit and to check that welfare facilities are in place. When the client's duties on domestic projects are delegated to the contractor, the contractor must take on responsibility for managing the safety of their own site.

What are the requirements to manage health and safety under the CDM Regulations?

Every client must take reasonable steps to ensure that the arrangements made for managing the project, including the allocation of sufficient time and resources, by persons with a duty under the CDM Regulations are suitable to ensure that

- construction work can be carried out so far as is reasonably practicable without risks to the health and safety of any person
- the requirements regarding site welfare facilities, as listed in Schedule 2 of the Regulations, are complied with in respect of any person carrying out construction work.

The client must also take reasonable steps to ensure that the arrangements are maintained and reviewed throughout the project.

The client must be satisfied that any contractor appointed to a project has made arrangements for managing the project.

The MHSWR provide a good guide to clients – or anyone else – on the steps that need to be taken to manage health and safety.

The client does not need to be an expert in health and safety nor construction processes but does need to know enough to be aware that a considered approach has been taken.

Minimum management standards would be

- the appointment of a site agent
- the implementation of a construction phase health and safety plan
- the implementation of a programme for appointing competent contractors
- site-specific hazard identification
- carrying out risk assessments and the provision of method statements
- the implementation of a personal protective equipment policy
- the provision of welfare facilities
- the implementation of procedures for communication across all those involved in the project (e.g. regular site progress meetings)

- the implementation of accident management procedures
- first aid provision.

Reasonable steps have to be taken to check that management arrangements are in place, maintained and reviewed.

This does not mean that the client has to go to site every day or week, but it could be that the client instructs their project manager or designer to note health and safety management arrangements at the site meetings.

The principal designer could visit the site and report on the facilities and procedures – such a visit could be combined with obtaining information for the health and safety file.

The CDM Regulations do not define 'reasonable steps', nor does the legal guidance. General interpretation of 'reasonable' is that as long as the arrangements put in place focus on the needs of the particular job and are proportionate to the risks arising from the work, then the arrangements will be reasonable.

Appendix to Chapter 15

Site safety checklist

1. Safe places of work

- Can everyone reach their place of work safely (e.g. are roads, gangways, passageways, passenger hoists, staircases, ladders and scaffolds in good condition)?
- Are there guardrails or equivalent protection to stop falls from open edges on scaffolds, mobile elevating work platforms, buildings, gangways, excavations and so on?
- Are holes and openings securely guardrailed, provided with an equivalent standard of edge protection or provided with fixed, clearly marked covers to prevent falls?
- Are structures stable, adequately braced and not overloaded?
- Are all working areas and walkways level and free from obstructions such as stored material and waste?
- Is the site tidy, and are materials stored safely?
- Are there proper arrangements for collecting and disposing of waste materials?
- Is the work adequately lit? Is sufficient additional lighting provided when work is carried on after dark or inside buildings?

2. Emergency/fire procedures

GENERAL

- Have emergency procedures been developed (e.g. evacuating the site in case of fire or rescue from a confined space)?
- Are people on site aware of the procedures?
- Is there a means of raising the alarm, and does it work?
- Are there adequate escape routes, and are these kept clear?

FIRE

- Is the quantity of flammable material on site kept to a minimum?
- Are there proper storage areas for flammable liquids and gases (e.g. liquid petroleum gas and acetylene)?
- Are containers and cylinders returned to these stores at the end of each shift?
- If liquids are transferred from their original containers, are the new containers suitable for flammable materials?
- Is smoking banned in areas where gases or flammable liquids are stored and used? Are other ignition sources also prohibited?
- Are gas cylinders and associated equipment in good condition?
- When gas cylinders are not in use, are the valves fully closed?
- Are cylinders stored outside?
- Are adequate bins or skips provided for storing waste?

- Is flammable and combustible waste removed regularly?
- Are the right number and type of fire extinguishers available and accessible?

3. Welfare

- Have suitable and sufficient numbers of toilets been provided, and are they kept clean?
- Are there clean wash basins, warm water, soap and towels?
- Is suitable clothing provided for those who have to work in wet, dirty or otherwise adverse conditions?
- Are there facilities for changing, drying and storing clothes?
- Is drinking water provided?
- Is there a site hut or other accommodation where workers can sit, make tea and prepare food?
- Is there adequate first aid provision?
- Are welfare facilities easily and safely accessible to all who need to use them?

4. Protective clothing

- Has adequate personal protective equipment (e.g. hard hats, safety boots, gloves, goggles and dust masks) been provided?
- Is the equipment in good condition and worn by all who need it?

5. Training and induction

- Is there suitable induction training and are records maintained?
- Are the personnel trained/competent?

6. Form F10/health and safety plan

- Is form F10 displayed?
- Is the construction phase safety plan on site?

7. Accidents

- Is the accident book kept on site?
- Are there any reportable accidents?

8. Risk assessments

- Have assessments been carried out for hazardous activities?
- Where appropriate, are records being kept?

9. Method statements

- Are method statements in place for hazardous or unusual activities?
- Are the method statements being followed?

10. Hazardous substances

- Have all harmful materials (e.g. asbestos, lead, solvents and paints) been identified?
- Have the risks to everyone who might be exposed to these substances been assessed?
- Have precautions been identified and put in place (e.g. protective equipment provided and used, and workers and others who are not protected kept away from exposure)?

11. Plant and machinery

- Is the right plant and machinery being used for the job?
- Are all dangerous parts guarded (e.g. exposed gears, chain drives and projecting engine shafts)?
- Are guards secured and in good repair?
- Is the machinery maintained in good repair, and are all safety devices operating correctly?
- Are all operators trained and competent?

12. Lighting

- Is there sufficient light in all working areas?
- Are hazardous areas/escape routes suitably lit?

13. Electricity

- Is the supply voltage for tools and equipment the lowest necessary for the job (e.g. battery-operated tools and reduced-voltage systems (110 V, or even lower in wet conditions))?
- Where mains voltage has to be used, are trip devices (e.g. residual-current devices (RCDs)) provided for all equipment?
- Are RCDs protected from damage, dust and dampness and checked daily by users?
- Are cables and leads protected from damage by sheathing, protective enclosures or by positioning away from causes of damage?
- Are all connections to the system properly made, and are suitable plugs used?
- Is there an appropriate system of user checks, formal visual examinations by site managers, and combined inspection and test by competent persons for all tools and equipment?
- Are scaffolders, roofers or others or cranes or other plant working near or under overhead lines? Has the electricity supply been turned off, or have other precautions, such as 'goal posts' or taped markers, been provided to prevent them contacting the lines?
- Have underground electricity cables been located (with a cable locator and cable plans), marked, and precautions for safe digging been taken?

14. Scaffolds

- Are scaffolds erected, altered and dismantled by competent persons?
- Is there safe access to the scaffold platform?
- Are all uprights provided with base plates (and, where necessary, timber sole plates) or prevented in some other way from slipping or sinking?
- Are all the uprights, ledgers, braces and struts in position?
- Is the scaffold secured to the building or structure in enough places to prevent collapse?
- Are there adequate guardrails and toe boards or an equivalent standard of protection at every edge from which a person could fall 2 m or more?
- Where guardrails and toe boards or similar are used
  - Are the toe boards at least 150 mm in height?
  - Is the upper guardrail positioned at a height of at least 910 mm above the work area?
  - Are additional precautions (e.g. intermediate guardrails or brick guards) in place to ensure that there is no unprotected gap of more than 470 mm between the toe board and the upper guardrail?
- Are the working platforms fully boarded, and are the boards arranged to avoid tipping or tripping?
- Are there effective barriers or warning notices in place to stop people using an incomplete scaffold (e.g. where working platforms are not fully boarded)?
- Has the scaffold been designed and constructed to cope with the materials stored on it, and are these distributed evenly?
- Does a competent person inspect the scaffold regularly (e.g. at least once a week), and always after it has been substantially altered, damaged or following extreme weather?
- Are the results of inspections recorded?

15. Demolitions

- Have risk assessments been carried out, and are method statements available?
- Are there procedures in place for restricted access and safe distances?
- Are noise/dust controls in place?
- Is personal protective equipment available?

16. Cranes and lifting appliances

- Is the crane on a firm, level base?
- Are the safe working loads and corresponding radii known and considered before any lifting begins?
- If the crane has a capacity of more than 1 tonne, does it have an automatic safe load indicator that is maintained and inspected weekly?
- Are all operators trained and competent?

- Has the banksman/slinger been trained to give signals and to attach loads correctly?
- Do the operator and banksman find out the weight and centre of gravity of a load before trying to lift it?
- Are cranes inspected weekly, and thoroughly examined every 14 months by a competent person?
- Are the results of inspections and examinations recorded?
- Does the crane have a current test certificate?

17. Excavations

- Is an adequate supply of timber, trench sheets, props or other supporting material made available before excavation work begins?
- Is this material strong enough to support the sides?
- Is a safe method used for putting in the support (i.e. one that does not rely on people working within an unsupported trench)?
- If the sides of the excavation are sloped back or battered, is the angle of batter sufficient to prevent collapse?
- Is there safe access to the excavation (e.g. by a sufficiently long, secured ladder)?
- Are there guardrails or other equivalent protection to stop people falling in?
- Are properly secured stop blocks provided to prevent tipping vehicles falling in?
- Does the excavation affect the stability of neighbouring structures?
- Are materials, spoil or plant stored away from the edge of the excavation in order to reduce the likelihood of a collapse of the side?
- Is the excavation inspected by a competent person at the start of every shift, and after any accidental collapse or event likely to have affected its stability?

18. Roof work

- Are there enough barriers and is there other edge protection to stop people or materials falling from roofs?
- Do the roof battens provide safe hand and foot holds? If not, are crawling ladders or boards provided and used?
- During industrial roofing, are precautions taken to stop people falling from the leading edge of the roof or from fragile or partially fixed sheets that could give way?
- Are suitable barriers, guard rails or covers and so on provided where people pass or work near fragile material such as asbestos cement sheets and roof lights?
- Are crawling boards provided where work on fragile materials cannot be avoided?
- Are people excluded from the area below the roof work? If this is not possible, have additional precautions been taken to stop debris falling on to them?

19. Powered access equipment

- Has the equipment been erected by a competent person?
- Is fixed equipment (e.g. mast climbers) rigidly connected to the structure against which it is operating?
- Does the working platform have adequate guardrails and toe boards or other barriers to prevent people and materials falling off?
- Have precautions been taken to prevent people being struck by the moving platform, projections from the building or falling materials (e.g. a barrier or fence around the base)?
- Are the operators trained and competent?
- Is the power supply isolated and the equipment secured at the end of the working day?

20. Ladders

- Are ladders the right means of access for the job?
- Are all ladders in good condition?
- Are they secured to prevent them slipping sideways or outwards?
- Do ladders rise a sufficient height above their landing place? If not, are there other hand-holds available?
- Are the ladders positioned so that users don't have to overstretch or climb over obstacles to work?
- Does the ladder rest against a solid surface and not on fragile or insecure materials?

21. Manual handling

- Has the risk of manual handling injuries been assessed?
- Are hoists, telehandlers, wheelbarrows and other plant or equipment used so that manual lifting and handling of heavy objects is kept to a minimum?
- Are materials such as cement ordered in 25 kg bags?
- Can the handling of heavy blocks be avoided?

22. Hoists

- Is the hoist protected by a substantial enclosure to prevent someone from being struck by any moving part of the hoist or falling down the hoistway?
- Are gates provided at all landings, including at ground level?
- Are gates kept shut except when the platform is at the landing?
- Are the controls arranged so that the hoist can be operated from one position only?
- Is the hoist operator trained and competent?
- Is the hoist's safe-working load clearly marked?
- If the hoist is for materials only, is there a warning notice on the platform or cage to stop people riding on it?

- Is the hoist inspected weekly, and thoroughly examined every 6 months, by a competent person?
- Are the results of inspection recorded?

23. Noise and dust

- Are breakers and other plant or machinery fitted with silencers?
- Are barriers erected to reduce the spread of noise?
- Is work sequenced to minimise the number of people exposed to noise?
- Are others not involved in the work kept away?
- Is suitable hearing protection provided and worn in noisy areas?
- Is dust being contained/kept to a minimum?

24. Protection of the public

- Is the public fenced off or otherwise protected from the work?
- When work has stopped for the day
  - Are the gates secured?
  - Is the perimeter fencing secure and undamaged?
  - Are all ladders removed or their rungs boarded so that they cannot be used?
  - Are excavations and openings securely covered or fenced off?
  - Is all plant immobilised to prevent unauthorised use?
  - Are bricks and materials safely stacked?
  - Are flammable or dangerous substances locked away in secure storage places?

25. Hazardous areas

- Are hazardous areas suitably signed?
- Is there restricted access to hazardous areas?
- Are hazardous areas protected/guarded?

26. Confined spaces

- Have any confined spaces been identified?
- Is a permit to work system in operation?

27. Hot works

- Are there fire precautions for hot works?
- Is a permit to work system in operation?

28. Traffic and vehicles

- Have separate pedestrian, vehicle access points and routes around the site been provided? If not, are vehicles and pedestrians kept separate wherever possible?
- Have one-way systems or turning points been provided to minimise the need for reversing?

- Where vehicles have to reverse, are they controlled by properly trained banksmen?
- Are vehicles maintained (e.g. the steering, handbrake and footbrake all work properly)?
- Have drivers received proper training?
- Are vehicles securely loaded?
- Are passengers prevented from riding in dangerous positions?

CDM 2015 Questions and Answers: A practical approach
ISBN 978-0-7277-6032-6

http://dx.doi.org/10.1680/cdmqa.60326.263

Chapter 16
Accident and incident management

What are the requirements of the Reporting of Injuries Diseases and Dangerous Occurrence Regulations 2013?

Where any person dies or suffers any of the injuries or conditions specified in regulations 4, 5, 6 or 7 of the Reporting of Injuries Diseases and Dangerous Occurrence Regulations (RIDDOR) 2013, or where there is a 'dangerous occurrence' as specified in Schedule 2, as a result of work activities, the 'responsible person' must notify the relevant enforcing authority.

Notification must be online by way of the Health and Safety Executive's (HSE's) accident reporting website, and confirmed in writing within 15 days.

Where any person suffers an injury not specified in the RIDDOR but which results in an absence from work of more than 7 calendar days, the 'responsible person' must notify the enforcing authority in writing, using the approved form F2508 and reporting the absence online.

The 'responsible person' may be the employer, a self-employed person, someone in control of the premises where work is carried out or someone who provided training for employment.

Where death of any person results within 1 year of any notifiable work accident, the employer must inform the relevant enforcing authority.

When reporting injuries, diseases (e.g. industrial diseases – those contracted as a result of work undertaken, such as Weil's disease or miner's lung) or dangerous occurrences, the approved forms must be used, either F2508 or F2508a.

Records of all injuries, diseases and occurrences that require reporting must be kept for at least 3 years from the date they were made.

Accidents to members of the public that result in them being taken to hospital as a result of the work activity must be reported.

Incidents of violence to employees that result in injury or absence from work must be reported.

Why must these type of accidents be reported?

National accident statistics are collated by the HSE in order to indicate the general state of health and safety across Britain. Fatalities, specified injuries and 'over 7 day' injuries are all recorded and allocated to industry-specific sectors so that the state of legal compliance, accident trends and so on can be judged.

However, the most important reasons for notifying accidents are

- it is a legal requirement
- so the enforcing authorities can investigate to establish whether the employer has contravened the law
- so serious incidents can be prevented from happening again.

Accident statistics for the construction industry for 2013–2014 in Britain are

- 42 fatalities in the construction sector
- 1900 major/specified injuries to construction workers
- 3293 'over 7 day' injuries to construction workers
- 4 fatalities to members of the public caused by construction activity.

If an accident is reported to the enforcing authority, will an investigation take place?

Not always. It depends on the severity of the accident and the approach of the enforcing authorities, in particular the HSE. Increasingly, the HSE is using accident notifications as a reason to carry out site inspections: this often allows them to invoke the fee for intervention scheme, as there is a high probability that they will encounter a material breach of health and safety regulations on the site.

Any accident that involves a specified injury is highly likely to be investigated, as it shows to the authority that something serious may have gone wrong with the employer's safety management system.

Sometimes, an enforcing authority will make a telephone investigation first and request further details of management systems so that it can assess the general attitude of the employer and their commitment to health and safety. If it finds information inadequate, it will make a site visit.

What are the consequences if I ignore the law on reporting accidents?

A failure to notify accidents, diseases and dangerous occurrences is an offence under the RIDDOR, and the number of prosecutions for non-compliance is rising. Fines are unlimited in the magistrates' court as well as in the Crown Court.

What are the type of accidents that have to be notified?

It is not actually accidents that have to be notified but the consequences of those accidents and the type of injuries that they cause.

Accidents and incidents that arise out of or in connection with work and that fall into the categories of

- fatality
- specified injury
- over 7 day injury

must be reported.

Also, any accident or incident that involves a member of the public or non-employee being sent to hospital needs to be reported. This is so that information can be gathered on how safe work practices are for members of the public using premises and so on.

Certain types of 'dangerous occurrence' must also be reported. These would be incidents that have the potential to cause specified or multiple injuries and that could affect large numbers of people (i.e. high-risk catastrophes).

Industrial diseases must also be reported within 12 months of the disease being identified.

What are specified injuries?

A specified injury must be reported to the enforcing authority. Any accident at work or caused by a work activity that results in the following is notifiable:

- any fracture of a bone (other than a finger, thumb or toes)
- any amputation
- dislocation of the shoulder, hip, knee or spine
- loss of sight (whether temporary or permanent)
- a chemical or hot-metal burn to the eye or any penetrating injury to the eye
- any injury resulting from an electric shock or electrical burn (including one caused by arcing) leading to unconsciousness or requiring resuscitation or admittance to hospital for more than 24 h

- injuries leading to hypothermia, heat-induced illness or unconsciousness, or requiring resuscitation or requiring admittance to hospital for more than 24 h
- loss of consciousness caused by asphyxia or by exposure to a harmful substance or biological agent
- acute illness requiring medical attention or loss of consciousness resulting from the absorption of any substance by inhalation, ingestion or through the skin
- acute illness requiring medical treatment where it may be caused by exposure to a biological agent, its toxins or infected material.

What are 'over 7 day' injuries?

When a person at work is incapacitated for *more than* 7 consecutive days from their normal work owing to an injury resulting from an accident at work, then the accident must be reported.

The day the accident happens does *not* count in calculating the 7 days. But any days that would not be normal working days (e.g. shift days, days off, holiday or weekends) *do* count in the 7 days.

If an employee remains at work but cannot carry out their usual work (e.g. they are put on 'light duties'), then the accident *must* still be notified.

Does it matter when accidents are reported or are there strict timescales?

As you would expect, there are strict timescales for the reporting of accidents:

- fatalities – immediately or as soon as possible after they happen
- specified injuries – immediately or as soon as possible after they happen
- over 7 day injuries – within *15 days* of them happening
- accidents to people who are not at work – immediately or as soon as possible after they happen
- dangerous occurrences – immediately or as soon as possible after they happen.
- diseases – as soon as apparent and without undue delay.

Any accident, disease or dangerous occurrence that is notified immediately – usually by telephone, email, fax or online – must be confirmed in writing on the appropriate form within *15 days*.

Where and to whom should accidents be notified?

Overall, the number of notifications of accidents is generally low, and there is serious under-reporting (hence the trend to prosecute for non-compliance).

In order to address this, the HSE has made the reporting of accidents much easier, and has provided a 'one stop shop' for all employers to report accidents and so on irrespective of whether their enforcing authority is the HSE or the local authority.

All notifiable accidents, dangerous occurrences and diseases can be notified by way of the HSE's RIDDOR website (http://www.riddor.gov.uk).

What forms have to be used for notification?

There is a standard form for use under the RIDDOR, known as F2508. Blank forms can be printed from the RIDDOR website. Online notifications also use the F2508 template. However, not having the correct form is not an excuse for failing to notify an accident.

Do records of accidents and so on need to be kept by the employer?

Yes. Records of injuries and dangerous occurrences must be kept by the 'responsible person' for at least 3 years.

Records must contain

- the date and time of the accident or dangerous occurrence
- for an accident suffered by a person at work
  - the full name of the person
  - the occupation of the person
  - the nature of the injury
- for an accident suffered by a person not at work
  - the full name of the person
  - the status (e.g. customer) of the person
  - the nature of injury
- the place where the accident or dangerous occurrence happened
- a brief description of the circumstances
- the date that the event was first reported to the enforcing authority
- the method by which the event was reported.

Keeping copies of the F2508 will suffice.

Many employers have an accident book, but the information contained in a standard accident book will not satisfy the detail of information needed if the accident is notifiable.

Responsible persons must also be aware that any information kept under accident reporting rules is subject to data protection rules, and must be kept secure and confidential.

Accident books should therefore have one page for each entry, and must not display personal data to anyone not authorised to read it.

The RIDDOR refer to the responsible person as being responsible for reporting accidents. Who is this?

The responsible person under the RIDDOR is

- the employer
- a self-employed person
- the person in control of the work premises where the work is carried out.

If self-employed persons are injured while at work in other people's premises, then the person in control of the premises (e.g. the employer, managing agent or the facilities management company) will need to notify the accident.

Is the principal contractor responsible for notifying site accidents?

No, unless the principal contractor is the employer of the person concerned or unless the person injured is a member of the public.

The employer of the person injured is responsible for notifying the HSE of the accident, or, if they are self-employed, they are responsible themselves for notifying the HSE.

Principal contractors may be the persons in control of the premises, and will be responsible for all common areas and so on and for the safety of persons who come on to the site. Members of the public obviously do not have an employer responsible for them as they will not be at work when visiting the site or when affected by its activities.

The principal contractor should always investigate site accidents and must ensure that the relevant person – either the employer, the self-employed person or themselves – completes the notification process.

The principal contractor should run through the correct site procedures with all contractors during site inductions.

What diseases need to be notified under the RIDDOR?

The number and type of diseases that need to be notified are many and varied, although the 2013 RIDDOR reduced the list quite significantly.

Generally, any disease that is caused by an activity at work or from being exposed to substances in use at work will be notifiable. The reportable diseases (and their associated hazards) are

- carpal tunnel syndrome (where the person's work involves regular use of percussive or vibrating tools)
- cramp of the hand or forearm (where the person's work involves prolonged periods of repetitive movement of the fingers, hand or arm)
- occupational dermatitis (where the person's work involves significant or regular exposure to a known skin sensitiser or irritant)
- hand–arm vibration syndrome (where the person's work involves regular use of percussive or vibrating tools, or holding materials subject to percussive processes, or processes causing vibration)
- occupational asthma (where the person's work involves significant or regular exposure to a known respiratory sensitiser)
- tendonitis or tenosynovitis (in the hand or forearm, where the person's work is physically demanding and involves frequent, repetitive movements).

What are dangerous occurrences?

The list of dangerous occurrences is quite long, and is included in a schedule to the RIDDOR. Indications of dangerous occurrences are

- the collapse, overturning or failure of load-bearing parts of lifts and lifting equipment
- the accidental release of biological agent likely to cause severe human illness
- the accidental release of a substance that may damage health
- the explosion, collapse or bursting of a vessel or associated pipework
- an electrical short circuit or overload causing fire or explosion
- an explosion or fire causing suspension of normal work for over 24 h
- the collapse of scaffolding.

Any dangerous occurrence that has the potential to cause significant harm must be checked to see whether it is notifiable. It is always better to err on the side of caution and report than not.

Does any injury that happens to a visitor to the site or to a member of the public have to be reported?

No. Only those accidents and resultant injuries that require the public to be taken to hospital as a result of the accident, and that were caused by the employer's *work activity*. Injuries that happen to the public or others that are due to carelessness or from something over which they have control will not be notifiable. The injury must result from an accident 'arising out of or in connection with work'.

Types of incident that would *not* be reportable if they caused injury to a person are

- acts of violence causing injury between fellow workers over a personal argument
- a person dying of a heart attack on the premises
- a visitor who tripped over their own bag or luggage
- acts of violence between customers or visitors.

Is there a legal duty to investigate accidents?

No, not at present – although it is good practice to do so.

Accident reporting is only one part of the process of health and safety management. When an accident or incident occurs, it is necessary to find out what caused it, what went wrong, and why and what can be done to ensure that it does not happen again.

There is an implied requirement and duty to investigate accidents because risk assessments have to be reviewed regularly and when circumstances change. An accident may be 'changed circumstances'.

Also, with the increase in civil claims, insurance companies are forcing employers to investigate the causes of accidents so that strategies can be put in place to prevent future occurrences. This will help bring down employers' liability insurance premiums, or at the very least prevent them from rising astronomically.

What are the key steps in an accident investigation?

Every employer should have an accident investigation plan as part of their health and safety management policy.

Accident investigation should be looked upon as identifying what happened and why, so that a reoccurrence can be prevented:

- Step 1: define the purpose of the investigation.
- Step 2: define the procedure.
- Step 3: define what equipment will be needed.
- Step 4: define how the investigation is to be carried out and what information will need to be gathered.
- Step 5: define the content of the report.
- Step 6: decide how recommendations will be implemented.

It is sensible to create an 'accident investigation kit' so that everything needed is in one place and valuable time is not lost in trying to find equipment.

Contents of an accident emergency investigation kit

- Report form.
- Routine checklist for basic questions or prompts to the investigator.
- Notebook, pad, paper and pen.
- Voice recorder for on-site comments or to assist at interviews.
- Camera – to take immediate photographs of the scene of the accident.
- Measuring tape (e.g. a surveyor's or builder's tape).
- Any special equipment regarding the work environment that could assist the investigation (e.g. noise meters or an air sampling kit).
- Witness forms for statements.

What issues will I need to consider when an accident investigation is carried out?

If there is an accident investigation procedure in place you should follow the specific guidelines.

If not, some pointers are

- Where did the accident happen (describe exactly)?
- Who was injured? Were they an employee, contractor or a member of the public?
- What were they doing?
- What equipment were they using?
- What time was it?
- What were the environmental conditions?
- What was the condition of the area or equipment?
- Were any defects noted, such as
  - maintenance issues
  - worn flooring/trip hazards
  - poor lighting
  - broken guarding?
- Was there a safe system of work in place?
- Were risk assessments available?
- Had assessments under the Control of Substances Hazardous to Health (COSHH) Regulations 2002 been completed?
- Who witnessed what happened?
- What did they *see* happen, or what did they hear?
- What actions were taken immediately after the incident?
- Was the accident notifiable?

- What needs to be done to prevent it happening again?
- Had equipment been routinely checked?
- Are maintenance records available?
- Had the work process been regularly reviewed and checked as part of safety monitoring?
- Did someone not do something they should have?
- Was an employee involved? If so, what training had they had?
- Were contractors involved? If so, had they had induction training and been made aware of any site specific hazards?

Appendix to Chapter 16

Accident and incident form

Premises			
Address			
Your name		Job title	
Date that the accident or incident was reported to you			
Name and address of person involved			
Was that person an	Employee/visitor/other:		
Male/female		Age	
If an employee, what is their job title?			
Details of what the person involved said happened			

Was the person involved injured? If so how	

Date and time of the accident or incident		If employee, are they expected to be off work or on light duties for more than 3 days?	

Where did the acident/incident take place?	

Did the person go to hospital?		Were the police involved?	

Signed: ______________________________

Date: ________________

CDM 2015 Questions and Answers: A practical approach
ISBN 978-0-7277-6032-6

ICE Publishing: All rights reserved
http://dx.doi.org/10.1680/cdmqa.60326.275

Chapter 17
Site welfare facilities

Is there a legal duty to provide welfare facilities on a construction site?

Yes. Schedule 2 of the CDM Regulations contains specific requirements for the provision of welfare facilities on all construction sites.

Generally, everyone who works on a construction site must have access to

- sanitary accommodation
- washing facilities
- warming facilities
- somewhere to eat their food
- somewhere to store clothes
- drinking water.

The CDM Regulations tend to be 'goal setting', which means they do not stipulate exact numbers of facilities for every site. 'Suitable and sufficient' is a term that is often used, and the employer has to determine what this might be.

The client, under duties imposed by CDM regulation 4, must ensure that welfare facilities are provided in respect of any person carrying out construction work both before and during the construction work. The client can specify what facilities they expect to see on the site, or whether existing facilities can be used, in the client brief that sets out the scope of the project and the expectations for managing the project safely. Such a brief should be included in any pre-construction information pack/document, as it is important that all persons involved in the project understand what is expected.

How is 'suitable and sufficient' or 'reasonable' determined?

To an extent, common sense should prevail.

There must be enough facilities for everyone to use them without excessive waiting and so on.

Guidance on appropriate numbers is provided by the Health and Safety Executive (HSE) or within other documents (e.g. British standards). Employers are expected to know about the existence of guidance, and would be expected to consult good practice guides.

Should an inspector from the HSE visit a site and declare that the facilities are inadequate or not suitable and sufficient, an improvement notice could be served, requiring the provision of additional facilities.

If the employer feels that the inspector is being unreasonable, they could appeal the improvement notice, and the matter would be heard at a tribunal. 'Suitable and sufficient' could be determined in this arena, or ultimately in a court of law.

Do facilities have to be provided on the site or could toilets available in the area be used?

In the majority of instances, it will be expected to provide welfare facilities actually on the construction site. It will be unreasonable to expect operatives to walk off site to find public conveniences, for example.

However, it may be reasonable to make proper arrangements with another employer to use their facilities elsewhere in the building for instance.

The HSE inspector would expect to see a proper arrangement covered in the construction phase health and safety plan or within information given to employees.

How many water closets, urinals and wash hand basins have to be provided?

The CDM Regulations state the facilities should be 'suitable and sufficient'.

Guidance on the actual numbers of facilities can be found in other documents, namely the Approved Code of Practice for the Workplace (Health, Safety and Welfare) Regulations 1992 and the British Standard BS 6465.

A good starting point for calculating the number of facilities is Table 17.1.

Do toilets always have to be plumbed in with running water?

No, not always, but it is preferable to have flushing toilets if at all possible.

If a water supply and drainage cannot be provided to the site welfare facilities, it will be acceptable to provide chemical closets.

Table 17.1 The number of facilities required (from BS 6465-1:2006)

No. of men at work	No. of water closets	No. of urinals	No. of wash stations
1–15	1	1	2
16–30	2	1	3
31–45	2	2	4
46–60	3	2	5
61–75	3	3	6
76–90	4	3	7
91–100	4	4	8
>100	An additional WC for every 50 (or part) men plus an equal number of additional urinals, plus an additional wash hand station for every 20 operatives		

Wash hand stations should be provided with adequate supplies of hot and cold running water. Water closets should preferably be wash-down water types

Suitable numbers of chemical closets must be provided, and they must have suitable mechanisms for maintaining the closets in a sanitary condition.

The site agent must ensure that greater attention is given to ensuring that chemical closets are kept clean, and regular cleaning schedules must be in place. Regular emptying of the sewage containers will be necessary, and plans will need to be made to deal with this.

If drains are available for discharging chemical closets, the question might be asked as to why flushable toilets cannot be used.

What hand-washing facilities would be acceptable if no running water is available?

Suitable facilities for washing hands and arms are essential on a construction site because the risk of hand-to-mouth infection is high.

If no running water is available, suitable containers of water must be provided, such as plastic containers with a tap usually associated with camping and caravanning.

If hot water cannot be provided, it is important to provide anti-bacterial soaps that work in cold water, or water-less hand gel that effectively sanitises the hands.

Cement dust in particular needs to be removed from the hands and arms so as to prevent skin diseases.

Do urinals have to be provided?

Not necessarily, as the overall number and accessibility of facilities is the most important factor. Urinals can be provided in addition to water closets, and, where they are, a slightly lesser number of water closets will be needed. Table 17.1 can be used to aid calculations.

Whose duty is it to calculate the number of sanitary facilities – the client or the principal contractor?

All construction projects fall under the CDM Regulations 2015, and the client may stipulate what they expect to see in respect of site facilities. Clients are expected to set the standards for improving the overall conditions on construction sites, and under the CDM Regulations they must approve the provision of welfare facilities before a project starts on site.

If a client has set the standards, these will be found either in the pre-construction health and safety pack or in the accompanying 'employer's requirements'.

The requirements listed in the pre-construction health and safety pack should be taken forward by the principal contractor and developed into the construction phase health and safety plan.

If a client has not stipulated any specific requirements, it will be assumed that the principal contractor or contractor will be responsible for ensuring that legal requirements in respect of facilities on the site are met.

The provision of site welfare facilities must be clearly covered in the construction phase health and safety plan. If the client feels that insufficient provision is made, they can prohibit start on the site of the works, as the construction phase health and safety plan will not be considered sufficient and CDM regulation 4(2)(b) relating to welfare facilities will not be met, causing the client to be in breach of their duty.

The client should give an indication to the principal contractor of the anticipated number of contractors or workers expected on the site, especially if any of these will be client direct appointments or nominated contractors. This should help the contractor to calculate the number of facilities.

If site welfare facilities are shared between all contractors, who is responsible for keeping them clean?

The construction phase health and safety plan should detail who is responsible for providing and maintaining welfare facilities. The bigger and more complex the project, the greater the requirement for detailed information on the welfare facilities to be provided.

Generally, the principal contractor is responsible for cleaning and maintaining the welfare facilities, although another contractor could be identified as being responsible.

The principal contractor is responsible for ensuring the co-ordination and co-operation of employers on a multi-occupied site, and there must be a clear indication of who is responsible for providing and maintaining site welfare facilities.

Where facilities are shared with a residual employer (e.g. refurbishment projects in occupied buildings), agreement must be reached between both sides regarding who is responsible for cleaning and maintenance or, indeed, whether the facilities can be shared. Sometimes an influx of site workers will cause the existing facilities to become inadequate, and the principal contractor may have to provide additional facilities.

If facilities are found to be in unsatisfactory condition (not clean, etc.) and an HSE inspector visits the site, it is likely that an improvement notice will be served under the Health and Safety at Work etc. Act 1974. Any action taken that could be deemed a material breach of the law will incur a fee for intervention charge.

Is it necessary to provide separate sanitary and washing facilities for women?

Men and women may use the same toilet provided it is in a separate room with a door that can be locked from the inside. Where possible, cubicle walls and door should be full height – floor to ceiling – so that the cubicle is totally enclosed.

Wash hand basins can be shared between the sexes for hand and arm washing. It would be good practice to have a water closet and wash hand basin in one cubicle, but if this is not practicable, communal wash hand basins in an ante-room to the water closets would suffice.

What provision needs to be made for clients and visitors?

There is no legal requirement for separate facilities to be provided for clients and visitors, although many sites do have separate facilities.

A client may stipulate that separate facilities are required, and this will be a matter of agreement between the contractor and the client. The HSE inspector will be concerned only that the number of welfare facilities on the site is suitable for the number of operatives working there.

Often, separate facilities are provided because they can be locked shut and kept in a more acceptable condition.

What provision needs to be made on a construction site for drinking water?

Schedule 2 of the CDM Regulations requires that a suitable supply of drinking water be provided on every construction site.

Drinking water is 'potable' water that meets the requirements of the drinking water regulations. It generally needs to be a mains piped supply, but adequate quantities of bottled water or water containers and dispensers will be satisfactory.

Every supply of drinking water must be conspicuously marked by an appropriate sign where necessary for health and safety reasons. This is especially important where there may be two or more supplies of water around the site and one is fit for drinking while the others are not.

Where a supply of drinking water is provided, there must also be provided a sufficient number of cups or other drinking vessels unless the water supply is from a purposely designed drinking fountain.

On large sites, a suitable supply of drinking water must be provided at readily accessible and suitable places. This could be within each floor if the construction site is a multistorey building, or within, say, every 100–200 m.

It would *not* be considered reasonably accessible for only one drinking water supply to be available in the canteen if several floors need to be climbed to get to the facility.

More drinking water points need to be provided in warmer weather than in winter. If the site has a 'no food and no drink on site' rule, it is imperative that the principal contractor makes adequate provision for drinking water supplies.

What facilities are required for the changing and keeping of clothing on site?

Schedule 2 of the CDM Regulations requires that 'suitable and sufficient' changing rooms be provided or made available for

- the clothing of any person at work on the construction site that is not worn during working hours
- special clothing that is worn by any person at work on the construction site but that is not taken home.

The facilities provided for keeping clothing on site must include suitable provision for drying clothing.

All operatives on a construction site have the right to keep their clothes safe during working hours, especially if they need to wear protective clothing that they put on when they reach site.

'Suitable and sufficient' may include

- coat hooks in the mess room
- coat hooks in an area adjacent to the sanitary accommodation
- a purposely designated hut with coat hooks, rails and so on
- individual lockers.

Coat hooks, lockers and so on can be shared.

Drying facilities will mean a room or an area where wet clothes can be hung to dry. A suitable heating device will be needed, but this should be safe to use (avoid liquid petroleum gas (LPG) heaters, open flames, etc.). Electrical heaters or drying rails will be ideal. Ensure that they are fitted with a safety, over-heating cut-off device.

Protective clothing and equipment must be kept safe so as to keep it suitable for use. Separate storage facilities may be required.

Is it necessary to provide separate accommodation for changing clothes for men and women?

All construction sites must provide suitable and sufficient accommodation for the changing of clothes where

- persons have to wear special clothing for work
- those persons cannot be expected for reasons of proprietary or health to change elsewhere.

It may thus be necessary for reasons of propriety to provide separate changing facilities for men and women working on the construction site.

Facilities could be a dedicated changing cabin or an integral part of the accommodation for storing clothes. Employers must consider all aspects of discrimination and must ensure that any female operatives are not harassed in any way.

The ante-room in which wash hand basins are located could double up as the changing room, provided it is for use by a single sex.

On large sites, a locker room and changing room are often combined. As more women join the professional trades, it is becoming more frequent for separate and quality accommodation to be provided.

Personal possessions of operatives should be able to be stored on site securely – lockable lockers are preferred. Alternatively, personal possessions could be kept in a secure site office.

What is meant by the term 'rest facilities'?

All employees, the self-employed and others 'at work' are entitled to have rest breaks from work. In fact, the Working Time Regulations requires categories of employees to be given formal rest periods during the working day. These usually include tea/coffee breaks, a lunch break and a mid-afternoon break. It depends on how long people are working before they are allowed a break.

If employees are expected to work for more than 6 h continuously, they are entitled to a minimum of a 20 min break. The CDM Regulations require suitable and sufficient rest rooms to be provided on all construction sites.

Rest facilities cannot be part of the construction site area. They must either be a designated room or area. 'Rest' is to be a period of recuperation from noise, dust, activities and so on.

Operatives can eat their food and drink in rest areas. These must be away from the risk of contamination of the foodstuffs.

On smaller sites, if the entire site stops at the same time and is cleared, an area within the construction site could be used as a rest area.

Suitable seating facilities must be provided, and preferably a table.

Rest rooms must be smoke free in order to comply with the law.

Rest rooms must also contain suitable provision for the eating of meals, including the preparation of meals. A microwave would be suitable, together with table and chairs. Some means of boiling water must also be provided – again, a microwave oven would suffice. Operatives must have the opportunity to make hot drinks if required.

Rest rooms or areas must be adequately ventilated. They should not be exposed to risks from dust, noise, trailing cables and so on.

Rest rooms or areas should not be used to store plant, equipment or materials.

HSE inspectors will expect to see adequate provision made on *all* sites for rest facilities. The details should be included in the construction phase health and safety plan.

Consideration needs to be given in *advance* whether the rest facilities need to be moved because of the progress of works on site. Plan in advance, and always ensure that some defined area is available. Future proposals for maintaining rest facilities during the progress of works must be incorporated into the construction phase health and safety plan.

What provision for heating has to be made of a construction site?

Regulation 34 in the CDM Regulations deals with the requirements for temperature and weather protection.

Indoor working temperatures have to be reasonable – there is no maximum or minimum temperature set down in the CDM Regulations (nor indeed in other health and safety legislation, although there is guidance).

Reasonable working temperatures are a matter of interpretation of a number of circumstances:

- What work needs to be done – how physical is it?
- How open is the site to the elements?
- How many people are on site?
- How easy is it for operatives to leave the work area to get warm?
- How much thermal clothing can be worn comfortably?

Generally, in order to arrive at a suitable decision as to what level of heating is required, a risk assessment will be necessary. The principal contractor should complete the risk assessment for all aspects of the site, and must communicate the findings to all contractors.

The hazards from too cold a working environment are

- increased risk of accidents
- lack of concentration
- increased risk of a heart attack
- hypothermia
- poor circulation of the blood.

Obviously, the biggest concern is the likely increase in accidents if people are too cold to hold tools effectively, mix materials and so on.

The principal contractor must identify what control measures can be put in place to reduce the hazards and risks associated with low working temperatures.

Space heating could be installed in suitable locations throughout the site. Heating appliances can be

- electric
- LPG
- gas.

Inadequately ventilated LPG and gas heaters could cause carbon monoxide gas to be produced, and this could lead to fatalities. Gas equipment may continually leak because valves have been left on. These potential hazards need to be weighed up when choosing a suitable heating source.

Electrical fan blowers purposely designed for construction sites and running on 110 V may be the safest option.

Case study

Three men died from carbon monoxide poisoning on a construction site. The men were using a steel container as their rest room, and were running a generator to provide the heating to keep warm. The steel container had no ventilation, and the three men were sleeping on the site. Carbon monoxide was produced due to the incomplete combustion in the generator: dangerous levels of carbon monoxide can accumulate within minutes, especially in a confined space.

All construction site operators must be aware of the risks. All flueless open-flame heaters fuelled by natural gas or LPG require an adequate supply of fresh air to prevent the formation of high levels of carbon monoxide.

Where work is undertaken outdoors, there must be suitable provision for protection from adverse weather, including facilities for keeping any protective clothing or equipment dry. In particularly cold or wet conditions, workers should be given the opportunity of warming themselves or drying their clothes at regular intervals during their shift.

What provision for ventilation has to be made for a construction site?

Regulation 33 of the CDM Regulations sets out the requirements for suitable and sufficient ventilation for a construction site.

All workplaces must be provided with adequate ventilation, which means a supply of purified air or fresh air. Construction sites are no exception, although, generally, there is little difficulty in providing adequate ventilation because of the open nature of many sites.

Particular attention has to be paid to the adequacy of ventilation when dust, chemicals, fumes or vapours are produced around the site.

Extract ventilation may be needed at certain times to eliminate dust or fumes. This requirement should be covered in the assessment for the Control of Substances Hazardous to Health (COSHH) Regulations 2002 completed for the work activity that would generate the dust or fumes.

It would be sensible to complete a risk assessment to determine the needs for ventilation. Certain areas of the site may be 'confined spaces', and additional provision may be required.

Increased ventilation may be needed in summer months – a constant review of conditions on site is necessary so that adaptations to ventilation requirements can be made.

What provision needs to be made for lighting on the site?

Regulation 35 of the CDM Regulations sets out the requirements for lighting on a construction site.

As expected, lighting on the site needs to be 'suitable and sufficient'. Lighting should also be, where practicable, natural light.

If artificial lighting is used, it must not cause any warning signs or symbols to be adversely affected by a change of colour.

Lighting must be suitable and sufficient and the only test is, 'Can you see where you are going around site?'

Is it possible to see the floor clearly, any small holes, drainage channels and so on? Can the task at hand be seen clearly without eye strain? Can operatives see what they are doing? Can they see the emergency exit signage? Can they see clearly when using steps and stairs, or are they at risk of falling? Is there a difference between outdoor sunlight and internal lighting that may cause temporary blindness? Is artificial lighting in the right place? Does it have enough brightness or lux?

Poor lighting increases the risk of accidents. Improve the lighting levels and the accident rate may reduce, causing the site to be much more efficient.

Early morning starts and dusk in winter need to be considered, and more lighting will be necessary at these times. Don't forget to consider entering a darker building from external sunlight – eyes need time to adjust, so make sure that there is more than enough lighting.

Is it clear what lighting is to be provided by whom? It should be clearly laid out in the construction phase health and safety plan. The principal contractor is responsible for background lighting and for lighting means of access and egress to the site. Contractors and sub-contractors may be responsible for providing task lighting for their individual trades.

The principal contractor should discuss the aspect of lighting in pre-start meetings, and should check the contractor's risk assessments to make sure that they have considered the need for task or background lighting.

The principal contractor will also be responsible for ensuring that adequate provision is made on the site for secondary lighting when it is considered there would be a risk to health and safety if the primary lighting failed.

Emergency lighting will be required on fire exit routes. The fire plan should deal with these issues.

CDM 2015 Questions and Answers: A practical approach
ISBN 978-0-7277-6032-6

http://dx.doi.org/10.1680/cdmqa.60326.287

Chapter 18
Asbestos

What are the main duties under the Control of Asbestos Regulations 2012?

Employers have duties under the Control of Asbestos (CAW) Regulations 2012 to protect their employees from exposure to asbestos-containing materials, as they may cause harm to health.

Under the CAW Regulations, employers are responsible for the health and safety of

- their employees
- other people's employees
- members of the public
- the self-employed

if they are or will be exposed to asbestos.

Employers must also

- provide information, instruction and training
- carry out risk assessments
- produce a written plan of work
- ensure asbestos types are identified
- prevent or reduce exposure to asbestos
- introduce control measures
- maintain effectively and control measures
- keep records of any tests and examinations
- provide suitable protective clothing
- provide changing facilities and clean clothing
- develop emergency procedures
- prevent or reduce the spread of asbestos
- clean equipment and premises after exposure to asbestos
- designate areas as 'respirator zones' or 'asbestos areas'

- display suitable hazard warning notices
- arrange for effective air monitoring
- keep records for 5 years, or for 40 years if to do with health surveillance
- provide health surveillance every 2 years to those exposed to asbestos
- remove asbestos waste under special waste provision.

Contravention of the CAW Regulations is an offence, and fines can be unlimited per offence in the magistrates' court or, for serious offences and breaches of improvement notices or prohibition notices, unlimited with prison sentences in the Crown Court.

What is asbestos and what are the risks from it?

Asbestos is a natural mineral fibre that has been used for decades as an effective heat insulator, fire-retardant material and general bonding material.

There are three main types of asbestos fibre

- chrysotile (white asbestos)
- amosite (brown asbestos)
- crocidolite (blue asbestos).

Historically, blue and brown asbestos have been considered the most dangerous, but research has indicated that all types of asbestos fibres are potential health hazards, so all types of asbestos have been banned for new building works and asbestos is no longer being imported or sold in the UK.

However, there is much residual asbestos in existence in buildings of all descriptions, in particular in those built up to about the mid-1980s. Brown and blue asbestos was banned in 1985, and white asbestos was banned in 1999.

Asbestos fibres cannot be easily visually identified by their colour, and laboratory analysis is needed to positively identify the type and volume of fibres.

Asbestos fibres can become airborne, and the microscopic fibres can be breathed into the lungs, where they lodge for many years and in so doing create an irritant and cause lung damage or more serious diseases.

Breathing in asbestos fibres can lead to the individual developing one of three fatal diseases:

- asbestosis – scarring of the lung tissue leading to shortage of breath and difficulty in breathing

- lung cancer
- mesothelioma – cancer of the lining around the lungs and stomach.

There are no cures for asbestos-related diseases.

People who smoke and who are exposed to asbestos fibres are at an even greater risk of developing asbestos-related diseases.

Exposure to minute amounts of asbestos fibres over prolonged periods of time will increase the risks of disease.

Asbestos-related diseases take from 10 to 60 years to develop from first exposure, and so it may be in retirement that the health effects of exposure are felt.

Are any employees or workers at greater risk than others?

Obviously, those who choose to work in the asbestos removal industry are exposed to a significant hazard, but the risk to which they are exposed may be quite minimal because the controls and checks placed upon the removal industry are considerable, and generally workers are well protected.

Most of the individuals suffering from asbestos-related illnesses work in the building and maintenance trades.

Employees (or the self-employed) who are most of risk are

- plumbers
- electricians
- builders
- carpenters
- roofing contractors
- gas fitters/service engineers.

How does asbestos get into the body?

Asbestos fibres can be taken into the body by way of

- breathing in fibres
- swallowing fibres.

Airborne fibres can be inhaled by way of both the mouth and nose. Larger asbestos fibres are generally filtered out by the lungs, but the microscopic fibres are not.

Ingestion occurs when hands become contaminated with asbestos fibres and the individual wipes their mouth, eats food or otherwise transmits the fibres by way of the 'hand to mouth' route.

Is asbestos cement as much of a worry as asbestos fibres?

No, not quite such a worry. However, it is classified as a hazardous substance because the fibres that are held in place with the cement can become loose and airborne.

What is the duty of care to manage asbestos?

The CAW Regulations 2012 include a duty for the management of asbestos-containing materials.

It is not illegal to have asbestos-containing materials in a place of work, because if asbestos is in good condition and not releasing fibres into the atmosphere it is generally quite safe. In fact, there is often less risk in leaving it in place than there is from removing it.

Regulation 4 of the CAW Regulations places legal responsibilities for managing asbestos onto duty holders.

A duty holder is defined in the CAW Regulations as

(a) every person who has, by virtue of a contract or tenancy, an obligation of any extent in relation to the maintenance or repair of non-domestic premises or any means of access to or from those premises; or
(b) in relation to any part of non-domestic premises where there is no such contact or tenancy, every person who has, to any extent, control of that part of those non-domestic premises or any means of access to or from those premises.

Where there is more than one duty holder, the relative contribution to be made by each person in complying with the requirements of the CAW Regulations will be determined by the nature and extent of the maintenance and repair obligation owed by that person.

A wide range of people will potentially have obligation under the CAW Regulations, including

- employers
- the self-employed
- the owner of premises
- managing agents.

The duty to manage asbestos does not extend to domestic premises but does apply to residential premises if they are let as a business (e.g. hotels, bed and breakfast establishments and caravan parks).

As a duty holder, what do I need to do?

Find out if asbestos-containing materials are present in your building or premises.

If the building was constructed before 1985, it is likely to contain some asbestos unless it has already been removed.

Buildings constructed up to 1999 may have asbestos cement materials.

What are the steps to be taken to establish whether asbestos-containing materials are present in the building?

Checking the building plans and other information such as operating manuals to see if any reference has been made to asbestos or asbestos materials.

Consult the design team, if possible, who undertook the building works, including any known contractors or sub-contractors, building services contractors and so on.

Carry out a full survey of the premises to identify likely asbestos-containing materials.

What types of surveys are there that can be used to identify asbestos-containing materials?

It is important to note that the CAW Regulations 2012 do not legally require surveys in order for a duty holder to comply with their duties under the regulations. However, as the CAW Regulations require asbestos-containing materials to be managed, you will have to know where they are, so a survey will be vital.

If the duty holder does not want to conduct a survey, they could make a 'presumption' that all suspect or likely materials will be considered to contain asbestos and act accordingly. But this could cause the duty holder to undertake costly preventative measures when lesser ones may be all that are required.

There are two types of survey usually undertaken to identify asbestos-containing materials in buildings. Whichever survey is chosen, the duty holder must always ensure that it is carried out by a competent person: their qualifications, accreditations and experiences should be checked. References should be sought from others.

Some survey reports are riddled with so many exclusion clauses that they are essentially meaningless. Asbestos identification may be a case where 'cheapest isn't always best'.

Management survey

This survey is the minimum survey that must be carried out if there is any likelihood that asbestos-containing materials are present in the building.

The survey identifies, as far as is reasonably practicable, the presence of asbestos in the building, and is often non-intrusive although samples of materials can be taken for analysis.

Asbestos materials can be 'presumed' present and recorded in an asbestos register so that, when more intrusive refurbishment works are required, note is made that further investigations are necessary.

The management survey can be used to indicate where more detailed and intrusive surveying is required.

A management survey could be used when asbestos materials are known to be present, with the purpose of identifying the condition of the materials and assessing any residual risk to health.

Refurbishment/demolition survey

Whenever refurbishment, demolition or intrusive works are required in any building where there is a likelihood of asbestos materials, a refurbishment and demolition survey must be carried out. This is an intrusive survey that removes equipment or the fabric of the building (e.g. ceiling panels and samples of material are taken for analysis).

A detailed survey report must be produced with clear information about risks from the asbestos materials and which material needs to be removed, encapsulated or managed,

A refurbishment/demolition survey cannot be undertaken in occupied areas as it involves destructive inspection and asbestos material disturbance. Should it be undertaken in occupied buildings, an environmental clean must be undertaken before the area is reoccupied, including a 'fit for occupation' certification.

If asbestos is present in the building, can it be removed immediately?

No. All work with asbestos-containing materials is potentially harmful and must be strictly planned, with risk assessments and method statements.

The removal of asbestos lagging, insulation, or asbestos insulating board (AIB) must be carried out by contractors licensed to carry out such tasks with the Health and Safety Executive (HSE).

Under the CAW Regulations 2012, some asbestos-containing materials no longer have to be removed by a licensed contractor, but these are usually the low-risk materials where asbestos fibres are imbedded in a 'matrix', such as asbestos cement, bitumen felt, vinyl floor tiles and rope gaskets in boilers/electrical switchgear.

There are strict controls on work procedures for the removal of all asbestos-containing materials and also for controlling any actual work on the material that does not necessitate removal (e.g. drilling holes in AIB to put up shelving). Further guidance must be sought from competent persons.

Asbestos-containing materials that need to be removed by a licensed contractor must be notified to the local HSE office by way of a *14 day notice*. No works of removal can take place until the notification period has expired.

What work with asbestos needs to be undertaken by a licensed contractor?

Most asbestos work must be undertaken by a licensed contractor.

Licensable work with asbestos is work

- where worker exposure to asbestos material is not sporadic and of low intensity
- where the risk assessment cannot show that the control limit will not be exceeded (i.e. 0.1 asbestos fibre per cubic centimetre of air)
- on asbestos coating
- on asbestos insulation or AIB where the risk assessment demonstrates that the work will last longer than 2 h within any 7 day period and that one person will work for longer than 1 h with the asbestos material (i.e. any work that is not deemed 'short duration').

An asbestos licence will be needed whether work is undertaken by a contractor or by the employer on their own premises using their own employees.

Some other works will need a licence, such as putting up and taking down scaffolding that provides access to licensable work.

What is notifiable, non-licensed work (NNLW) with asbestos?

Some non-licensed work, where the risk of fibre release is greater, must

- be notified to the enforcing authority
- be properly recorded with details of employees who have worked on NNLW.

Persons who undertake non-licensable work must be subjected to medical examinations, and suitable records must be kept. Medical records/health surveillance records must be kept for 40 years.

NNLW with asbestos is a matter of judgement, and a suitable risk assessment should be completed.

Examples of types of work that will not need a licensed contractor but that could be notifiable to the enforcing authority are

- minor, short-duration maintenance work with asbestos insulation
- minor removal of AIB
- entry into a confined space (e.g. the ceiling void above AIB ceiling), when no decontamination or cleaning has taken place
- removal of substantially degraded asbestos cement
- steaming or scraping of textured coatings.

Notification must be made on line only, to the relevant enforcing authority.

Works can start as soon as notification has been made, unlike licensable work, which must be notified 14 days in advance and not started until the 14 days have ended.

What should be done if asbestos materials are uncovered or damaged?

If asbestos-containing materials are discovered unexpectedly, they may not cause an immediate health and safety risk, provided they have not been damaged and fibres have not been released into the atmosphere.

Once asbestos-containing materials in good condition have been found, the asbestos register should be updated, the material labelled and a management procedure put in place.

If damaged asbestos materials are found, or damage has been caused to existing but previously good-condition asbestos, then the following steps should be adhered to

- stop work immediately
- decide who must do the work – a licensed contractor may be needed
- minimise the spread of contamination to other areas
- keep exposure levels as low as possible (e.g. damp down the material)
- clean up the contamination
- keep people out of the area
- display suitable warning signs.

Samples of the material may need to be sent for analysis: until the results come back, no work in the area must be undertaken.

If the material is not asbestos lagging or AIB, it can be cleaned up by competent, trained employees or contractors who do not need to be licensed.

If the material is lagging or AIB, all further work on it, including removing debris and cleaning up must be carried out by a licensed contractor.

Where in a building might asbestos-containing materials be found?

It helps to know where asbestos-containing materials might be, so start with a survey in these high-risk areas:

- ceiling voids – may have sprayed asbestos lagging or coating to structural timbers for fire protection, or loose asbestos material may be packed into voids as fire breaks
- pipework – sprayed asbestos coating or lagging is often found around heat pipes as both insulation and fire protection
- boilers – asbestos gaskets and seals are used for insulation and fire protection
- electrical switchgear – asbestos gaskets and so on are used as for boilers
- ductwork, structural steels, firebreaks and so on – sprayed asbestos coating or lagging is used for insulation and fire protection
- insulating boards – asbestos cement insulation is contained in many boards for fire protection and insulation (on soffit boards, ceiling panels, partition walls, etc.)
- ceiling boards – lay-in grid ceiling tiles can contain asbestos
- roofing felt – may contain asbestos
- guttering and rainwater goods – may contain asbestos cement
- roof coverings – may contain asbestos cement
- water tanks – may be made of asbestos cement
- floor tiles – may contain asbestos for both insulation and fire protection
- fire doors – often contain AIB for fire proofing.

Assume materials contain asbestos unless evidence is available in the contrary.

Do not disturb and break into material to see if it contains asbestos – you often cannot tell by looking at it, and if it is asbestos you expose yourself and others to health risks from fibres.

If the initial survey indicates that materials are likely to contain asbestos, a full destructive survey will need to be carried out whereby samples of the material are taken for analysis. Only trained and competent persons should undertake sampling.

What procedures need to be taken if asbestos-containing materials are found?

Assess its condition by considering the following:

- Is the surface of the material damaged, frayed or scratched?
- Is any part of the material peeling or breaking off?
- Is it detached or loose from the structure it was applied to (e.g. falling away from pipes or structural steel)?
- Are protective coatings and coverings damaged?
- Is asbestos debris or dust evident in the area?

What steps need to be taken if asbestos material is in poor condition?

Asbestos in poor condition is a serious health hazard. The duty holder must do one of the following:

- remove it
- repair it
- encapsulate it
- seal it.

Licensed contractors will need to be appointed to work on most asbestos materials. The duty holder must agree an action plan with them, and must ensure that all legal requirements are met. Any asbestos-containing material in poor condition is best dealt with by a competent, licensed contractor.

What steps need to be taken where asbestos-containing materials remain in the building?

An asbestos management register must be created to help ensure all asbestos is located and regularly inspected.

Hazard warning signs must be displayed on all residual asbestos, or some other form of identification.

Permit to work procedures for maintenance works, refurbishment works and so on must be in operation.

It is not illegal to have asbestos-containing materials on the premises, provided they do not cause a health hazard. It is safer to have asbestos in good condition in situ and manage the potential risk than to remove it and create high health risks.

What are some of the steps for managing asbestos?

Make regular checks of the condition of the asbestos material to ensure that it is not deteriorating.

Carry out risk assessments for any work in the area: for example, could anything puncture the asbestos material, releasing fibres?

Keep records up to date and show that an asbestos management plan is in place.

Introduce a permit to work system for all contractors, maintenance engineers and others so that the duty holder knows where they will be working, why, on what and what the hazards and risks are. If they are to work near asbestos material, this can be highlighted, and safety precautions stipulated.

How often do I need to check the condition of asbestos material?

Once every 12 months will usually be sufficient, but if the asbestos is in the area that is at risk of damage, then more frequent inspections will be required.

How do I dispose of asbestos waste?

Asbestos comes under the hazardous waste regulations (e.g. Hazardous Waste (England and Wales) Regulations 2005). Regulation is similar in England, Wales, Scotland and Northern Ireland.

Asbestos can only be removed to a licensed waste disposal site by a licensed carrier. *It* must be double bagged and sealed in heavy duty polythene bags and clearly labelled with a recognised asbestos label.

Failure to comply can lead to hefty fines.

Case study

Two companies and two directors were prosecuted under the CAW Regulations 2012 for failing to remove asbestos-containing materials safely from a demolition site.

Steel pipework was to be removed from the building by the contractors. No assessments/surveys for asbestos-containing materials were carried out, even though it was very apparent that the pipes were lagged.

No measures were put in place to prevent any release of fibres, and both workers and members of the public were put at risk. Lagging was found strewn on the ground, and an operative was using a mechanical excavator to move the steel pipes into a skip.

The company that owned the building and that had commissioned the pipe removal (i.e. the duty holder) had failed to provide information to the contractor by not having any survey work carried out. There was an absolute contravention of the duty to provide information about asbestos-containing materials to those who may need it. The contractor removing the pipework was charged with failing to carry out any assessments for asbestos materials and with exposing workers to risk.

Fines and costs amounted to approximately £30 000.

CDM 2015 Questions and Answers: A practical approach
ISBN 978-0-7277-6032-6

http://dx.doi.org/10.1680/cdmqa.60326.299

Chapter 19
Hazardous substances

What are the legal requirements about the use and exposure to hazardous substances?

There are a number of legal requirements governing the use and exposure to substances, liquids, vapours, gases, fumes and biological agents that may be 'hazardous to health' – the potential to cause harm to the health of a person.

Harm to health can be by

- ingestion
- inhalation
- absorption through the skin.

The main pieces of legislation are

- Control of Substances Hazardous to Health (COSHH) Regulations 2002
- Control of Asbestos Regulations 2012
- Control of Lead at Work Regulations 2002.

Generally, an employer must not expose employees or others to substances hazardous to health unless a risk assessment has been carried out – usually known as a COSHH assessment.

Good practice in respect of managing exposure to hazardous substances makes reference to 'the hierarchy of risk control', namely

- eliminate the use of hazardous substances
- substitute for lesser-risk substances
- control substances at source
- provide personal protective equipment (PPE)
- monitor and review controls.

Employers must remember that when two or more 'harmless' substances are mixed together they may combine to form a hazardous substance.

What are the health effects of hazardous substances?

Examples of the effects of hazardous substances include

- skin irritation and dermatitis
- asthma
- respiratory conditions
- loss of consciousness as a result of being overcome with toxic fumes
- cancer
- infection from bacteria.

When is a substance classed as hazardous under the COSHH Regulations?

If a substance or mixture of substances is classified as dangerous under the Chemicals (Hazard, Information and Packaging for Supply) Regulations, then the COSHH Regulations apply to it.

Hazardous substances can be identified by their warning label, and the supplier must provide safety data sheets for all hazardous substances. Warning symbols can be of the international type – white background with red borders, pictogram symbols and no words – or of the traditional orange square with black symbols and words.

The Health and Safety Executive (HSE) publishes lists of the most commonly used dangerous substances.

Substances that have an occupational exposure limit are classified as hazardous. These are substances that can be considered 'safe' up to certain exposure limits.

Any kind of dust in concentrations specified in the COSHH Regulations is deemed to be harmful.

The COSHH Regulations do *not* cover asbestos and lead, as these are dealt with under specific legislation.

What types of substances in use on a construction site could be considered as hazardous?

The following are commonly in use on construction sites and could be considered hazardous substances under the COSHH Regulations:

- solvents
- silica
- cement
- mineral oils

- carbon monoxide gas
- carbon dioxide gas
- adhesives
- paints
- wood dusts
- welding fumes
- carbon deposits/soot
- acid cleaners
- detergents/degreasers
- pesticides.

Details from the suppliers or manufacturers must be obtained for all substances planned to be used on site, or brought in by others to be used on site.

Remember, while the operatives in the immediate vicinity of the hazardous substance may be protected by PPE, others working some distance away could be affected by the substances, fumes, vapours, mists or dust.

Case study

Two lift engineers were working in a lift shaft, welding the metal frame. They were adequately protected by PPE and had a local exhaust ventilation system that removed the welding fumes. However, unbeknown to them, the exhaust extract was faulty because it had not been subjected to regular inspection, and the welding fumes were leaking into the roof void above the lift shaft in which two other operatives were laying electrical cables. Both operatives in the roof void were overcome by fumes, and were found when the site agent checked on progress of works.

How do hazardous substances affect health?

Hazardous substances will affect the body in different ways, depending on how the substance enters the body, known as the route of entry.

Inhalation

Inhalation of a hazardous substance accounts probably for the majority of deaths associated with using substances.

Inhalation of dusts, fumes, mists and vapours will affect the health of an individual in different ways and at different speeds.

Acute exposure refers to the immediate onset of symptoms: that is, the exposure is so great that an immediate adverse reaction happens (e.g. unconsciousness, acute respiratory failure or a heart attack).

Chronic exposure is exposure to a substance over a prolonged period of time, where the harmful substance accumulates in the body over that period of time – frequently through little and often exposure.

Lung diseases often occur because of persistent and long-term exposure to the hazard (e.g. miners and coal dust).

Ingestion

Substances once swallowed will enter the stomach and intestines and pass into the blood stream. When this happens, the toxic substances are transported around the body to all major organs.

Ingestion mostly occurs from 'hand to mouth' contact (i.e. eating or drinking with contaminated hands). Also, if drops of the harmful substance, dust and so on are accidentally transferred on to the mouth and the person wipes their mouth, they could transfer the substance onto the tongue and then swallow it.

Absorption through the skin

Certain substances and micro-organisms can pass through the skin, which acts as a protective barrier to underlying tissue and the blood stream.

Substances can therefore enter into tissue and the blood stream and then travel around the body to vital organs. Some substances are so hazardous that only a minute amount needs to come into contact with the skin (e.g. ricin, which has been used as a terrorist chemical weapon) in order to cause multiple organ failure and death.

Any cuts or grazes to the skin, abrasions and so on will increase the risk of absorption of a chemical through the skin.

Sometimes the skin does not allow complete transference of the substance into underlying tissue and the blood stream, and an allergic reaction takes place. This manifests itself in visible skin irritation, rashes, blisters, burns and so on.

The most common parts and organs of the body affected by hazardous substances are

- the lungs
- the liver
- the bladder
- the skin
- the brain
- the central nervous system

- the circulatory system
- the reproductive system
- the urogenital system.

The most common illnesses suffered by construction workers are

- lung diseases (e.g. pneumoconiosis)
- lung cancer
- silicosis
- dermatitis
- asthma
- leptospirosis
- Legionnaires' disease.

What are the workplace exposure limits?

Workplace exposure limits (WEL) are British occupational exposure limits and are set in order to help protect the health of workers. WELs are concentrations of hazardous substances in the air, averaged over a specified period of time, referred to as a time-weighted average (TWA). Two time periods are used

- Long term – 8 hours
- Short term – 15 minutes

Short-term exposure limits (STELs) are set to help prevent effects such as eye irritation, which may occur following exposure for a few minutes.

Workplace exposure limits are listed in a document called EH40 and employers are recommended to consult this document in order to establish whether specific exposure limits exist for certain substances. However, the absence of a substance from the list does not indicate that it is safe.

We have safety data sheets for all our substances. Do we still have to carry our COSHH assessments?

Yes. A good safety data sheet (SDS) is not a substitute for carrying out and recording a COSHH risk assessment. Gathering SDSs is only the first stage in the assessment process. The SDS will provide information on the hazardous properties of the substances you are using, any health effects associated with its use, how likely it is to get into the air or onto the skin, and what risk reduction measures you should use to control exposure to an acceptable level. However, it will not be specific to your workplace and cannot take into account the particular environment you work in.

What is health surveillance?

Health surveillance is required under certain circumstances, and requires the employer to assess the health of their employees regularly. If employees are exposed, for instance to a substance that causes skin irritation, then it may be necessary to check the condition of hands and arms by visual examination from time to time.

Health surveillance allows an employer the opportunity to monitor the effectiveness of the control measures in place.

If employees are exposed to breathing in fumes or dust, then routine lung tests or blood tests can be used.

Health surveillance can be carried out by a medical doctor or occupational nurse, or an employer can carry out simple assessments and refer to experts for advice.

Does the principal contractor have to arrange for health surveillance for operatives on site?

If the operatives on site are employees of the principal contractor, then the principal contractor may well have the duty to organise health surveillance. However, if the operatives are employees of other contractors or are self-employed, then they themselves will be responsible for complying with the COSHH Regulations and arranging health surveillance as necessary.

Principal contractors may be responsible for arranging for health surveillance for all operatives on site if the potential health hazards are site-wide.

Principal contractors should assess the procedures that contractors have for health surveillance as part of their competency and assessment investigations.

Occupational ill health is a common problem on construction sites, and a principal contractor could implement a site-wide scheme to carry out health checks on all operatives for skin dermatitis. This could be a simple process of checking the hands and forearms of all workers leaving the site on one day.

Information should indicate to the principal contractor whether appropriate PPE is being worn and whether it is working well.

If operatives are displaying signs of skin irritation, then the principal contractor is in a good position to investigate the use of materials, the application practices, the use of PPE and so on.

Pro-active health surveillance by whoever undertakes it will always highlight trends in potential health risks, and good, preventative action can then be taken.

What are the requirements for a COSHH assessment?

Regulation 6 of the COSHH Regulations requires employers to carry out a 'suitable and sufficient' assessment of the risks to health from using the hazardous substance.

'Suitable and sufficient' does not mean perfect, but the guidance that supports COSHH regulation 6 lists a number of considerations that must be taken into account during the process.

The first step in providing a suitable and sufficient COSHH assessment is to ensure that the person carrying it out is *competent*. COSHH regulation 12(4) requires any person who carries out duties on behalf of the employer to have suitable information, instruction and training.

A competent person does not necessarily need to have qualifications as such, but they should

- have adequate knowledge, training, information and expertise in understanding the terms 'hazard' and 'risk'
- know how the work activity uses or produces substances hazardous to health
- have the ability and authority to collate all the necessary relevant information
- have the knowledge, skills and experience to make the right decisions about risks and the precautions that are needed.

The person carrying out the assessment does not need to have detailed knowledge of the COSHH Regulations, but does need to know who to contact and where to go to obtain more information. They need to be able to recognise when they need more information and expertise.

The COSHH assessment process can be broken down into *seven steps*, as follows.

Step 1: assess the risks

Identify the hazardous substances present on the site, or intended to be used on the site.

Consider the risks these substances present to your own employees and all others.

List down all of the substances likely to be used. If the substances are on site, read the labels and look for the hazard warning symbols.

The symbols used on products will either be orange in colour with black symbols and writing or will be the newer plain white diamond-shaped boxes with red edging and black symbols. The new symbols have been introduced to harmonise the use of hazardous substances symbols across the world using the Globally Harmonized System of Classification and Labeling of Chemicals (GHS). The old and new hazard symbols are shown in the appendix to this chapter.

Obtain safety data sheets from the supplier or manufacturer.

A safety data sheet must be provided by the manufacturer or supplier, and it must give information on, among other things, common usage, constituent ingredients, exposure limits, PPE, emergency procedures, first aid and spillage precautions.

If sub-contractors are supplying the hazardous substances, then they must be required to submit manufacturers data sheets together with their COSHH assessments: *all* COSHH assessments must have product safety data sheets attached to them.

Consider the risks of the hazardous substances identified to people's health – all people, not just employees: the COSHH Regulations clearly state that the effect on health of a hazardous substance to 'other persons' must be clearly considered.

The assessing of risk to health from using the hazardous substance is one of judgement. Sometimes it is not possible to know for definite what level of exposure will be harmful. In these cases, it is preferable to always err on the side of caution.

Some questions to consider are

- How much of the substance is in use?
- How could people be exposed?
- Who could be exposed to the substance and how often?
- What type of exposure will they have?
- Could other people be exposed to the harmful substance?

Step 2: decide what precautions are needed

The first responsibility for an employer is to eliminate the risk of using a harmful substance. This is an effective precaution to take but may not always be possible.

Next, consider whether an alternative substance could be used that is less hazardous than the original substance. If there is something on the market that does the job more safely, then it should be used.

If the substances cannot be eliminated or substituted, then it must be used with *suitable controls* so as to reduce any level of exposure to an acceptable limit.

Suitable controls may be

- changing the way that the work is done (e.g. painting instead of spraying)
- reducing concentrations of the substance
- modifying the work process (e.g. to reduce the exposure time)
- reducing the number of employees and others exposed to the substance
- adopting maintenance controls and procedures
- reducing the quantities of substances kept on site
- controlling the overall working environment (e.g. increasing the ventilation or damping down dust)
- appropriate hygiene measures (e.g. easily accessible hand/arm-washing stations to prevent skin absorption)
- enclosing the work activity.

Step 3: prevent or adequately control exposure

Having decided what controls are required, the next step is to implement the controls on the site.

The very nature of a construction site makes it quite difficult to implement some controls, but ones to consider will be

- substituting substances
- excluding workers not needed from the area
- introducing local exhaust ventilation
- changing work processes (e.g. reducing spraying in favour of brush application)
- creating zoned work areas
- undertaking tasks in the open air, where practicable.

Construction sites are particularly prone to a 'cocktail of substances' whereby different substances may combine to form a new and harmful substance. The principal contractor should co-ordinate the use of these different substances and should complete the COSHH assessment for any 'communally' produced hazardous substance.

If there are no effective alternatives, it is permissible to issue operatives with PPE.

PPE should be considered as the last resort – all other control measures must have been effectively considered and introduced if at all possible.

PPE will include

- face masks
- respiratory masks
- gloves or gauntlets

- goggles
- safety boots or shoes that are chemical resistant.

The control of harmful substances *must* be to a level that most workers can be exposed to day after day without adverse effects on their health.

Adequate control of harmful substances can be defined by referring to the WEL of a substance. An WEL is set at a limit that is not likely to damage the health of workers who are exposed to the substance day after day.

If a substances has an WEL and the level of exposure is kept to the limit stated, then, as an employer, you will be deemed to have adequately controlled the risk.

Short term, infrequent exposure to higher levels than the WEL is permissible, but exposure must be the exception rather than the norm. Usually, such higher exposure will be because of an emergency (spillage, substance release, failure of exhaust ventilation, etc.).

Any substance with an WEL must be reduced to a level that is below the WEL, and employers are required to ensure exposure to the substances is as *low as is reasonably practicable*.

Step 4: ensure that control measures are used and maintained

Employees and others must make proper use of the controls that an employer puts into place.

The principal contractor on a multi-occupied site must ensure that contractors follow their own COSHH assessments and that others affected by the works also adhere to the controls identified as necessary.

An employer has a duty to provide PPE to employees and to ensure that it is suitable and sufficient for their needs and is properly maintained.

Employees must report defects to PPE to the employer, or, if PPE has been provided by the principal contractor, to the principal contractor or their representative.

All controls introduced onto the site to reduce the exposure to harmful substances must be adequately maintained so as to ensure that they are effective.

Engineering controls and local exhaust ventilation equipment must be regularly inspected, tested and examined, and records must be kept. Local exhaust ventilation must be inspected every 14 months.

Records and test certificates must be kept for at least 5 years. The principal contractor should keep these for any communal controls introduced into the site.

Step 5: monitor exposure

If the risk assessment determines that there is a serious risk to health if people are exposed to a substance, then health surveillance must be considered.

If any of the following apply, health surveillance will be essential:

- if control measures fail or deteriorate
- if exposure limits could be exceeded
- if control measures are not maintained adequately.

Air monitoring must be carried out if there is a risk of exposure to the harmful substance due to inhalation.

Air monitoring is either site-wide with environmental air monitoring, or can be achieved by giving exposed workers personal air monitors.

Records of air monitoring should be kept for at least 5 years.

Step 6: carry out health surveillance

Health surveillance is defined in the COSHH Regulations as

> an assessment of the state of health of an employee, as related to exposure to substances hazardous to health, and includes biological monitoring.

Health surveillance will be necessary in the following circumstances:

- any exposure to lead fumes
- exposure to substances that cause industrial dermatitis
- exposure to substances that may cause asthma
- exposure to substances of recognised systemic toxicity (i.e. substances that can be breathed in, absorbed through the skin or swallowed and which affect parts of the body other than where they enter).

Any health surveillance carried out should be recorded, and records should be kept for at least 40 years. It is for the *employer* to decide whether health surveillance is necessary.

Step 7: information, instruction and training

All employees expected to use or to come into contact with substances hazardous to health must receive suitable information, instruction and training.

The site agent should ensure that all contractors have adequate records for training their operatives in the health and safety risks of using hazardous substances.

The site agent should ensure that the site induction training covers the use of hazardous substances on the site and the control measures to be followed.

All operatives, visitors, contractors and so on are entitled to see the COSHH assessment and the attached safety data sheet.

Information may be by way of safety notices, toolbox talks, site rules and so on.

Instruction generally covers one-to-one exchanges on how to do something, what to use, what PPE to wear and so on. The health effects of using the hazardous substance should be clearly discussed.

It is useful to keep records of any exchange of information on the COSHH Regulations in the site records book.

What are the health hazards associated with lead?

Lead is a major health hazard, and is controlled by its own legislation: the Control of Lead at Work Regulations 2002.

Managing the risks from lead on site can be quite complex although the general principles of the COSHH Regulations broadly apply.

Any exposure to lead, lead fumes or vapours, dusts and so on must be subject to a 'suitable and sufficient' risk assessment.

Specific controls are required if employees are going to be exposed to 'significant exposure', which is when

- the exposure exceeds half the OEL for lead
- there is substantial risk of an employee ingesting lead
- there is a risk of the employees' skin coming into contact with lead alloys or alkyls.

If exposure is deemed to be 'significant', then an employer must

- issues employees with protective clothing
- monitor lead in air concentrations
- place employees under medical surveillance

- use risk assessments to determine control measures
- identify any other measures to reduce exposure and comply with the Control of Lead at Work Regulations.

How is the level of silica dust determined?

If a significant level of stone cutting, cement drilling/cutting and so on is to be undertaken on site, it will be highly likely that the airborne dust levels of silica will be high.

The principal contractor should arrange for air monitoring to be undertaken by a competent person so that levels can be determined.

Results can be analysed in a laboratory, and, in some instances, immediate indication can be given from electronic sampling equipment.

Levels in excess of the MEL will need to be reduced. Generally, levels greater than 0.1 mg/m^3 can be regarded as harmful and significant, and control measures will have to be put in place to reduce levels of exposure.

Air monitoring by competent persons should also indicate what type of control measures will reduce the dust levels (e.g. local exhaust ventilation).

What are the hazards associated with silica?

Crystalline silica is present in substantial quantities in sand, sandstone and granite, and often forms a significant part of clay, shale and slate. It is also found in chalk, limestone and other rocks and stones.

Crystalline silica can cause lung cancer and other lung diseases.

The health hazards of silica come from breathing in the dust. Operatives and members of the public can be exposed to inhaling the dust.

Not only could operatives undertaking activities directly involving crystalline silica be exposed to the hazards but so also could operatives in the vicinity.

Dust travels around a site, and can be inhaled by all persons, although obviously concentrations are reduced the further away the dust travels from its source.

Activities that can expose workers or others to exposure to dust are

- stone masonry
- façade renovations

- demolition
- sandblasting or stone cleaning
- concrete scabbling
- concrete cutting
- concrete drilling
- tunnelling.

The use of power tools will increase the volume of dust that can be created during the activity (Table 19.1).

Exposure is intensified in confined spaces – the dust cannot dissipate, and greater quantities are inhaled.

Breathing in fine particles of silica dust can cause silicosis. Dust particles will be microscopic.

What are the symptoms of silicosis?

Silicosis is the scarring of the lung tissue, which leads to breathing difficulties.

Exposure to very high concentrations over a relatively short period of time can lead to acute silicosis, resulting in rapidly progressive breathlessness and death within a few months of onset.

Usually, constant exposure to relatively low levels of silica dust results in progressive silicosis. Time lengths vary but can be several years. Prolonged exposure causes fibrosis – a hardening or scarring of the lung tissue – which results in the loss of lung function. Victims are likely to suffer shortage of breath, and will find it impossible to walk even short distances once the disease has taken hold.

The condition is irreversible, and continues to develop even when exposure to the dust has ceased.

Heart failure is a common cause of death.

Sufferers are often unable to enjoy retirement, and are probably not aware of the damage they are doing to their health during their working years.

Are there any safe limits for exposure to silica dust?

Silica has been assigned a maximum exposure limit (MEL) of 0.3 mg per cubic metre of air.

Table 19.1 Examples of typical levels of silica exposure in some common construction activities

Activity	Control measures	Exposure	Improvements required[a]
Drilling in a poorly ventilated undercroft	■ No dust suppression ■ No extraction ■ No forced ventilation ■ Inadequate respiratory protective equipment (RPE)	High – 300 times the MEL	■ Fit water suppression or dust extraction to drilling equipment ■ Provide appropriate PPE ■ Ensure correct use of RPE
Drilling onto brickwork under an arch blocked at one end	■ Primitive extraction by a fan and airbag ■ Disposable face masks worn	High – 5 times the MEL	■ Fit water suppression or dust extraction to drilling equipment ■ Provide appropriate RPE ■ Ensure correct use of RPE
Using jackhammers to break out concrete in a large open indoor area	■ Limited ventilation ■ No dust suppression ■ No local exhaust ventilation ■ No RPE in use	Medium – within the MEL but double toe level regarded as reasonably practicable	■ Wet down concrete and rubble
Chasing out cracks in a screeded cement floor in a large open indoor area	■ PRE provided but not worn properly ■ Breathing zone of a worker crouching over a grinder very close to the source of the dust	High – 6 times the MEL	■ Attach dust extraction to the grinder ■ Wet down ahead of chasing ■ Provide appropriate RPE ■ Ensure correct use of RPE
Chasing out mortar between bricks prior to re-pointing	■ Ineffective extraction fitted to a hand-held electric grinder ■ RPE correctly worn but not to the correct standard	High – 21 times the MEL	■ Attach dust extraction to the grinder ■ Provide appropriate RPE ■ Ensure correct use of RPE
Cutting paving kerb (33% silica) in an open area	■ Petrol-driven saw not fitted with a water spray or local exhaust ventilation	High – 12 times the MEL	■ Provide effective water suppression system to the saw

Table 19.1 Continued

Activity	Control measures	Exposure	Improvements required[a]
Cutting blue brick (32% silica) in an open area	■ Petrol-driven saw not fitted with a water spray or local exhaust ventilation	High – 5 times the MEL	■ Provide effective water suppression to the saw
Cutting breeze block (3% silica) in an open area	■ Petrol-driven saw not fitted with a water spray or local exhaust ventilation	High – twice the MEL	■ Provide effective water suppression to the saw
Cutting window openings in a concrete wall with a wall saw/cutting concrete with a floor saw	■ Water suppression on a saw used	Low – well below the MEL and also below the level regarded as significant	
General clearing and removing rubble	■ Hand sweeping with a brush	High – twice the MEL	■ Damp down rubble before clearing ■ Use mechanical means to sweep up ■ Provide appropriate RPE ■ Ensure correct use of RPE
General clearing and removing rubble	■ Use of a mechanical sweeper with rotating brushes and vacuum extraction	Medium – within the MEL but double the level regarded as significant	■ Provide appropriate RPE ■ Ensure correct use of RPE
Concrete crushing from a demolition job for use as hard core	■ Machine with an enclosed cab ■ Water jets fitted	Low – well below the MEL and also below the level regarded as significant	

Reproduced from the HSE's *Construction Information Sheet No. 36: Silica*
[a] To reduce exposure to below the MEL and so far as is reasonably practicable

Exposure to respiratable silica dust must be reduced to as low as is reasonably practicable and always to limits below the MEL.

Exposure even at limits below the MEL should be short and infrequent.

Is a COSHH assessment necessary for silica?

Yes. It will generally be for the principal contractor to complete the COSHH assessment for exposure to silica dust across the whole site.

If the works to be undertaken are carried out by a specialist contractor and are contained in a localised area, then the specialist contractor could complete the COSHH assessment.

The COSHH assessment must be reviewed by the principal contractor in order to assess whether site-wide preventative or control measures need to be implemented.

The COSHH assessment should evaluate the risk of exposure, describe the precautions necessary to control the exposure and set out how the control measures are to be monitored, supervised and maintained.

What are the health hazards associated with cement?

Cement can cause ill health mainly by

- skin contact
- inhalation of dust
- manual handling of sacks/bags.

Wet cement causes skin irritation and/or burns.

Skin irritation is usually termed dermatitis, and the symptoms are red scaly skin, itching, soreness and cracking of the skin.

Some people can be allergic to the properties of cement. Allergic reactions are usually demonstrated by the same skin symptoms as irritation.

People can become sensitised to cement over a period of time – particularly older workers.

Cement dust is a major health hazard, and breathing in the fine dust can irritate and clog up the lungs. In the short term, or for short exposures, the nose and throat will be irritated.

Cement dust can also contain silica – which is itself a health hazard.

Manual handling issues are associated with the weights of cement, although the manufacturing industry has generally limited the weight of bags to 25 kg. However, manual handling issues can be caused when sites use the on-site mixing silos and generate their own wet cement mix to be moved around the site. There may be a tendency to overload wheelbarrows, carry bags and son so that the mix is moved quickly.

What are some of the control measures that need to be put in place to control exposure to cement?

The first step is to look at the possibilities of eliminating the need for cement or substituting the need for cement by using a less hazardous substance – which is easier said than done! So, the control measures that will be practicable are ones that limit exposure to the hazardous substance.

Avoid direct skin contact with both dry and wet cement.

Issue gloves to be worn by all operatives.

Provide easily accessible wash stations with clean, warm running water, soap and towels.

Ensure operatives wear long-sleeved garments and long trousers. Reduce exposure to the skin.

If on-site silos are used, ensure that there is no need to directly touch either the dry material or the wet prepared cement.

Avoid the risk of anyone breathing in dust when mixing cement. Issue suitable respiratory masks to individuals, or, better still, mix the cement in one controlled area and provide local exhaust ventilation.

Manual handling injuries are prevented by providing appropriate mechanical aides such as wheelbarrows, sack trucks, trollies and so on.

Ensure your supplier is providing bags that are no heavier than 25 kg.

Remember: the health hazards of cement accumulate over time – exposure that is little and often but over prolonged periods (i.e. years) is more harmful than a brief burst of airborne dust.

What are the health hazards from solvents?

Solvents are chemical substances used as carriers for surface castings, such as paints, varnishes, adhesives, fungicides, pesticides and perhaps other specialist cleaning products.

Common solvents are

- white spirit
- xylene (found in paints, lacquers and adhesives)
- isobutanol (found in resins and paints).

Many products will contain mixtures of solvents.

Solvents can cause ill health through

- Inhalation of vapours as the solvent dries.
- Skin contact – some solvents are absorbed through the skin. Prolonged or repeated contact with liquid solvents can cause industrial dermatitis.
- Eye contact – causing irritation, inflammation and, potentially, severe eye damage.
- Ingestion through contaminated food, drink and smoking. 'Hand to mouth' contact is a common route – solvent on the hands then eating something can result in ingestion of the solvent.

It is not uncommon for people to drink solvents – often inadvertently because the liquid has been put in an empty drinks bottle that has not been relabelled.

Poorly labelled hazardous products on a construction site are a major issue, and indicate overall poor safety management practices.

The common symptoms of exposure to solvents are

- skin irritation
- eye irritation
- lung irritation – difficulty in breathing
- headache
- nausea
- dizziness and light headedness
- impaired co-ordination.

Once people are exposed to the hazardous vapours of the solvent they may become more accident prone because co-ordination has deteriorated – for instance, the inhalation of a vapour cloud of solvent while working from a ladder could cause the person to fall off the ladder, sustaining more serious injury from the fall than from the exposure to the vapour.

Loss of concentration is common, and reactions slow down. Feelings of intoxication can be experienced (the reasons why some people 'glue sniff' – which is inhaling solvent vapours).

High exposures to solvent vapours can have an immediate impact on consciousness, and can even cause death.

Solvent exposure is obviously increased when the work area is confined or poorly ventilated.

Large spillages can be a cause of wide spread vapour inhalation.

How is exposure to solvents controlled?

Work with solvent-based products is governed by the COSHH Regulations, and exposure to the hazardous substance has to be assessed, prevented, reduced or controlled.

Following the hierarchy of risk control takes us first to step 1:

- Prevent exposure. Does the solvent-based substance have to be used at all? Is there a non-harmful equivalent?

If step 1 isn't possible, then consider step 2:

- Substitute for a lesser hazardous product. Maybe a solvent-type product is necessary, but is there one that can be used watered down? Perhaps a water-based product can be used?

If the above fail, try step 3:

- Control exposure to the hazard.

What are some of the controls that can be implemented on site to control exposure to solvents?

If solvent-based products have to be used, then ensure that the area is well ventilated. Open the doors and windows, and install local exhaust ventilation if possible.

Avoid spraying solvents – fine mist and vapours are more likely to be inhaled, as greater quantities of the substance get into the air.

Paint solutions on the surface with a brush or dip the product into the solvent solution.

Store solvents safely with fully fitting lids.

Label all containers clearly with the correct contents. Ensure the hazard warning symbols are displayed on the container.

Dispose of solvent waste carefully. Use sealed and labelled containers, and dispose of as special waste.

Solvent-soaked rags, brushes and so on must also be safely disposed of – place them in sealed containers.

Consider the possibility of spillages, and have an emergency plan ready for immediate action. Spillages can generate toxic clouds of vapours that have very high concentrations of hazardous substances, and these 'clouds' can drift quickly over large areas on the air currents.

Emergency procedures will be listed on product data sheets, and the controls necessary should be incorporated into the COSHH assessment.

Remember, too, that many solvents are flammable and could significantly contribute to the fire risk on a site. Hazardous substances that are flammable should be included in the fire risk assessment, and, if necessary, the emergency services will need to be advised so that they are properly prepared to tackle any emergency situation without undue risk to themselves or others.

What PPE will be necessary if working with solvents cannot be avoided?

PPE is the last resort in controlling safety hazards and risks. But if the risks cannot be controlled by any other methods, then PPE must be issued.

For work with solvents, the following PPE would be a minimum:

- protective overalls
- gloves – suitable for use with solvents (e.g. some may disintegrate if exposed to contact with the solvent)
- face shields or full-face visors (these are better that goggles because they leave no part of the skin on the face exposed)
- respiratory equipment where ventilation is poor.

What other precautions are necessary for controlling solvent use and abuse?

Good operative training and the provision of information and instruction are essential.

People have the right to know what they are using during their working day, and they need to be aware of the hazards and risks to their health. They also need to know how those risks will be controlled.

Personal hygiene is important – good arm and hand washing, and not eating or drinking when the hands could be contaminated.

Contaminated overalls must be disposed of safely – they can be safely laundered, but the laundry must be advised of the risk of fumes and vapours to their own employees. Clothes should be aired before being re-worn.

What is Weil's disease?

Weil's disease is the common name for leptospirosis, a bacterial disease that affects humans and animals. The bacteria can be found in rats, cattle, dogs, horses and pigs.

Outbreaks of leptospirosis are usually caused by exposure to water contaminated with the urine of infected animals. To become infected, the person needs to either swallow the contaminated water or have a sore, cut or their eyes/mucous membranes come into contact with the contaminated water.

The illness takes approximately 2 days to 4 weeks from exposure to the infected water to develop, and it can last from a few days to several weeks.

Symptoms of Weil's disease are similar to flu, including high fever, headache, aches, chills, muscle pains, vomiting and, possibly, jaundice. If the disease is not treated, complications could develop, including meningitis, kidney damage and liver failure.

What duties does the principal contractor or any other employer have regarding Weil's disease?

All employers have the duty to protect their workers and other people from the hazards and risks associated with their work activity.

Where there is a risk of contracting Weil's disease, the principal contractor should complete a risk assessment, as should any other employer.

The risk of contracting the disease on a construction site will depend on

- the proximity to watercourses
- the likelihood of rodent infestation of the site.

Rats, a known carrier of the leptospirosis bacterium, are ever increasing and are often a major problem on construction sites due to readily available food sources, lack of pest-proofing to the site, disturbance of drains and so on.

If rats are known to be a problem, the client and/or the CDM co-ordinator should provide information to this effect, and should advise the principal contractor that there is a potential risk – no matter how small.

Principal contractors must then evaluate the likelihood of operatives being infected, and put in place control measures. Such control measures could be

- pest control by competent persons – either by putting controlled poison baits down or proofing the site from access
- eliminating pools of stagnant water
- keeping drains covered
- providing adequate supplies of plasters – waterproof dressings and so on
- providing information to site workers – the HSE publishes a useful pocket guide
- providing training to operatives
- ensuring good provision of welfare facilities, running water and so on
- ensuring appropriate PPE is worn by all those at risk of exposure (e.g. gloves and boots).

What are the risks from sunburn and heat stress?

In the building and construction industry, operatives will often have to work outdoors on building sites and in partially constructed buildings where there is no protection from the sun, or where work conditions are extremely hot. For employees such as bricklayers, concrete employees, roof carpenters and tillers working outdoors there is a risk of heat stress, sun stroke, sunburn and skin cancer from prolonged exposure to ultraviolet radiation from the sun.

The effects of exposure to the sun are cumulative – the longer the skin is exposed, the greater the risk of skin cancers, regardless of tan or skin pigment. Short-term risks include sunburn, blistering and peeling, acute skin reactions with certain drugs, ointments and creams, and sore, gritty, swollen eyes that are sensitive to bright light. Long-term risks include skin cancers, premature ageing, wrinkling, wasting skin tissues, excessive pigmentation, and clusters of tiny blood vessels and cataracts in the eye.

What duties does the principal contractor have to manage sunburn and heat stress?

The principal contractor should assess whether the day's tasks could cause heat stress or heat stroke, and consider ways of eliminating or reducing the risks by considering factors such as the weather forecast, the availability of shade, knowledge of the job ahead and an awareness of individual heat tolerance. Where possible, reschedule heavier work for cooler times of the day (or wait for cooler days), and rotate work so that workers spend less time on heavier tasks.

All operatives should be trained in safe work procedures for working in the sun and in hot areas of the workplace. Training should spell out the action required if an employee shows symptoms of heat stress or sun stroke.

What are the symptoms of heat stress?

The effects of heat stress range from simple discomfort to life-threatening illnesses such as heat stroke. Heat stress makes it difficult to concentrate on the job, which may be hazardous in itself. Signs of heat stress include tiredness, irritability, inattention and muscular cramps.

If you believe someone may be suffering the effects of heat stress, rest them in a cool, airy area and give them cool (rather than cold) fluids. Report the problem immediately to your supervisor or first aid officer.

Heat stroke

Heat stroke is common. A person suffering from heat stroke will stop sweating, and their body temperature will be high. Skin will be hot and dry. Confusion and loss of consciousness may occur.

Heat stroke is life threatening, and urgent treatment by a doctor is very important. While waiting for medical help to arrive, cool the patient as quickly as possible. Soaking the person's clothing with cold water and increasing air movement by fanning can do this. If the person is conscious, give them water to drink.

What can the principal contractor do to manage sunburn and heat stress?

The principal contractor and all other employers should make sure that all operatives are trained in ways to reduce the risk of sunburn, heat stress and heat stroke. Some of these are

- drinking lots of water, juice or soft drinks
- taking rest breaks in a cool place
- wearing cool, protective clothing such as a shirt with a collar and long sleeves and long trousers
- protecting the head – wear a hard hat
- applying SPF30+ sunscreen before exposure to sunlight as well as on overcast days – noses, lips, ears, necks and backs of hands need extra protection, sunscreen reapplied regularly, and communal dispensers provided

- if possible, working in shaded areas in the high-risk hours between 11 am and 3 pm
- not working near reflective surfaces such as water, cement, shiny metal or white-painted sheds in strong sunlight
- protecting the eyes from glare.

What are the hazards and risks associated with welding fumes?

There are many different types of welding processes, and the hazards and risks depend on what activity is being carried out. Nearly all the processes will produce metal fumes and gases. Some of the metal fumes will be

- aluminium
- cadmium
- chromium
- copper
- fluorides
- iron
- manganese
- lead
- molybdenum
- nickel
- tin
- titanium
- vanadium
- zinc.

Some of the gases will be

- ozone
- oxides of nitrogen
- carbon monoxide.

The health hazards associated with welding fumes can be

- asthma
- lung damage
- emphysema
- skin irritation
- lung cancer
- metal fume fever
- respiratory irritation
- siderosis (an accumulation of iron oxide in the lungs)
- lead poisoning

- unconsciousness
- death (especially from carbon monoxide).

The greater the exposure to the welding fumes and the higher the concentration of those fumes, the greater the risk of ill health.

What controls should be put in place on site to control the risks of exposure to welding fumes?

A wide range of controls can be implemented, and where exposure to fumes is a site-wide issue, the principal contractor must take the collective measures necessary to protect the majority.

Controls could be

- control access to the work area (allow only those who need to be there)
- locate the work away from openings, doors and windows (unless needed for ventilation, but proper planning is needed)
- provide good ventilation to the work area (approximately 5–10 air changes per hour)
- provide an air speed of approximately 1 m/s so as to clear fumes effectively
- provide local extraction equipment and ensure that it is working (check regularly and monitor with a manometer)
- discharge clean air outside of the building
- do not let fumes re-enter the building or discharge to other areas where people are working
- remove grease and any surface coatings from the materials to be welded
- ensure that the operative is in the correct position and that their head can be above any welding fumes generated
- extinguish the torch during rest periods so as to prevent nitrous fumes
- check the equipment to ensure that it is in good working order (e.g. all hoses and clips are in place)
- ensure operatives wear the correct PPE
- keep records of checks.

All welding work must be the subject of pre-planning and general discussion between all those undertaking it and those who may be affected by it. This will ensure that everyone is familiar with what works are to be undertaken, what controls will be in place, who will check for what and, most importantly, who will monitor standards and authorise any remedial or improvement works.

Appendix to Chapter 19

Figure A19.1 Hazardous substances symbols

Former European symbol	New GHS symbol	Description of risks
		Unstable explosives Explosives of divisions 1.1, 1.2, 1.3, 1.4 Self-reactive substances and mixtures, types A, B Organic peroxides, types A, B
		Flammable gases, category 1 Flammable aerosols, categories 1, 2 Flammable liquids, categories 1, 2, 3 Flammable solids, categories 1, 2 Self-reactive substances and mixtures, types B, C, D, E, F Pyrophoric liquids, category 1 Pyrophoric solids, category 1 Self-heating substances and mixtures, categories 1, 2 Substances and mixtures, which in contact with water, emit flammable gases, categories 1, 2, 3 Organic peroxides, types B, C, D, E, F
		Oxidizing gases, category 1 Oxidizing liquids, categories 1, 2, 3
No symbol		Gases under pressure: ■ Compressed gases ■ Liquefied gases ■ Refrigerated liquefied gases ■ Dissolved gases
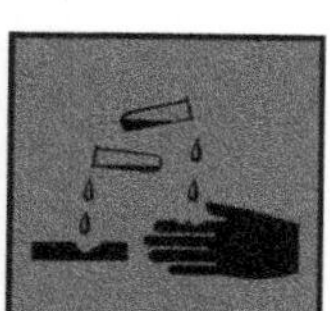		Corrosive to metals, category 1 Skin corrosion, categories 1A, 1B, 1C Serious eye damage, category 1

		Acute toxicity (oral, dermal, inhalation), categories 1, 2, 3
		Acute toxicity (oral, dermal, inhalation), category 4 Skin irritation, category 2 Eye irritation, category 2 Skin sensitisation, category 1 Specific target organ toxicity – single exposure, category 3
 		Respiratory sensitisation, category 1 Germ cell mutagenicity, categories 1A, 1B, 2 Carcinogenicity, categories 1A, 1B, 2 Reproductive toxicity, categories 1A, 1B, 2 Specific target organ toxicity – single exposure, categories 1, 2 Specific target organ toxicity – repeated exposure, categories 1, 2 Aspiration hazard, category 1
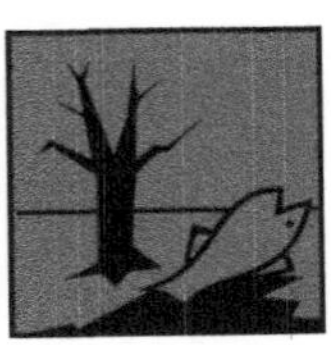		Hazardous to the aquatic environment ■ Acute hazard, category *1 ■ Chronic hazard, categories 1, 2

COSHH assessment

Company:
Address:
Contact:

Product:

Job task:

Application:

Equipment:

Safety data sheet attached: Yes No

Risk identification	
Hazardous component(s): Hazardous nature of component(s): Health hazards (known): Persons affected: Duration of exposure: Level of exposure:	
Risk category:	

Control measures

For users:

For persons in location:

Training:

Health surveillance:

Re-assessment:

Date of assessment/revision:

COSHH Assessment carried out by:

Company:

CDM 2015 Questions and Answers: A practical approach
ISBN 978-0-7277-6032-6

http://dx.doi.org/10.1680/cdmqa.60326.329

Chapter 20
Safe places of work

What are the duties of the site agent in respect of safe means of access to site?

Under Section 2 of the Health and Safety at Work etc. Act 1974, an employer is responsible for ensuring that employees and others have safe means of access and egress to their place of work.

A 'place of work' can be anywhere where an individual is expected to perform their duties, and can include buildings, rooms, open spaces, working platforms, roofs and scaffolding.

The site agent has a duty on behalf of their employer to ensure that employees have safe means of access and egress. In addition, the site agent has to ensure that persons other than those in their employ have safe means of access and egress: that is, access and egress without risk of injury or harm.

There should be no risk of being run over by vehicles, no risk of tripping or falling over materials, plant and so on, no risk of falling from any height, and no risk of objects falling on to people using the access and egress route.

Safe access and egress needs to be considered in relation to whether tools and materials need to be carried to and from the place of work. Vehicles colliding with pedestrians is a common accident cause on construction sites.

Under the CDM Regulations, the principal contractor has the duty to prevent unauthorised access to site, and where no principal contractor is appointed the contractor has that duty.

The CDM Regulations apply to all construction projects, and the site agent must include in the construction phase health and safety plan the details of safe access and egress to the site (the preferred pedestrian route, how and where materials and plant will be delivered, vehicular access, routes out of the site, emergency exits, etc.).

What is the requirement for safe access and egress under the CDM Regulations?

Regulation 17 of the CDM Regulations states that

> (1) There must, so far as is reasonably practicable, be suitable and sufficient safe access and egress from—
> (a) every construction site to every other place provided for the use of any person whilst at work; and
> (b) every place construction work is being carried out to every other place to which workers have access within a construction site.
> (2) A construction site must be, so far as is reasonably practicable, made and kept safe for and without risks to the health of any person at work there.
> (3) Action must be taken to ensure, so far as is reasonably practicable, that no person uses access to or egress from or gains access to any construction site which does not comply with the requirements of paragraph (1) or (2).
> (4) A construction site must, so far as is reasonably practicable, have sufficient working space and be arranged so that it is suitable for any person who is working or who is likely to work there, taking account of any necessary work equipment likely to be used there.

A 'place of work' is defined as any place that is used by any person at work for the purposes of construction work or for the purposes of any activity arising out of or in connection with construction work.

Places of work can therefore be

- the whole construction site
- specific areas of the site
- a work platform
- a roof
- a ladder
- a crane
- an elevating platform
- a trench or excavation
- a confined space
- the mess room
- the site office.

Can pedestrians and vehicles use the same access and egress routes?

It is important to keep vehicles and pedestrians as separate as possible, as a high number of accidents, including fatalities, occur when vehicles and pedestrians share access routes.

Regulation 27 of the CDM Regulations requires that construction work be so organised that, where practicable, pedestrians and vehicles can move safely and without risks to health.

In particular, traffic routes will not be considered safe if

- steps have not been taken to ensure that pedestrians, when using traffic routes, can do so without danger to their health and safety
- any door or gate opens directly into the traffic route without any clear view of approaching traffic or vehicles
- pedestrians cannot have a place of safety to view oncoming vehicles
- there is no adequate separation between pedestrians and vehicles.

So, pedestrians and vehicles can use the same traffic route but *strict* safety rules apply and, generally, a risk assessment will be necessary that identifies the hazards and risks.

What precautions can be taken to separate pedestrians and vehicles, or manage the risks to their health and safety?

A number of safety precautions can be taken to ensure the safety of pedestrians

- designate a safe walking area by painting hatch markings on a coloured walkway
- put up guardrails to delineate the walkway
- use a banksman to direct vehicles and pedestrians
- use mirrors so that drivers and pedestrians can see routes clearly
- ensure all vehicles have audible warning devices – not only for when they are reversing
- have a policy for all vehicles to ensure that lights and hazard lights are on during all hours – not just at dawn and dusk.

What will constitute a safe means of access to upper work levels?

Where practicable, permanent means of access to upper levels will be expected: for example, installation of a permanent staircase is preferable.

If the permanent structure cannot be installed, than a purposely designed temporary staircase is preferable.

When determining safe means of access, remember to consider what people have to carry to their place of work (e.g. materials and tools).

Ladders are not necessarily classed as a safe means of access, especially for longer-term projects. Ladders give immediate access to higher levels – but they are not necessarily safe.

Safe means of access to upper levels may be by way of hoists and lifts.

Mobile elevating platforms are safer for accessing high-level works (ductwork, ceiling works, etc.). They also provide a safe working platform.

What other precautions need to be considered for safe access and egress to the place of work?

Access and egress routes must

- be clearly lit at all times
- be free of obstructions and trailing cables
- have clearly defined steps or slopes
- have handrails if there are significant changes in level
- be of adequate size for the number of operatives using them
- be protected from hazards
- not to be underneath activities being carried out at height (e.g. under a crane sweep)
- be on stable ground
- be clearly visible
- be designed to be away from material delivery points and so on
- be identified in the pre-construction health and safety pack
- consider any other access or egress routes into the building (e.g. if the employer is still operating out of the building)
- not create crush points during clocking on/off times
- be adequately signed so that people know where they are expected to go
- have designated crossing points if vehicle/traffic routes need to be crossed
- be highlighted on a plan at the entrance to the site.

Is it necessary to provide a security point and signing in station to a construction site in order to manage safe access?

It is good practice to have a site control point because it is necessary to know who is and is not on site at any one time for fire safety purposes.

As a minimum, a designated signing-in place is essential at the site entrance. A site log for the signing in and out of the site is essential. This log can act as a reference log for fire safety purposes, and will help to check whether everyone is accounted for in the event of a fire.

On complex sites it is good practice to have a full-time security person operate the control point. This allows not only strict control of persons onto and off the site but also allows for checks on deliveries, skip removals and so on, and may help with the problem of materials and equipment thefts.

The pre-construction health and safety pack should indicate whether the client expects the principal contractor to provide a full-time operated access control point.

The principal contractor or, where one is not appointed, the contractor has to take responsibility under the CDM Regulations for preventing access of unauthorised persons to the site. A properly controlled access control point will discharge this responsibility, and even if someone should gain access to site and put themselves in danger, the principal contractor/contractor would then have a defence.

There has been a great deal of development over the years in computerised security systems. These are based on a swipe card and computer database, and record access and exit times. Many systems also combine photo-identity cards, and are used for training record purposes.

Does the construction site have to have more than one exit point?

The number of exit routes from a construction site depends on the number of operatives working on the site and the size and complexity of the site.

Regulation 31 in the CDM Regulations stipulates that a suitable and sufficient number of exit routes must be provided that can be used in an emergency. Any person on the site must be able to reach a place of safety quickly and without hindrance.

It is always good practice to have more than one exit route. The access or way in can also double up as the exit route, and often this will be the most popular route, as it is well known. There should be alternative exits from places of work if the travel distance to an exit route is more than 45 m if it is a clear run, or more than 30 m if the route is less direct.

On large sites, several exits from the place of work will be needed.

An exit route cannot be counted if it leads people back into the building or site or leads them to a dead end. An exit route must lead to a place of safety.

Do emergency exit routes have to have emergency lighting?

If emergency exit routes are to be used in poor daylight conditions (e.g. dawn or dusk), or daylight is generally poor (as in winter), emergency lighting will be essential on construction site emergency exit routes. Emergency lighting should come on if any artificial lighting fails or if visibility is low. Such emergency lighting must be automatic.

Emergency lighting is particularly essential on staircases leading out of the building – poor lighting could lead to people falling on the stairs, causing potential bottlenecks and crushing hazards.

Emergency signs should also be clearly visible, and the emergency fire signs should be illuminated. At the very least, photo-luminescent signs should be used.

What are the legal requirements for a safe place of work?

The CDM Regulations require that every place of work, so far as is reasonably practicable, be made and kept safe for, and without risks to health to, any person at work there (CDM regulation 17(2)).

Also, every place of work must, so far as is reasonably practicable, have sufficient working space and be so arranged that it is suitable for any person who is working or who is likely to work there, taking account of any necessary work equipment present (CDM regulation 17(4)).

The principal contractor has responsibility for the construction site on all sites that have more than one contractor, and where there is only one contractor, the contractor generally will have responsibility for the sites. It is therefore the responsibility of 'contractors' to keep the site safe and to ensure that adequate space is provided for operatives.

Many accidents happen on site because the site is congested and inadequate planning has contributed to poor work flows, poor siting of materials, restricted work areas and so on.

'Housekeeping' may not be the most exciting of terms, but it encompasses everything that needs to be done to keep a site safe.

Key areas for attention are

- correct storage of materials in dedicated areas
- clearing away of debris
- removal of trailing cables and so on
- removal of trip hazards
- control of airborne contaminates
- ensuring level surfaces.

A major contributor to workplace accidents is 'over-manning' of the site, especially at that point in many projects when they near completion and the site is flooded with operatives or tradesmen all working on top of one another to get the job done.

A site will soon become unsafe when operatives are working above other operatives (e.g. ceiling finishers are working in the area where the floor finishers are). Everyone wants to get their ladders or other equipment to their place of work, and soon there will be inadequate space to work – operatives will then take shortcuts or work unsafely.

Proper sequencing and planning should help to reduce the need for all trades to be present at the same time.

On projects that have more than one contractor, Health and Safety Executive (HSE) inspectors may want to see how the final phase of the project will be managed, and will expect to see the matter covered in the construction phase health and safety plan.

What are some recommendations for site set-up?

To achieve a high standard of health and safety within a construction area, as well as compliance with the relevant legislation and approved codes of practice, requires the careful set-up of site. To achieve this, the following areas must be addressed and appropriate action taken.

Statutory notices

It is mandatory that the following statutory notices are displayed in a prominent position within the site to which they relate (Figure 20.1):

- notification of the project on form F10 where applicable
- the Health and Safety Law poster
- the employer's liability insurance
- hazard warning signs and safety signs as deemed appropriate by any risk assessment.

Additional signage and notices are to be displayed in a prominent location where applicable, such as

- hazard warning signs
- mandatory signs
- prohibition signs.

Figure 20.1 Site safety notices (HSE Health and Safety Law poster: http://www.hse.gov.uk/pubns/books/lawposter.htm)

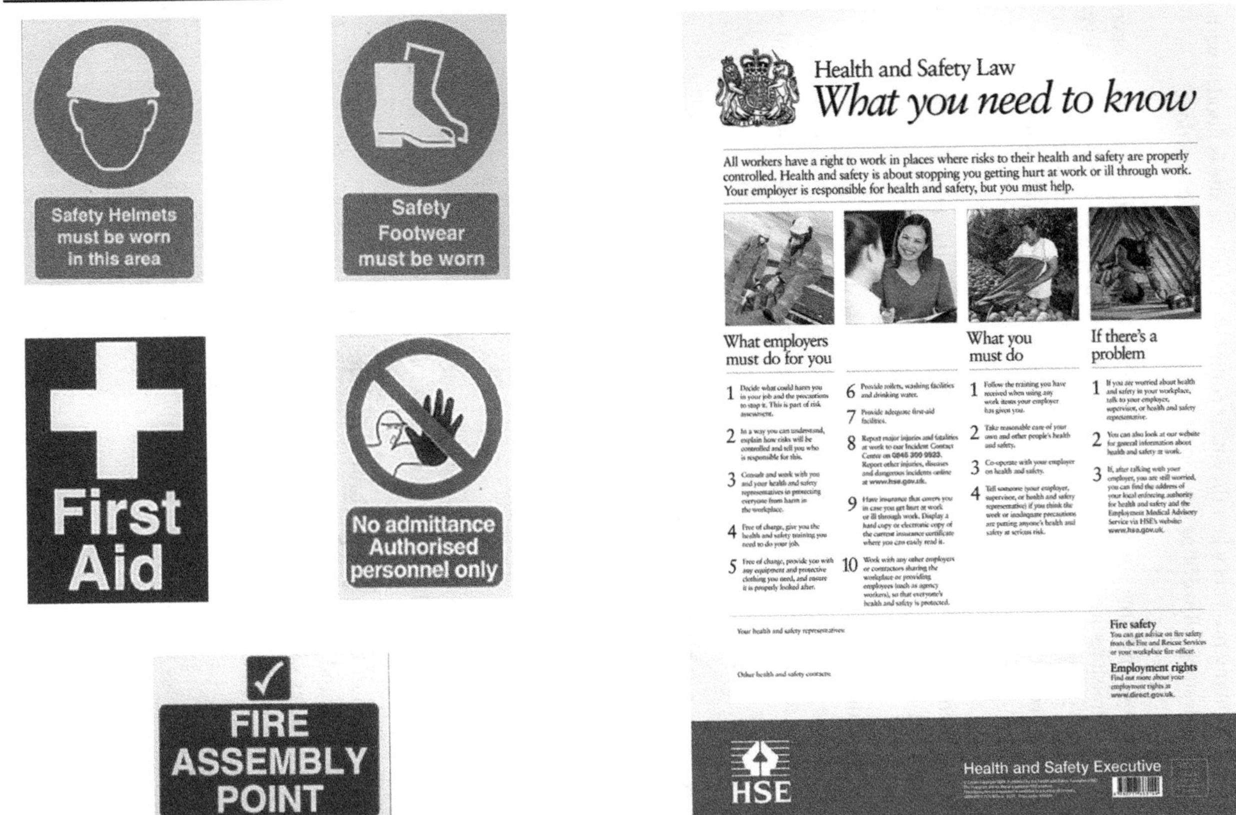

Table 20.1 Example equipment content of a health and safety station

Description	Quantity	Unit
Carbon dioxide or dry powder fire extinguisher	1	No.
Water or hydrospray fire extinguisher	1	No.
Large first aid kit	1	No.
Ear plugs (disposable)	1	Box
Face/dust masks (disposable) FFP.2 (or suitable type)	1	Box
Hard hats	3	No.
Safety goggles/glasses	2	Pair
Rubber gloves	2	Pair
Riggers gloves	4	Pair
Disposable gloves	1	Box
Waterproof plasters	1	Box
Accident book	1	No.
Eyewash (500 ml)	2	No.

Site safety station

The site safety station should be located in a prominent location close to the site entrance. This contains first aid, personal protective equipment (spare), and emergency equipment for use by site operatives and visitors (Table 20.1).

Segregation

To reduce the risk of theft, vandalism and unauthorised access to the site, the site must be secure. Adequate fencing and barriers to prevent access, but allowing emergency exit, must be in place. Signage stating ‘No unauthorised access’ should be displayed.

First aid

A large first aid kit, additional plasters and at least 1 litre of sterile eyewash should be located within the site safety station.

Signage showing the location and the name of the first aider should be displayed. It is expected that if five or more persons are present on site, a qualified first aider is present. For less than five personnel, an appointed person must be present.

Personal protective equipment (PPE)

Although PPE should be used only as a last resort to protect employees from hazards in construction areas, it is often the most practicable solution. Hard hats, safety boots, gloves and so on should be made available to operatives as required, and identified in the risk assessment. The health and safety station should contain additional PPE for use by

Table 20.2 The number of sanitary facilities required (from BS 6465-1:2006: sanitary installations)

No. of men at work	No. of water closets	No. of urinals	No. of wash stations
1–15	1	1	2
16–30	2	1	3
31–45	2	2	4
46–60	3	2	5
61–75	3	3	6
76–90	4	3	7
91–100	4	4	8
>100	An additional WC for every 50 (or part) men plus an equal number of additional urinals, plus an additional wash hand station for every 20 operatives		

Wash hand stations should be provided with adequate supplies of hot and cold running water. Water closets should preferably be wash-down water types

visitors and additional equipment for site crew if required. The station should also contain hard hats, ear plugs, dust masks, safety goggles and protective gloves.

Signs informing of the need for hard hats (remember that the regulations requiring the wearing of hard hats were repealed in 2014) and protective footwear should be displayed.

Welfare arrangements

Sanitary accommodation should be provided and must be available for use at all times. It must be kept clean, well lit and in working condition. The number of facilities and wash stations to be provided must comply with Table 20.2.

What are some of the steps that a principal contractor needs to take to manage the hazards and risks associated with slips and trips?

The principal contractor is responsible for ensuring that the construction site is managed effectively from a health and safety perspective and that hazards and risks to those who work on the site or manage the site are eliminated or reduced to acceptable levels.

Slips and trips are a key hazard on a construction site, and account for 27% of all major accidents (HSE annual accident statistics 2013–2014).

Principal contractors can significantly improve the incident rate of slips and trips by instigating a 'zero-tolerance' campaign for poor housekeeping, trailing cables, uneven surfaces and so on.

Some simple things to do are to

- keep storage areas tidy
- plan deliveries to minimise the amount of materials on site
- delineate footpaths around the site and ensure that they are free of obstructions, level and so on
- grit any icy surfaces in winter
- lay stones on muddy ground
- keep walkways clear
- tie up loose cables and so on
- instigate good housekeeping
- put fences or barriers around holes – or fill them in
- remove waste regularly from the work areas
- train the workers – implement a series of toolbox talks
- provide bins for waste
- ensure there is an adequate number of labourers on site to clear up
- clean up spillages
- highlight kerb edges or shallow changes in the level.

Add the category of slips and trips to any site-specific safety audit checklist.

Take action when hazards are spotted – as the HSE says, *see it: sort it*!

CDM 2015 Questions and Answers: A practical approach
ISBN 978-0-7277-6032-6

http://dx.doi.org/10.1680/cdmqa.60326.341

Chapter 21
Vehicles and transport

Vehicles and mobile plant are in common use on construction sites. What are the main hazards?

Vehicles, mobile plant and pedestrians don't go well together, and vehicles and mobile plant are the causes of many accidents on construction sites, many with fatal consequences.

The key hazards when using vehicles or mobile plant are

- moving vehicles running over operatives
- overturning vehicles or mobile plant
- reversing vehicles
- vehicles or plant too close to excavation edges
- vehicles or plant positioned on unstable ground
- vehicles or plant coming into contact with overhead power lines
- vehicles or plant coming into contact with buried services
- restricted access to the site
- restricted vision of vehicle and plant operatives
- untrained operatives driving vehicles
- leaving vehicles unattended, with keys in the ignition
- overloading and therefore overbalancing of mobile plant
- inadequate operating space when using vehicles or mobile plant.

How can hazards associated with vehicles and pedestrians be reduced to a manageable level on site?

Good site planning will help reduce hazards on the site. This could start at the design stage of the project and should involve the principal designer, where appointed under the CDM Regulations, so that all designers and others on the project are aware of what vehicles and plant may be needed on the site.

Provide safe entry and exit points with adequate turning space and good visibility for drivers.

Ensure that there is good lighting and visibility in areas close to pedestrians. Provide additional 'street lights', avoid blind corners and obstructions. Keep a good sight line for drivers of vehicles.

Plan to keep vehicles and pedestrians separate by having different entrances and exits for vehicles and people. Make sure each entrance and exit is properly signed, with clear text or pictogram signs that can be seen from all areas.

Provide separate, barriered walkways for pedestrians.

Where vehicles in particular have to be used in close proximity to pedestrians, provide a banksman.

What procedures need to be in place for reversing vehicles?

More accidents are caused by reversing vehicles than by any others, and the safety record of any site can be significantly improved by managing reversing vehicles.

Consider a one-way system for the site. This should be done at the planning stage, and the principal designer needs to co-ordinate with the design team and any contractors appointed any alterations to the site layout so that a one-way system can be accommodated. It may be necessary to slightly relocate the position of a building so that adequate access is made available.

If vehicles need to reverse around the site, ensure that they are fitted with audible reversing alarms and lights.

In any highly populated area, a banksman should be appointed to oversee the movement of vehicles, and the site manager should ensure that the person so appointed wears a high-visibility jacket and is properly trained in the tasks to be undertaken.

Prevent persons from crossing or moving in the area until a vehicle has completed its manoeuvres.

What are the good practice guidelines regarding routeways around the site?

It is sensible to set out clear and signed routeways across the site.

Avoid blind corners, sharp bends, narrow gaps, and places with low head room.

Avoid step gradients, adverse cambers, shafts and excavations.

Provide a temporary road surface – this not only helps to delineate the roadway but also helps to provide a level, stable surface on which vehicles travel.

Introduce a regular inspection and maintenance programme for all of the routeways. Pot holes create hazards that could cause vehicles to overturn or loads to dislodge. Regular repairs need to be instigated.

Keep vehicles away from temporary structures – especially scaffolds, as an accidental knock of a scaffold pole could cause the scaffold to collapse.

Erect speed limit signs (e.g. for a 5 mph *maximum* speed limit) throughout the site. Enforce the rules!

Erect directional signs to the entrance, exit or materials/delivery area.

Reduce the amount of mud transferred around the site and ultimately onto the highway by installing wheel washers at key locations (e.g. entrances and exits).

Protect any excavations with barriers that both highlight the excavation guard to prevent vehicles falling into the void.

Do not allow other vehicles to park on routeways, nor for routeways to be used for the storage of materials or plant.

What precautions are needed for vehicles that carry loads?

Make sure that vehicles that carry loads have been designed to carry loads – do not improvise and adapt vehicles.

Loads should be securely attached, and any loose materials (e.g. bricks or timber) must be secured with netting, tarpaulin or similar. Materials blowing off vehicles create major health and safety hazards, and falling material can cause serious injuries to site operatives and pedestrians.

Vehicles must not be overloaded, as they will often become unstable. Vehicles used for carrying or lifting loads will have a *safe working load* limit – make sure that this limit is understood and weights are within the limits. Overloaded vehicles are also difficult to steer and their braking efficiency is impaired.

Operatives who drive vehicles must be over 18 years of age.

All operatives must be properly trained to drive vehicles.

Case studies

A labourer riding as an unauthorised passenger on a dumper, fell and struck his head on the road. He died of head injuries.

An untrained labourer thought he could drive a dumper truck to collect some debris. The keys had been left in the ignition. He drove it over uneven ground, lost control and turned it over on the edge of an excavation. He was crushed to death as it overturned.

A poorly maintained dumper overturned into an excavation, causing the driver to be trapped underwater. The braking system had failed due to lack of routine servicing.

Are there any special precautions in respect of health and safety to be taken when using dumper trucks?

Compact dumpers, the official name for site dumpers, are responsible for approximately one-third of all construction site transport accidents.

The three main causes of accidents involving dumper trucks are

- overturning on slopes and at the edges of excavations, embankments and so on
- inadequately maintained braking systems
- driver error due to lack of experience and training (e.g. failure to apply the parking brake, switch off the engine and remove the keys before leaving the driver's seat).

The way to avoid such accidents is to pay attention to the following:

- ensure that all dumper trucks in use have 'roll over protective structures' (ROPSs) and seat restraints
- ensure that where there is a risk of drivers being hit by falling objects or materials, dumper trucks are fitted with 'falling object protective structures' (FOPSs).

Both ROPSs and FOPSs are legally required on all mobile equipment. If the equipment is hired, the hire company must ensure that the equipment complies with the law.

Also,

- ensure that a safe system of work operates when using dumper trucks
- provide method statements and risk assessments for using dumper trucks
- ensure that a thorough maintenance check is carried out and, in particular, that the braking system is checked

- check drivers' training records, driving licence and general competency to drive vehicles
- operate a strict no alcohol or drugs policy and prohibit anyone from driving who appears unfit to do the job
- plan any additional precautions necessary for using dumper trucks in inclement weather, and restrict their use in icy conditions
- provide adequate stop blocks to prevent dumper trucks falling into excavations and so on when tipping.

What are the key requirements for operating fork lift trucks safely?

There are a few simple measures that can be taken to manage the use of fork lift trucks on a construction site, namely

- manage lift truck operations using safe systems of work
- provision of adequate training for operators, supervisors and managers
- using suitable equipment for the job to be done
- laying out premises in such a way as to ensure that lift trucks can move safely around
- ensuring that lift trucks are maintained safely
- ensuring that the premises and site in which they are to be used are maintained in safe conditions (e.g. pot holes in access roads infilled).

What legislation applies to the use of fork lift trucks?

The main legislation includes the following:

- Health and Safety at Work etc. Act 1974
- Management of Health and Safety at Work Regulations 1999
- Provision and Use of Work Equipment Regulations 1998
- Lifting Operations and Lifting Equipment Regulations 1998
- Workplace (Health, Safety and Welfare) Regulations 1992.

Breaches of sections of the Health and Safety at Work etc. Act 1974 can incur unlimited fines, and fines for the other regulations can also be unlimited per offence.

How do I know if a fork lift truck operator is properly trained?

All fork lift truck drivers and operators must be able to demonstrate that they have received proper training from a competent instructor and from a recognised training body.

Operatives should have a valid certificate of training issued by one of the following:

- Association of Industrial Truck Trainers
- Construction Industry Training Board
- Lantra National Training Organisation Ltd
- National Plant Operators Registration Scheme.

Operators should be able to show the three stages of training as being completed:

1. basic
2. specific job
3. familiarisation.

Basic training should cover the skills and knowledge required to operate a fork lift truck safely and efficiently.

Job training should be site specific, and should include

- knowledge of the operating principles and controls of the fork lift truck to be used, especially where these relate to handling attachments specific to the job
- knowledge of any differing controls on the machine to be used, as this may be different to the one the operator trained on
- routine servicing and maintenance of the fork lift truck in accordance with the operator's handbook, as may be required to be carried out by the operator (e.g. pre-start safety checks, visual checks, and oil and brake checks)
- information on specific site conditions (e.g. slopes, overhead cables/beams, excavations, one-way vehicle routes, confined spaces, designated exits, speed limits and site rules).
- details of the tasks to be undertaken, the type of loads to be carried, hazardous areas, materials, unloading and unloading areas, and so on.

Familiarisation training should take place on the site, with the operator being supervised by a competent person. A full 'walk through' the site is recommended so that hazardous areas and the site layout can be explained. Also, emergency procedures must be covered during site familiarisation.

Even when operators have formal certificated training qualifications in operating fork lift trucks, it is sensible to record on a site training record the subjects covered in the site familiarisation stage and have the operator sign acknowledgement.

Safe driving practices: tips

- Check tyres, brakes, operating systems on all vehicles at the beginning of every day.
- Wear protective clothing and equipment (e.g. ear defenders, high-visibility jackets).
- Use dumper trucks with ROPSs and FOPSs.
- Use vehicles with reversing alarms and adequate lighting.
- Follow site speed limits.
- Check that any loads are evenly distributed.
- Do not overload vehicles.
- Know the characteristics of the vehicle in all weather conditions.
- Ensure that all drivers are trained and competent.
- Separate pedestrians from vehicles.
- Do not stand on vehicles when they are being loaded or unloaded.
- Select neutral gear, switch off the engine and remove the keys when stopping and leaving any vehicle.
- Keep to designated routeways.
- Make sure stop blocks are used.
- Do not use vehicles on steep inclines, adverse cambers, etc., without planning the job safely.
- Consider the unexpected and have a plan of action ready.
- Ensure good visibility at all times – do not overload dumpers, etc., so as to restrict vision.
- Check the area by means of mirrors, shouting, etc., before moving off from an area.
- Do not drive any vehicle if unfit to do so for any reason.

Case studies

An employer had to pay out nearly £30 000 in fines and costs for an accident involving the operator of a fork lift truck.

The employee who was driving the truck was crushed to death when it overturned.

A hole had been dug in the yard of a new building being constructed to lay sewer pipes. On the day of the accident, the employee was using the fork lift truck to load goods onto the back of a lorry. He reversed the lift truck across the yard, but one of the wheels slipped into the hole, and the vehicle toppled over. The driver was trapped beneath the vehicle, and suffered severe crushing injuries and died at the scene.

The employer failed to ensure that the surface of the yard was suitable for a fork lift truck. Everyone knew about the hole, but it was getting bigger due to water erosion. The employer had no system in place to monitor the condition of the yard and had not required a safe system of work, nor had he carried out risk assessments.

Prosecutions were brought under regulation 12 of the Workplace (Health, Safety and Welfare) Regulations 1992.

CDM 2015 Questions and Answers: A practical approach
ISBN 978-0-7277-6032-6

http://dx.doi.org/10.1680/cdmqa.60326.349

Chapter 22
Excavations and demolitions

What are the legal requirements governing excavations?

Regulation 22 of the CDM Regulations sets out the requirements for excavations.

First, though, regulation 2 of the CDM Regulations must be consulted in order to define the term 'excavation'.

In CDM regulation 2 (Definitions) an excavation is deemed to include

- earthworks
- trenches
- wells
- shafts
- tunnels
- underground working.

CDM regulations 22(1) to 22(3) set out the following:

> (1) All practicable steps must be taken to prevent danger to any person, including, where necessary, the provision of supports or battering, to ensure that—
> (a) no excavation or part of an excavation collapses;
> (b) no material form a side or roof, or adjacent, to any excavation is dislodged or falls; and
> (c) no person is buried or trapped in an excavation by material which is dislodged or falls.
> (2) Suitable and sufficient steps must be taken to prevent any person, work equipment, or any accumulation of material from falling into any excavation.
> (3) Suitable and sufficient steps must be taken, where necessary, to prevent any part of an excavation or ground adjacent to it from being overloaded by work equipment or material.

The final requirement in relation to excavations is to ensure that all supports and battering are properly inspected by a competent person. CDM regulations 22(4) and 22(5) state that

> (4) Construction work must not be carried out in an excavation where any supports or battering have been provided in accordance with paragraph (1) unless—
>
> (a) the excavation and any work equipment and materials which affect its safety have been inspected by a competent person:-
>
> (i) at the start of every shift in which the work is to be carried out;
>
> (ii) after the event likely to have affected the strength of stability of the excavation; or
>
> (iii) after any material unintentionally falls or is dislodged; and
>
> (b) the person who carried out the inspection is satisfied that construction work can be safely carried out there.
>
> (5) Where the person carrying out an inspection has informs the person on whose behalf the inspection is carried out of any matter about which they are not satisfied (under regulation 24(1)), construction work must not be carried out in the excavation until the matter has been satisfactorily remedied.

What does CDM regulation 22 actually mean?

The CDM regulation 22 states that excavations are hazardous and steps must be taken to prevent danger and harm to people.

Working in excavations requires a planned approach that identifies risk assessment, safe systems of work, monitoring and constant review of procedures.

The safety of all people associated with the excavation and those who may be in close proximity must be considered.

Regulation 22 doesn't define 'suitable and sufficient steps', and it is for the employer or person in control of the site to determine what needs to be done to comply with the 'suitable and sufficient' requirement.

Regulation 22 implies that a risk assessment should be carried out for excavation works. It is only when this has been done that the precautions necessary to prevent danger can be identified.

Once the hazard and risks have been identified on the risk assessment, control measures need to be implemented to reduce the risks to acceptable levels. The method statement will formulate the controls and 'system of work' that will need to be taken.

When must an inspection be carried out?

The guidance for inspections can be found in regulation 24 of the CDM Regulations 2015.

Excavations must be inspected

- before any person carries out work at the start of every shift
- after any event likely to have affected the strength or stability of the excavation or any part of it
- after any accidental fall of rock or earth or other material.

What information must be included in the report?

Regulation 24 of the CDM Regulations states that a report of inspection must include the following particulars:

- the name and address of the person on whose behalf the inspection was carried out
- the location of the place of work or part of that place inspected (including any plant and equipment or materials, if any)
- the date and time of the inspection
- details of any matter identified that could give rise to the health and safety of any person
- details of any action taken as a result of any matter identified that could cause safety issues
- details of any further action considered necessary
- the name and position of the person making the report.

There is no formal format for a report, but guidance and templates are suggested by the Health and Safety Executive. An example of a report template is given in the appendix at the end of this chapter.

What needs to be considered when planning work that involves excavations?

The key to safety in working in excavations is planning. Planning starts with knowing what ground conditions exist on the site and obtaining information. Before digging or starting any excavation, it is essential to consider and plan against the following:

- the collapse of sides
- materials falling onto people working in the excavation
- people and vehicles falling into the excavation
- people being struck by plant

- undermining nearby structures
- contact with underground services
- access to the excavation
- fumes
- accidents to members of the public.

It is important to ensure that materials needed to protect excavations are readily available on site *before* works to excavations start.

Materials required may include

- props
- trench sheets
- baulks
- timber planks
- ladders
- guardrailings
- signage.

The risk assessment should have identified what materials and equipment are necessary.

What needs to be done to prevent excavations collapsing?

Ground movement causes trench or excavation sides to collapse. Often, the weight of the removed spoil placed adjacent to the excavation imposes an increased loading that exacerbates instability of the ground. The ground type must first be identified. The first rule about excavation collapses is

- do not believe that any excavation is safe from collapse.

Excavations in semi-rigid soils, rock and so on may look safe but they may stand unsupported from 30 s to 30 days. But there is no knowing when the ground may move and the sides collapse.

Spoil heaps also move, and can cover and infill the excavation or surrounding areas.

There is no legal minimum height or depth of an excavation that must be shored or propped – there is often a depth of 2 m used as a guide for when excavations need propping. However, this is erroneous, and an excavation of only 1 m depth may be unsafe and need propping.

Any excavation where there is a risk of falling into the void must legally be guarded with edge protection.

Excavations are prevented from collapsing by proper shoring and propping of the side walls.

The sides to excavations could be made safe by battering them to a safe angle, but, generally, a safer option is to shutter the sides with timber planks and props, trench sheets specially designed for the job or specialised proprietary propping systems.

The risk assessment must determine at what intervals the propping takes place (e.g. continuous or at, say, 1.0 m intervals). If propping is not continuous, there may be a danger of partial collapse.

Propping or shuttering of an excavation must be properly planned by a competent person – for large and complex excavations a structural engineer may be necessary to design the system because the timber props, panels and so on will need to resist horizontal lateral loads and other forces.

Sheeting, vertical props and so on must be sunk deep enough into the ground to ensure stability. Many accidents have happened because the trench or excavation shuttering has collapsed because of inadequate design and installation.

Consideration must also be given to the risk of water ingress into the excavation or the water logging of adjacent ground, as this will weaken trench or excavation sides and increase the likelihood of collapse.

Water itself exerts great pressure, and could cause any shuttering installed to fail.

Excavation collapses are also prevented if there is an 'exclusion zone' placed around the excavation, as this will prevent vehicles and people walking too near the sides or on spoil heaps and so on, thereby undermining the stability.

Case studies

A ground worker was buried up to his neck as he worked in a trench of approximately 2 m depth. Trench sheets had been used but only at 900 mm intervals. They had not been properly secured. A heavy breaker was being used to remove boulders from the ground. The vibrations caused the shoring to give way, and the trench collapsed, trapping the ground worker.

Three ground workers were buried alive when the excavation they were working in collapsed. The trench was approximately 4 m deep, and because the operatives believed the ground to be semi-rigid as it was mudstone, they did not heed advice and shore up the sides of the excavation. With no warning of impending collapse, the trench side gave way, trapping all four workers. Three died but one survived with multiple fractures.

A labourer suffered multiple fractures to his head and upper body, as the brick wall situated next to the 900 mm trench he was digging collapsed into the excavation because he had undermined the walls foundations without realising. Even though his trench depth was only approximately 1 m deep, the foundations of the wall were higher.

What steps can be taken to prevent people or equipment and vehicles falling into excavations?

The easiest and most effective safety measure to take is to protect the opening of any excavation so as to prevent people and equipment falling into it.

Any excavation that could cause someone to fall 2.0 m or more *must* have adequate edge protection and guarding.

Guardrails must be substantial and withstand impact loads if people fall against them. They must have a top rail height of approximately 1000 mm, a mid-rail at 450 mm and a toe board at least 150 mm high.

Ensure that an excavation is properly signed with hazard warning signs. Also, ensure barriers and so on that are used to guard the void are highly visible.

Protect the excavation while operatives are working in it, so as to prevent persons or equipment falling in and potentially crushing the operatives.

When work in an excavation is finished for the shift, it might be safer to full board over the hole so as to prevent any risk of persons falling. Consideration must obviously be given to span widths and so on, so as not to create an even greater hazard from unsafe surfaces.

Keep all people, vehicles, plant and equipment away from excavations. Create an 'exclusion zone' around the excavation. Barrier off the excavation well beyond the edges. Erect hazard warning signs. Make sure that the excavation can be clearly identified. Consider any hazards such as poor lighting, restricted access, traffic routes and walkways.

Complete risk assessments for any tipping activity into the excavation, and ensure that stop blocks are used to prevent vehicles over-running. Consider whether hand balling infill spoil and so on will be safer, and stop the lorry away from the excavation edge.

Remember: health and safety is about *accident prevention*, so introduce safety procedures over and above the bare minimum (e.g. excavations shallower than 2.0 m may need edge protection).

What are some of the other hazards relating to excavations that need to be identified?

Excavations often involve the use of excavators, diggers, etc., and, often, more accidents are caused because people are killed by the vehicle or plant than by falling into or being crushed by the excavation.

Keep workers separate from moving plant. Where this is not possible, employ a banksman to guide the excavator and to protect people. Not all operatives on a site will be familiar with the hazards associated with excavations.

Develop safe systems of work that manages when, how, where and by whom excavation work is undertaken, and decide how such work may impinge on the safety of others.

Structures often become undermined because of excavation works, and adjusting buildings or structures can collapse into the excavation or in close proximity to it.

Structural engineers should be consulted when excavations are to take place adjacent to buildings or structures. Details of foundation depths and so on should be given to the principal contractor by way of the pre-construction health and safety pack and/or the principal designer.

Appendix to Chapter 22

Excavation checklist

Location of the excavation

1.	Have all underground pipes, utilities, etc., been identified and located?	Yes No N/A
2.	Has an adequate supply of timber, trench sheets, props and other supporting material been delivered to the site prior to excavation work beginning?	Yes No N/A
3.	Have Risk Assessments and Method Statements been provided?	Yes No N/A
4.	Is the material and method of support chosen suitable for supporting the sides?	Yes No N/A
5.	Is the method chosen for putting in the timbering safe (i.e. negates the need for persons to work in an unsupported trench)?	Yes No N/A
6.	Is the angle of any slope or batten sufficient to prevent collapse or land slip?	Yes No N/A
7.	Is there safe access and egress to the excavation (fixed ladders, tiered steps, etc.)?	Yes No N/A
8.	Are there barriers all around the excavation to stop people falling in?	Yes No N/A
9.	Does the excavation affect the stability of any neighbouring buildings?	Yes No N/A
10.	Are materials stacked safely, away from any edges and not likely to cause collapse of the excavation sides or slip into the excavation?	Yes No N/A
11.	Is any plant or material stored near the excavation sides so as to destabilise the excavation walls?	Yes No N/A
12.	If vehicles tip into the excavation, are proper vehicle stops, banksmen or other safe methods of working used to prevent the vehicle tipping into the excavation?	Yes No N/A

13.	Are there any risks of fumes or other noxious or hazardous fumes, vapours or mists drifting into the excavation causing operatives to be affected?	Yes No N/A
14.	Is there an emergency plan for evacuation or dealing with unexpected events?	Yes No N/A
15.	Are all operatives trained and competent to work in the excavation?	Yes No N/A
16.	Other matters – describe:	

Actions required

Indicate the steps that need to be taken to ensure that the excavation can be used safely.

Signed: ______________________________

Position: ______________________________

Company: ______________________________

Date: ______________________________

Time of inspection: ______________________________

CDM 2015 Questions and Answers: A practical approach
ISBN 978-0-7277-6032-6

http://dx.doi.org/10.1680/cdmqa.60326.359

Chapter 23
Temporary works

What are temporary works?

'Temporary works' is a widely used expression to describe an 'engineered' solution used to support or protect an existing structure or permanent works during construction or to support an item of plant or equipment or the vertical sides or side slopes of an excavation, or to provide access.

The construction of most types of permanent works will require the use of some form of temporary works.

What are some examples of temporary works?

- Earthworks:
  - trenches
  - excavations
  - temporary slopes and stockpiles.
- Structures:
  - formwork
  - falsework
  - propping
  - façade retention
  - needling
  - shoring
  - edge protection
  - scaffolding
  - temporary bridges
  - site hoarding and signage
  - site fencing
  - cofferdams.
- Equipment/plant foundations:
  - tower crane bases supports
  - anchors and ties for construction hoists

- ground works to provide suitable locations for crane bases/plant erection
- piling rigs.

Why does temporary work need to be managed as a separate topic?

Temporary works, if not managed effectively, have the potential to cause major collapse of a structure, which could result in multiple fatalities and major injuries.

The accident statistics for fatalities and major injuries caused by temporary works are unacceptable, and the Health and Safety Executive (HSE) has identified the subject of temporary works as being a major inspection topic across all sectors of construction.

Are there any legal requirements to manage temporary works?

Yes. The CDM Regulations and the associated HSE guidance document are directly applicable to the design and management of temporary works.

The definition of a structure in the CDM Regulations includes 'any formwork, falsework, scaffold or other structure designed or used to provide support or means of access during construction work'.

Designers under the CDM Regulations include temporary works engineers, including those designing auxiliary structures such as formwork, falsework, façade retention schemes, scaffolding and sheet piling.

What exactly is falsework or formwork?

Falsework is any temporary structure used to support a permanent building or structure while it is not self-supporting, either in new construction or refurbishment.

Any failure of the falsework could lead to the collapse of the permanent structure.

Formwork is the term given to either temporary or permanent moulds into which concrete is poured.

Usually, falsework supports the moulds or formwork.

Formwork can be timber, engineered moulds, plastic moulds, permanent plastic moulds or fibre-reinforced plastic.

How is it best to manage temporary works?

Systems and procedures for the management of temporary works should follow the BS 5975:2008 + A1:2011 standard, 'Code of practice for temporary works procedures and the permissible stress design of falsework'.

Site-specific arrangements for temporary works should be included in any construction site health and safety plan.

The temporary works plan should be included alongside any demolition phase plan.

BS 5975:2008 + A1:2011 recommends specific arrangements for managing temporary works, namely that

- a competent temporary works co-ordinator is appointed
- a temporary works supervisor is appointed
- a temporary works file is created
- designs and calculations are recorded for all temporary works
- arrangements are put in place to monitor and review all temporary works and designs as they are being installed
- arrangements are put in place to inspect any temporary works prior to loading
- regular monitoring and inspection of all works is carried out
- proper planning, review and monitoring is undertaken when striking any temporary works.

What role does the principal designer, appointed under the CDM Regulations, have to play in managing temporary works?

Designers include temporary works designers. Arrangements should be included in construction phase plans for the stability of structures, including temporary structures, and works of excavation are included as requiring management.

The principal designer has to take reasonable steps to ensure co-operation between permanent and temporary works designers and that arrangements are in place to ensure that designs are compatible and that permanent works can support any loadings from temporary works.

What should the principal designer be doing on projects that involve temporary works?

The principal designer should ensure

- competence of the principal contractor's temporary works designers
- that the initial construction phase plan includes arrangements for controlling significant site risks such as temporary works
- adequate designer co-ordination and co-operation when there is interaction between temporary works and permanent works
- adequate information, instruction and training for all those involved in the project has been considered and information shared among the team

- that permanent works designers minimise the need for temporary works
- that the permanent design considers the design and erection of temporary works
- provision of information on their designs to inform the temporary works designers.

Who enforces health and safety in respect of temporary works?

The HSE is the enforcing authority for temporary works, as they fall within the definition of construction works.

The HSE has been so concerned about the safety of temporary works that it has made them a specific inspection topic and focus and has included site inspections as a priority in its annual work plans.

The HSE's aims for temporary works are

- to promote awareness and knowledge of the importance of managing temporary works effectively
- to improve contractors' management arrangements for temporary works
- to increase the competence of those involved in temporary works management and design
- to reduce accidents arising from temporary works failures.

HSE inspectors will target temporary works when carrying out site inspections, and also carry out programmed inspections of projects where temporary works are indicated or implied on the works descriptions on the F10 notification. In particular, they will target medium/large principal contractors to review in more detail their temporary works management arrangements and 'track back' to designers and clients.

What happens when temporary works go wrong?

Generally, when temporary works go wrong, a structure or building collapses with often fatal consequences – people get killed by falling masonry, scaffolding, walls and so on or are crushed under the weight of the structural collapse.

Temporary works are often inadequately planned and designed, and the main causes of something going wrong are

- lack of adequate structural stability
- inadequate foundations
- overloading
- unavailability or inappropriate parts
- lack of competent temporary works designers or erectors
- poorly constructed works.

Investigations into temporary works incidents have revealed

- no or inadequate temporary works procedures
- no or an inadequate temporary works co-ordinator appointed
- no or inadequate temporary works design
- no or inadequate investigation of ground conditions, underground services and the structural condition of existing buildings
- unauthorised changes to the temporary works design
- lack of a competent contractor/builder/developer.

Are there any specific areas of temporary works design that require more attention in order to prevent failure?

There are *five* well-defined causes of failure in temporary works

1. incorrect estimation of the loads to be supported
2. design error or loading programme changes after the design has been completed
3. inadequate detailing and/or execution of points of load transference
4. inadequate horizontal lacing and diagonal bracing to resist lateral loads
5. inadequate foundations.

When loads are being calculated, it is not just the weight of the permanent structure that is required but also the temporary weight of the materials (e.g. wet concrete). Often, such calculations get overlooked.

The loading sequence is important, as are

- plant and material loads
- the competency of the contractor.

My employer has told me that I have to manage temporary works on the site. I don't feel competent. What do I need to know?

Before being able to manage temporary works you will need to be able to demonstrate that you are competent to do so.

Competency isn't just having information – it is having the experience, knowledge and training to understand the hazards and risks associated with the task and being able to manage the health and safety of those tasks in an efficient and capable manner.

You will need to know exactly what role your employer is expecting you to take on and what responsibilities you will have. There will be a vast difference between managing compliance with a written management procedure for temporary works and being expected to design temporary works.

Designing temporary works is a specialist task, and must only be undertaken by competent persons – as defined in the CDM Regulations. Any designer of temporary works is a designer under the CDM Regulations, and will have statutory responsibilities.

If you are being asked to supervise temporary works, you may be undertaking the duties of a temporary works supervisor.

The temporary works supervisor assists the temporary works co-ordinator in the supervision and checking of the temporary works. This will include the supervision of the erection, use, maintenance and dismantling of temporary works.

Checks will need to be carried out of the temporary works during the construction process.

Any modifications or adaptations to the temporary works will have to be assessed and reported to the temporary works co-ordinator.

So, in order to adequately supervise, you will need to have a thorough understanding of temporary works, including an understanding of

- loads
- falsework and formwork
- scaffolding
- underpinning
- propping
- ground conditions
- the effects of water on ground conditions
- structural stability
- piling
- earthworks
- excavations
- the influence of the site environment on temporary works (access, services, adjacent paths/roads, drainage, etc.)
- cranes and load lifting
- the effects of wind conditions and wind loadings on structures.

Are excavations temporary works?

The excavation itself may not necessarily be temporary works, but the support and shuttering required to ensure that the excavation sides do not collapse are temporary works.

The CDM Regulations specifically cover excavations, and state that all practicable steps must be taken, where necessary, to prevent danger to any person, including, where necessary, the provision of supports or battering to ensure that

- any excavation or part of an excavation does not collapse
- no material from a side or roof of, or adjacent to, any excavation is dislodged or falls
- no person is buried or trapped in an excavation by material that is dislodged or falls.

Steps should also be taken to prevent persons falling into an excavation.

Who should design the temporary support for excavations?

The temporary works designer should design the supports for excavations, as a technical assessment of loads, ground conditions, potential for heave and so on will need to be undertaken.

The range of supports available will include

- hydraulic walling frames
- manhole shores
- trench boxes
- drag boxes
- slide rail systems
- traditional raker shores
- self-supporting driven sheet piles.

Deciding on which system to use requires competency (i.e. knowledge, experience and training).

What would a risk assessment for temporary works design for excavations include?

The following table outlines a risk assessment for temporary works for excavations.

Hazard associated with activity	Precautions to be taken to reduce the risk	Comments
Inadequate design	Appoint a suitably qualified and experienced designer.	Essential to provide the designer with a comprehensive design brief, including a representative borehole or trial pit

Hazard associated with activity	Precautions to be taken to reduce the risk	Comments
Lack of co-ordination and communication between parties	Appoint a responsible person as a temporary works co-ordinator to ■ compile the design brief ■ check the submitted temporary works design ■ assess the risk and monitor ■ ensure compliance with the temporary works design ■ ensure kit is assembled and installed according to the manufacturer's instructions ■ monitor changes in the site conditions when compared with the design brief	Check frames and struts are positioned as specified Check the required 'toe-in' of sheets/piles Check that the excavated ground and the groundwater regime are as assumed in the design Monitor activities
Excavation is relocated after the design has been submitted	Check if the appropriate borehole log has been used Check if the surcharge details have changed Check if the depth has changed Refer back to the temporary works designer to re-work the design	Check if access and lifting requirements have changed
Contaminated ground	Check with the principal designer if contaminated ground is to be expected in the location of the excavation Continually monitor by sight, smell and use of gas detection equipment the excavated profile and excavated material for possible contamination If the excavation is known to be in contaminated land, check with the shoring supplier that structural integrity of the shoring equipment will not be compromised through contact	Method of work to be specified to accommodate contamination If suspected contaminated ground is encountered, cease work immediately and inform the safety co-ordinator

Hazard associated with activity	Precautions to be taken to reduce the risk	Comments
Soil profile encountered different to that used in the temporary works design	Continually monitor the soil profile Temporary works co-ordinator to check the actual profile against the design profile	If the actual profile varies from the design profile, immediately inform the temporary works designer, so they can check the design stability If the variance in profile is deemed to be significant, stop work until the design has been rechecked
Groundwater characteristics and control	Regularly monitor and record groundwater characteristics (e.g. rate of flow and strike levels) Temporary works co-ordinator to check the actual groundwater regime is consistent with the design assumptions Ensure the proposed method of groundwater control is as per that used in the 'basis of design' (e.g. do not use a sump pump when well point de-watering has been specified)	If the actual regime varies from the design, immediately inform the temporary works designer, so they can check the design stability If the variance in profile is deemed to be significant, stop work until the design has been rechecked
Change in the depth of the excavation	If the depth is to be varied, immediately inform the temporary works designer to re-work the design based on the altered depth	Do not exceed the design depth without the design being reworked
Change in the plan dimensions of the excavation	If the depth is to be varied, immediately inform the temporary works designer to re-work the design based on the altered dimensions	Do not exceed the stated dimensions without the design being reworked
Change in the surcharge	Ensure surcharge assumptions are correct, e.g. ■ the weight of excavator ■ the position of spoil ■ the position of adjacent roads and batters	Immediately inform the temporary works designer of change so the design can be reworked

Hazard associated with activity	Precautions to be taken to reduce the risk	Comments
Introduction of a new surcharge	Monitor if new surcharges are introduced, e.g. ■ a new haul road adjacent to the dig ■ a large plant positioned near the dig ■ spoil dumped near the excavation	Immediately inform the temporary works designer of change so the design can be reworked
Unknown structures/ services encountered	Note the position and nature of the structure and services, and inform the temporary works designer to assess the impact on the design	
Ground reduction details	Ensure all ground reduction details (batters, etc.) as specified in the temporary works design are complied with – if not, inform the temporary works designer to re-work the design	
Stability of the shoring system during use	Ensure that the system has been installed as per the temporary works design Continually monitor equipment for signs of overloading (e.g. deflection and deformation)	If in doubt, refer to the equipment supplier Do not use the shoring system as a work platform for the storage of materials Keep shoring equipment clear of spoil and debris
Stability of adjacent structures and batters	Continually monitor adjacent structures and batters for movement If the excavation is in or around embankments or slopes, carry out a global stability analysis (e.g. slip circle checks)	If in doubt, refer to the temporary works designer
Instability of excavation during extraction of the shoring system	Work to an approved method of work to ensure the stability of the excavation during the extraction of equipment Identify where 'short-term stability' is being assumed when considering the stability of the excavation during extraction	Do not remove any cross struts unless specifically allowed for in the temporary works design

Hazard associated with activity	Precautions to be taken to reduce the risk	Comments
Temporary works design checked by an external organisation	Prior to commencing work, ensure all relevant external organisations are issued with the temporary works design for checking purposes	
Change of the method statement	If method statements incorporating temporary works designs are amended, ensure that the original design assumptions are not compromised	

What is a temporary works register?

It is useful for a temporary works register to be prepared for any project. This is a written record of the works and actions taken in respect of temporary works. It should contain a list of all the identified temporary works items associated with the project.

Appropriate headings would be

- Design brief number
- Date issued
- Short description of temporary works
- Date required
- Category of temporary works
- Designer
- Design checker
- Date design complete
- Date design checked/approved
- Erection complete and checked, Permit to load issued or Permit to strike/dismantle issued.

What is the design brief?

A design brief should be designed for each item of temporary works so that all decisions can be assessed in the context of what the design intent was.

All data relevant to the design of the temporary works should be included (ground conditions, use of the structure, permanent works proposed, environmental conditions, proposed temporary works schemes, objectives, etc.).

What should the temporary works co-ordinator for a project do?

- Co-ordinate all temporary works activities.
- Ensure that various responsibilities have been allocated and accepted (e.g. designers, design checkers, erectors and site supervisors).
- Ensure that risks identified at the design stage, as well as assumed construction methods and loading constraints, are incorporated into the temporary works design brief.
- Ensure that the temporary works design is satisfactory.
- Ensure that a design check is carried out, covering the concept, structural adequacy and compliance with the design brief.
- Ensure that the design is made available to relevant parties.
- Register or record all drawings, calculations and other relevant documents relating to the final design.
- Ensure that those responsible for on-site supervision receive full details of the design, including any limitations and guidance notes.
- Ensure that risk assessments and guidance notes are prepared covering the safe erection and dismantling sequences.
- Make checks at appropriate stages during construction of temporary works.
- Ensure that appropriate maintenance is carried out to temporary works (e.g. to façade retention structures).
- After a final erection check, issue the permit to load.
- Once the permanent works have attained adequate strength, issue formal permission to dismantle the temporary works and specify any relevant sequence.
- Ensure that the temporary works are dismantled in accordance with a defined procedure.

Are there any temporary works that are considered lower risk and that could be designed and managed by a competent contractor without the need for specialist designers?

There are some temporary works that could be classed as simple or with potentially lower risks, such as

- standard scaffold
- formwork less than 1.2 m high
- hoarding and fencing up to 1.2 m high
- simple propping schemes – one or two props
- internal hoarding systems and temporary partitions not subject to wind loading
- shallow excavations less than 1.2 m deep/high.

All the above have the potential to create considerable hazards and risks, and fatalities and/or major injuries could result. *All* temporary works have to be properly planned and managed, no matter how simple they first appear.

Risk assessments and method statements are essential.

What type of temporary works would be medium risk?

- Falsework up to 3.0 m high.
- Formwork for columns and walls up to 3.0 m high.
- More complex propping schemes – multiple props at a single level.
- Needling of structures up to two storeys high.
- Excavations up to 3.0 m deep/high.
- Safety net systems fixed to robust primary members.
- Simple designed scaffold.
- Temporary roofs.

And what type of temporary works are considered high risk?

- Falsework and formwork over 3.0 m high.
- Trenchless construction.
- Working platforms for cranes and rigs.
- Tower crane bases.
- Façade retention schemes.
- Flying and raking shores.
- Complex propping schemes.
- Needling of structures greater than two storeys.
- Ground support schemes greater than 3.0 m deep.
- Complex designed scaffold.
- Cofferdams.
- Bridge erection schemes.
- Jacking schemes.
- Complex structural steelwork.
- Pre-cast concrete erection schemes.
- Hoarding and fencing over 3.0 m high.

All of the above will need to be designed by a *competent* designer, and the works should be under the supervision of a temporary works co-ordinator and a temporary works supervisor.

What checks should be carried out before a temporary works scheme is erected?

Before erection commences, the temporary works design should be checked for

- the design concept
- strength and structural adequacy (including foundations and lateral stability)
- compliance with the design brief.

The design check should be carried out by an independent, competent person. Where temporary works are complex, it is imperative that the design checker is independent and not swayed by team/project pressures to speed up the checking process and so on.

Safe erection, use and dismantling of formwork: a good practice guide

Falsework is any temporary structure used to support a permanent structure while it is not self-supporting, either in new construction or refurbishment. Any failure of falsework may lead to the collapse of the permanent structure. This could cause injury or death to those working on or near to it, as well as loss of time and money.

The causes of many past failures were foreseeable and could have been prevented by proper consideration when planning, erecting, loading or dismantling the falsework. Investigations into falsework collapses have identified a lack of co-ordination between the various trades and suppliers of falsework as a major cause.

Failures often occur on fairly simple structures erected by smaller falsework contractors, who may not employ design staff.

Contractors' responsibilities include

- preventing the falsework collapsing under load
- ensuring that those constructing and dismantling falsework can carry out their work safely, with particular regard to preventing falls from height
- minimising risks to the health and safety of others who may be working on, or passing by, the construction activity – risks could arise, for example, from falling materials, wind-blown plywood or scaffold boards, noise and dust.

On sites where there is a principal contractor and a number of contractors, the principal contractor is responsible for the safe co-ordination of all activities on site (including liaison with specialist proprietary suppliers).

Management

The law requires falsework to be erected and dismantled only under the supervision of a competent person. As early as possible, a person should be appointed for each site as a falsework co-ordinator, with responsibility for co-ordinating the various items and stages of use of the falsework.

- The falsework co-ordinator is commonly known as the *temporary works co-ordinator*.
- On a large contract, the appointed co-ordinator might be a suitably qualified engineer, whereas on a small building contract the role might be taken on by the site agent or foreman appointed by the contractor.
- Whoever it is, the appointed co-ordinator is responsible for ensuring that correct falsework procedures are followed and that operations are carried out safely.

Falsework should be constructed in accordance with BS 5975:2008 + A1:2011, 'Code of practice for temporary works procedures and the permissible stress design of falsework'.

Planning

All concerned should contribute towards the preparation of a design brief, which should serve as the starting point for subsequent decisions, design work, calculations and drawings. Initial planning should cover

- what needs to be supported, and how it should be done
- how long the falsework will be in use.

Design

All falsework should be designed. This will vary from the use of simple standard solution tables and graphs to site-specific design and supporting drawings. Designs should be checked. The designer of the temporary works and the person interpreting the standard solutions are commonly known as *temporary works designers*.

The term 'designer' has a broad meaning, and includes

- anyone who specifies or alters a design, or who specifies the use of a particular method of work
- contractors carrying out design work as part of their contribution to a project
- temporary works engineers, including those designing formwork, falsework, scaffolding and sheet piling.

Designers have duties under the CDM Regulations, and must

- identify the hazards
- eliminate the hazards, if feasible

- reduce the risk by design
- provide the information necessary to identify and manage the remaining risks.

Standard solutions for scaffolding, falsework and so on that comply with recognised codes of practice are often used. Such solutions, when used with the recommended procedures, will normally meet the risk control requirements of the CDM Regulations. However, where such solutions are adapted, consideration needs to be given to whether the risk is still effectively controlled.

Particular consideration should be given to the following:

- stability requirements, lateral restraint and wind uplift on untied decking components
- designing falsework that can be erected, inspected and dismantled safely, including how striking will be achieved (it may be craned into position in one piece but could have to be removed piecemeal)
- selecting adequate foundations or providing information to ensure adequate foundations are used
- providing the information that the temporary works co-ordinator will need to manage the interface between the falsework and the permanent structure safely.

Materials

Falsework should be constructed, or adapted, so as to be suitable for the purpose for which it is used:

- it should be strong enough and stable in use
- damaged components should not be used
- different proprietary components should not be mixed, unless expressly approved by the designer.

Erecting the falsework

Before erection begins, a risk assessment should be carried out and a safe system of work developed. A method statement that includes how all the hazards are to be managed should be prepared. This should be read and understood by those doing the work.

To ensure safety, falsework should be stable at all stages of erection and be regularly checked. Only 'working drawings' and not 'preliminary drawings' should be used. Erectors should know

- where to start
- whether the equipment supplied is the same as that ordered

- at what stage checks or permits are required
- whether checks and permits have already been carried out or issued.

Loading

Once complete, all falsework should be inspected and certified as ready for use (a written permit-to-load procedure is strongly recommended). The frequency of subsequent inspections will depend on the nature of the temporary works. They should be carried out frequently enough to enable any faults to be rectified promptly.

Striking and dismantling

The temporary works co-ordinator should agree the time of striking for each section of the falsework (a written permit-to-strike procedure is strongly recommended).

During dismantling, ensure that workers can work safely and cannot be injured by falling objects. A sequence for dismantling should be agreed and detailed.

Training

Temporary works co-ordinators, and those erecting and dismantling falsework, should be competent and trained in the safety of falsework.

(Source: HSE Information Sheet No. 56)

CDM 2015 Questions and Answers: A practical approach
ISBN 978-0-7277-6032-6

http://dx.doi.org/10.1680/cdmqa.60326.377

Chapter 24
Underground and overground services

What are the hazards from underground services?

Every year a significant number of workers are injured through contact with underground services.

Underground services include

- electricity mains
- gas mains and distribution pipes
- water mains
- telecommunications cables
- sewers and foul drainage systems.

All have the potential to cause fatal injuries both to those carrying out the excavation and others in the vicinity of the work. Non-fatal injuries include severe electrical burns, multiple fractures, head injuries and the effects of breathing toxic fumes.

Information on underground services should be available *prior* to commencing any excavation. Investigations should be carried out to confirm the information, or to ascertain the information.

Check utilities service drawings and plans.

Look around for obvious signs on site (valve covers, road repairs, inspection covers, etc.).

Use pipe and cable locators, and mark the ground accordingly.

If there is any possibility that unexpected services will be encountered, follow safe digging practices:

- dig trial holes with hand tools
- use spades or shovels

- do not throw or spike spades or shovels into the ground – use gentle foot pressure
- use any picks, forks and similar tools with care to break up larger clumps of soil, stone and so on
- try to excavate alongside the anticipated service pipe or cable so as to avoid a 'direct hit'
- use an 'air digging' tool, which will remove soil by air pressure and prevent impact damage to services
- assume all services are live until a proper disconnection and permit to work system indicates otherwise
- when buried services have been uncovered, wait for identification before proceeding
- support any disturbed pipes or cables
- do not use pipes and so on as foot or hand holds to exit a trench
- label any services found.

Ensure that a comprehensive emergency plan and procedure has been drawn up in case of an accident, explosion, fire and so on.

Who is responsible for detecting underground services?

Information on the presence of services, whether underground or overground, should be provided by the client as part of the pre-construction information. If the client does not have specific information, they must commission surveys.

The principal designer should take responsibility for finding out all relevant information about services on the site. If necessary, site exploration works should be commissioned and accurate site plans drawn up.

Where pre-investigative works have not been possible, the client must ensure that the principal contractor has been given adequate time to carry out site surveys and obtain information before any major construction works are commenced.

The principal contractor, and indeed all contractors, have a duty to share information with those who need it, and so, if any contractor finds out about underground services, they must share that information by way of the principal contractor. The principal contractor must include the information in the construction phase health and safety plan for the project.

In order to ensure that future risks are managed, the information of residual underground services and any overhead services should be included in the health and safety file for the project.

What are the duties placed on designers?

Designers have a duty to reduce or 'design out' the risks arising from damage to underground services. Having reduced the risks to a level as low as reasonably practicable by design, information should be provided to those doing the work about the risks that remain. In most cases, the best way of informing contractors and individuals doing the work is by providing this on working drawings.

The principles of prevention should be considered when designing structures in areas where there could be underground services. The main principle of prevention is to avoid the risk whenever possible. You will need to know if there are underground services present so that you can amend the design to avoid them where possible.

For building work, re-siting the services away from the work is often a reasonably practicable means of avoiding the risk. Ask the service owner/operator to do this, and include adequate notice.

Other options to re-siting the services may include

- repositioning or redesigning structures or parts of structures, to ensure that services are avoided during the work
- arranging for the supply to be disconnected during the work.

If neither of the above is possible, a method should be selected that avoids the services: for example, by using ground beams to bridge or span the services.

For electricity cables more than other services, there may be a need to make them dead for the work to proceed safely. Contact electricity companies as early as possible to allow them to isolate supplies. Plan project schedules to allow sufficient time for this to happen.

If the cable cannot be made dead, an alternative safe way of doing the work will be required. Permanent structures such as buildings should generally not be built over services, nor should services be encased in concrete, as this may introduce additional risks to construction workers and can prevent future access to the services. If it is not possible to avoid building structures over any service, make arrangements with the utility to relocate the services in a duct or something similar.

Consider the location of underground gas pipelines when planning building, excavation, landfill or other such work. Such activities may either cause damage to the pipelines or deny access to them for maintenance purposes. Make suitable arrangements for future access and maintenance before undertaking the work.

Consider ancillary work, including the erection of perimeter fencing and walling, or the position of temporary and permanent roadways onto the site that may affect underground services at the site perimeter. Early identification and planning are essential to control risks during the entire construction phase of the project, including enabling works.

What steps should be taken for identifying exposed services?

Once underground services have been uncovered, failure to identify them correctly is a common cause of accidents. A wide variety of materials and colours have been used for services over the years. Some services may be very similar in appearance, and some services run in ducts made of various materials, including asbestos cement, making them difficult to identify. Adopt the following approaches until you have positively confirmed the identity of the service.

Water pipes, electricity cables and telecommunication cables may be covered in black plastic. If any black plastic service is found, assume it is a live electricity cable. Iron and steel water pipes and gas pipelines may appear very similar. If any such pipe is uncovered, treat it as if it were a gas pipe.

Always treat continuously welded steel pipes as if they contain a hazardous or high-pressure fluid. At collieries, beware of electricity cables, some of which are yellow or blue and may be mistaken for other services.

On some building sites, beware of electricity cables being placed in yellow service pipes or blue water pipes. Where there is any doubt about the identity of an exposed service, treat it as an electricity cable or gas pipe until proved otherwise.

Most underground cables are laid in trenches between 450 mm and 1 m deep. Some high-voltage cables will be deeper. However, never assume depths: cables may be found at shallower depths.

In most cases, there will be no permanent surface marker posts or other visible indication of the presence of an underground cable. Even if no cables are shown on plans, or detected by a locator, there may still be cables present that could be live, and a close watch should be kept for any signs that could indicate their presence.

A cable is positively located only when it has been safely exposed. Even then, digging should still proceed with care, as there may be other cables and services adjacent or lower down. In addition, some lines of 11 kV or greater can be laid out as separate single-phase cables, spread out up to 600 mm across, particularly near cable joints. Where it is clear there is a risk of damage to a cable during the course of any work, the owner may wish to be present on site.

What steps need to be taken when working near overhead power lines?

Many deaths and serious injuries are caused when operatives come into contact with overhead power lines.

Work must be carefully planned to avoid accidental contact.

Wherever possible, the electrical power supply to the overhead cables should be disconnected and isolated. Often, contact with overhead power lines is inadvertent because the operative is unaware of the risks – often the assumption is made that the lines are dead.

Carrying long scaffold tubes, handling long metal roof sheets, carrying long ladders, and using lifting plant or mobile elevating work platforms – all can contribute to the likelihood of hitting an overhead power line with fatal consequences.

A permit to work system should operate where overhead power lines exist.

Work should, where possible, be away from the lines.

The lines may be temporarily switched off.

Plan what works need to be carried out in the area of overhead power cables – measure the height of the cables and specify that any long pieces of material or equipment shall be less than the height of the cables, including a safe tolerance area.

Prevent vehicle access to overhead power lines by creating barriers and restrictive areas.

What are the steps for detecting services?

There are different levels of survey:

- desktop study
- desktop study and site investigation
- physical identification of the services.

Desktop study involves requesting and considering the service drawings from the owners of underground services. This should be done for all projects that involve excavation or penetrating the ground.

Desktop study and site investigation involves using the information from the desktop study to assist a physical inspection of the site (looking for physical signs such as inspection hatches, reinstated excavations, street lights and telecoms boxes) and a survey using detection tools.

In addition to the above, physical identification of the services involves taking steps to detect and identify the underground services through trial holes to verify their location, depth and identity. It may also involve passing a tracing device through a pipe or tunnel. The level of survey needed will depend on the nature of the work site. Some congested urban locations will require a more detailed survey than some brown and green field sites. The decision on the necessary level of survey should be informed by an assessment of the likelihood of underground services being present, based on the information obtained for the work site. The principal designer should make this decision at the planning stage in consultation with the client, contractor and surveyor. The results of the survey should be shared with the designers, and recorded in a clear, usable format on working drawings to be shared with those working on the site and, where possible, marked out on site.

Those doing the survey need to have sufficient knowledge and experience in the use of survey equipment and techniques. They will need to understand the limitations of the equipment, the effect of differing ground conditions on the survey results, how to survey a given area effectively, and to appreciate the limitations of plans and drawings provided by the service owners. The position of any services in or near the proposed work area should be pinpointed as accurately as possible using a detecting device in conjunction with up-to-date service plans and other information that provide a guide to the possible location of services and help to interpret the signal.

Take account of any indications that underground services exist, such as the presence of lamp posts, illuminated traffic signs, gas service pipes entering buildings, pit covers, pipeline marker posts, evidence of reinstated trenches and so on. However, if there are no such indications, this does not mean that there are no underground services. Plans do not normally show the position of gas service connections, and their existence should be assumed. It may be possible to estimate the probable line of the service connection pipe from the gas meter position, or from the point of entry into the premises. Gas plant may be shown by valve boxes, pits and housings. However, covers for valve boxes and pits will sometimes not show clearly whether gas is the service present: if in doubt, contact the gas transporter.

Anyone selecting detection tools and survey methods must understand the range of methods and tools and their limitations. In particular, they need to be aware of the potential for false readings or signals in certain techniques, as they may lead to inaccurate information being included in the plan of work and, in turn, lead to a false sense of security.

Emergency work is required. How should I manage the risks from underground services? (HSE, 2014)

There will be occasions where it is necessary to do emergency work to repair damaged services in order to make them safe or restore them following damage. Often, this emergency work may just require a temporary fix before a permanent repair can be done, though circumstances may mean it is appropriate to make a permanent repair immediately.

Routine work that does not arise from a safety-critical situation, and could be planned, should not be done as emergency work.

Emergency work still requires planning and assessment of the risks before and while carrying it out. Attempts must be made to obtain information about underground services in the area. The principal designer should take the responsibility for co-ordinating the surveys and information.

Inevitably, there will be greater emphasis on the work on site to locate services and excavate safely. The work must be done by individuals with sufficient knowledge and experience, and be reviewed as it is undertaken.

Those managing emergency work must balance the risk of potential damage to underground services against the continuing risk from the emergency situation. For example, the risk of damage to cables while isolating a gas pipe to stop a leak must be weighed against the risk of fire and explosion from the continued leak. In all cases, make sure that the overall risk to safety is not increased.

Case studies

Failing to have a safe system of work

An operative using a powered pneumatic breaker lost both his arms when he hit a live 11 kV mains cable. The cable had been identified within a general area and the operative set about digging the trench. He had dug down only approximately 0.5 m when he hit the mains cable.

The investigation showed that no safe system of work was in operation at the site. No detailed investigation had been done to locate the exact position of the cable, and no safe digging techniques were being followed.

The employer of the operative was prosecuted for failing to have a safe system of work, and so was the principal contractor. Other contraventions included

- no risk assessment
- no training record.

Negligence – failing to advise

A construction worker was hired by a principal contractor to help install modular buildings as part of a city's redevelopment project. The construction project involved the converting and redeveloping of a former naval base into a mixed-use development. A construction worker was working in a trench when a drill that he was operating hit an underground energised electrical line.

The construction worker suffered extensive physical injuries, including an amputation of his left arm, lower right leg, and transtibial amputation of his left foot. He also suffered extensive third-degree burns over approximately 20% of his body that required skin grafting, and a brain injury that left him with severe neurological problems.

At the time of the accident, employees of the utility company were present and observing operations at the construction site to identify the locations of underground electric utilities and to try to avoid the risk of injury to the construction workers. The construction worker had been told by an employee of the utility company that he was drilling in a safe location.

The construction worker sued the principal contractor for negligence, claiming that he failed to advise him of the dangerous condition created by the underground energised electrical lines.

CDM 2015 Questions and Answers: A practical approach
ISBN 978-0-7277-6032-6

http://dx.doi.org/10.1680/cdmqa.60326.385

Chapter 25
Electrical safety on the site

Is it against the law to use mains electricity on a construction site?

No. The law does *not* state that you cannot use 230 V electrical supply on a construction site. You could do so if you manage the risks adequately.

However, 230 V electricity is a major hazard and has great potential to kill and cause injury to people if used unsafely or in conditions that are unsafe – and managing health and safety is about eliminating or reducing hazards to acceptable levels.

Requiring contractors to use 110 V power tools or 12 V power tools reduces significantly the likelihood of injury from the electrical supply. It is therefore a safe system of work to use 110 V or 12 V power tools.

What safety steps must I take if 230 V is the only available power on site?

If power tools have to operate on 230 V, then the following are required as an absolute minimum:

- a detailed risk assessment
- all equipment and cables/leads to be in good repair
- use of residual current devices (RCDs) – also called 'trip' devices – that operate at 30 mA with no time delays
- regular daily checks
- avoidance of hazardous environments (eliminate damp conditions, water, dust, use of chemicals, etc.).

All equipment must be connected to RCDs or, if agreed by a competent electrician, an RCD can be fitted to the main power supply so that it protects the whole of the electrical system.

Risk assessments must be constantly reviewed, and updated should conditions on the site change.

Doubly insulated equipment should be used where possible.

RCDs must be installed with great caution, as a poorly installed one will create more hazards than it solves.

RCDs must be

- kept free of moisture
- kept from dirt
- protected from vibration
- protected from mechanical damage
- properly installed and enclosed
- properly protected with sealed cable entries
- checked daily by using the test button
- considered as a safety device and not the complete eliminator of a hazard.

Tools must be connected only to sockets protected by RCDs.

Cables and leads supplying equipment running on 230 V must

- be protected from damage
- kept at high levels
- protected inside impact-resistant conduit.

For additional safety, use armour-protected flexible cables/leads.

What should the site agent do if a contractor comes onto the site and wants to use 230 V equipment when the site operates on 110 V?

A well managed and planned construction project should have processes in place to pre-advise contractors and sub-contractors about the rules of the site so that they are aware of what is expected before they arrive on site. If a contractor or sub-contractor arrives on site ignoring information about site safety it should immediately throw up questions about their competency.

The first response will be to refuse them access to the site on the grounds that they will not be following the site rules. However, a more realistic approach needs to be taken, as turning the contractor away may affect the programme, which in turn could create more serious safety issues.

A full review with the contractor is necessary:

- What needs to be done?
- Who needs to do it?
- Where does it need to be done?
- Where are the risk assessments?
- Where are the method statements?
- What work may affect others in the area?
- What controls for safety does the contractor propose using (e.g. RCDs)?
- How does the contractor propose to manage the hazard of 230 V tools and cables?
- What permit to work methods does the contractor propose?
- Can the work area be isolated from others so as not to expose them to unnecessary hazards?

If the risk assessments and method statements demonstrate that a safe system of work can be operated, that the risks to *all* operatives can be controlled and that the work is of relatively short duration, then it *may* be possible to allow the work to go ahead.

Using 230 V on a construction site is a serious hazard that poses great risk of electrocution, electric shock, electric burns and fire, but it is *not* against the law, and, provided the risks can be managed, it may be permissible. However, best practice and all safety guides advocate the use of 110 V power tools or cordless 12 V power tools.

Does all electrical equipment have to have an inspection and test certificate?

The Electricity at Work Regulations 1989 require the 'duty holder' (usually the employer) to ensure that electrical equipment is: 'maintained in a safe condition'. The regulations do not specifically require elaborate testing and inspection, but if these steps are needed to ensure that the equipment is safe, then they must be carried out.

A simple and, usually, effective, procedure to check electrical equipment for faults and damage is visual inspection. Approximately 95% of faults or damage can be detected by carrying out a thorough visual inspection.

All operatives should be trained on what to look for when carrying out a visual inspection, and everyone should be encouraged to use the techniques *every* time they use electrical equipment.

Before an electrical hand tool, RCD, lead or cable extension, and so on is used, check that

- no bare wires are visible
- the cable covering is not damaged

- the cable cover is free from cuts and abrasions
- the plug is in good condition
- the plug pins are not bent, missing, the casing cracked and so on
- there are no joined cables with insulating or other tape – any cable connectors must be proprietary
- the outer cable covering (sheath) is gripped properly where it enters the plug or equipment (e.g. no coloured wires showing)
- the equipment itself is in good condition with no parts missing, screws missing, cracks in the casing and so on
- there are no overheating marks, burn or scorch marks around the plug, socket or equipment
- the RCD is working correctly.

Some equipment will require more definitive evidence that it has been formally inspected, and a certificate is a good record.

Equipment that has a high risk of becoming unsafe through misuse should be inspected and tested, usually using a portable appliance test (PAT). The PAT will check the earth continuity.

If equipment is going to be used in hazardous areas (e.g. confined spaces), it should be inspected and tested regularly, with suitable records kept.

If equipment is tapped to earth, it should be regularly checked to ensure that the earth wire is intact and connected. Abuse of 110 V tools by multi-users will often cause wiring to come loose. This may not be seen by visual inspection. A formal process of combined testing would be sensible.

Table 25.1 lists common types of equipment and the inspection and testing required.

What precautions which need to be taken when using electrical equipment in hazardous areas?

Hazardous areas include

- confined spaces
- areas with flammable atmospheres (e.g. paint shops)
- chemical plants
- areas subject to dust and particle accumulation.

Wherever possible, electrical equipment should not be used in hazardous areas.

Table 25.1 Tips: inspection and testing of electrical equipment

Equipment	Voltage	User checks	Formal visual inspection	Combined inspection and testing
Battery-operated power tools	Less than 25 V, usually 12 V	No	No	No
Torches	Less than 12 V	No	No	No
25 V portable hand lamps	25 V, powered from a transformer	No	No	No
50 V portable hand lamps	50 V, secondary winding, centre tapped to earth	No	No	Yearly
110 V portable hand tools, extension leads, site lighting	Maximum of 55 V, tapped to earth	Weekly	Monthly	Before first use on site, then 3 monthly
230 V portable and hand-held tools	230 V mains supply	Daily, every shift	Weekly	Before use and then monthly
230 V portable floodlighting, extension leads	230 V	Daily	Weekly	Before first use, monthly
230 V 'fixed' equipment (e.g. lifts, hoists and permanent lighting)	230 V	Weekly	Monthly	Before first use and then 3 monthly
RCDs – portable		Daily, every shift	Weekly	Before first use and then monthly
RCDs – fixed		Daily, every shift	Weekly	Before first use and then 3 monthly
Mains equipment in site offices (e.g. fax machines and photocopiers)	230 V	Monthly	6 monthly	Before first use, then yearly

Source: Health and Safety Executive (2013), *Maintaining Portable and Transportable Electrical Equipment*, 3rd edn (HSG 107)

Before any work is carried out in these areas, a *permit to work* should be issued. This will detail all of the safety precautions to be taken for working in the area, including which type of electrical power tool is appropriate.

In any hazardous environment, 12 V portable tools are safer to use than any other.

Are there any special precautions to take in respect of health and safety when commissioning and testing of fixed plant and equipment takes place prior to project completion?

Equipment will need to be commissioned during the course of the development works, and this is likely to involve the use of 230 V mains power supply.

There is a health and safety risk when using a 230 V supply on a construction site, and the following procedures should be adopted.

- A permit to work on electrical equipment should be implemented across the site.
- The principal contractor should be required to produce an agreed plant commissioning programme that will incorporate information about the site and from other sub-contractors/nominated contractors who require plant and equipment to be commissioned
- All sub-contractors must be informed about the commissioning of plant and equipment, and the principal contractor must display a series of notices on either the individual equipment or at the entrance to the area that advises everyone that the equipment is now live and connected to 230 V.
- Secure means of isolation must exist for each part of the installation on which work is carried out. Padlocks and keys should be clearly identified and held by the principal contractor.
- Circuits that are not in use (e.g. used for commissioning and thereafter not required until the site is handed over) should be locked off, and doors to switch rooms, fuse boxes and so on should be locked shut with keys kept by the site agent.
- All electricians should be suitably trained to work with high voltages and the Electricity at Works Regulations 1989 must be complied with.
- Each employer will be responsible for their own employees, and must ensure suitable information, instruction and training has been provided, that their employees are competent to do the job and so on.
- The principal contractor will be responsible for ensuring that site safety rules are followed and that all contractors and sub-contractors are kept informed of all site safety matters.

Mechanical plant and equipment (e.g. air handling units, lifts and escalators) will be operating from mains voltage while the site is still a construction site. The hazards and risks of 230/240 V on site must therefore be properly and safely managed.

All cables, ductwork and so on carrying 230/240 V power must be labelled with a hazard warning sign (yellow and black) as follows:

WARNING: LIVE 230/240 V POWER
DO NOT TOUCH

or similar.

All electrical cables running adjacent to work areas *must* be encased in armour-plated conduit (or other recognised material) and sufficiently and visibly labelled

LIVE ELECTRICS

The principal contractor must ensure that *all* operatives on the site are aware that the mechanical and electrical equipment may be operational. Toolbox talks and notices displayed in mess rooms will provide timely reminders.

Essential steps for safe isolation

- Step 1: check with all interested parties that isolation of the equipment/supply is required.
- Step 2: identify the supply system.
- Step 3: locate and identify the circuit/supply to be isolated.
- Step 4: select the approved voltage indicator device and verify on a known supply and on a proving unit.
- Step 5: verify that the circuit/supply is functional using an approved voltage indicator device. If not operational, dead testing will be required to verify the circuit/supply.
- Step 6: identify suitable means for isolation.
- Step 7: isolate the circuit/equipment by switching off, using circuit breakers, removing fuses and disconnecting.
- Step 8: fit appropriate lock-off devices and locks, and retain the key or security number for padlocks.
- Step 9: fit warning labels/signage for the isolation and works being undertaken.
- Step 10: verify that the circuit/system/equipment is isolated. Use an approved voltage indicator to prove the circuit is dead.

- Step 11: recheck that the approved voltage indicator device is still functional on the same known supply as in step 4 and on the same proving unit as in step 4.
- Step 12: the circuit/equipment should now be safe to carry out works. Recheck using the voltage indicator regularly and on leaving and returning to the works/task.

Case study

An employee, an electrical fitter, received 33 000 V burns when he climbed live electrical apparatus in a substation when he believed it to be a safe working zone.

His injuries resulted in the amputation of both arms.

The company did not have an adequate demarcation for safe working zones, and no monitoring of any working areas was carried out. Staff received inadequate training about working in the atea and how to set out safe working zones.

A prosecution was brought under section 3 of the Health and Safety and Work etc. Act 1974, and a fine of £50 000 was levied on the employer.

CDM 2015 Questions and Answers: A practical approach
ISBN 978-0-7277-6032-6

http://dx.doi.org/10.1680/cdmqa.60326.393

Chapter 26
Safe systems of work

What is a safe system of work?

There is no legal definition of what constitutes a safe system of work, and it will be a matter of 'fact and degree' for the courts to determine.

Precedence was, however, set in the Court of Appeal in the 1940s, when the then Master of the Rolls said

> I do not venture to suggest a definition of what is meant by system. But it includes, or may include according to circumstances, such matters as physical lay-out of the job, the sequence in which work is to be carried out, the provision … of warnings and notices and the issue of special instructions.
>
> A system may be adequate for the whole course of the job or it may have to be modified or improved to meet circumstances which arise: such modifications or improvements appear to me to equally fall under the heading of system.
>
> The safety of a system must be considered in relation to the particular circumstances of each particular job.

This means that a system of work must be tailored for each individual job.

What is the legal requirement for safe systems of work?

The Health and Safety at Work etc. Act 1974 sets out specifically in section 2 that the employer is responsible for

> the provision and maintenance of plant and systems of work that are, so far as is reasonably practicable, safe and without risks to health.

The Confined Spaces Regulations 1997 also require employers to establish a safe system of work if work and entry into confined spaces cannot be avoided.

What are the provisions for a safe system of work?

Generally, developing a safe system of work will involve

- carrying out a risk assessment
- identifying hazards and the steps that can be taken to eliminate them
- designing procedures and sequences that need to be taken to reduce exposure to the hazard
- considering whether certain things or actions need to be completed before others
- designing permit to work or permit to enter systems
- writing down the procedure
- training employees and others.

Is a method statement the same as a safe system of work?

Generally, the two are similar, and a method statement is a written sequence of work that should be followed by the operator in order to complete the task safely.

This is the same as a 'safe system of work', which is a sequence of events needed in order to reduce or eliminate the risks from a hazard that in itself cannot be eliminated.

Method statements are common in the construction and maintenance industries, and they are often required under construction laws: for example, demolition works must always be accompanied by method statements.

The Control of Asbestos Regulations 2012 require all work with asbestos to be supported by a work plan or plan of work (i.e. a method statement or safe system of work).

When employees have to undertake hazardous tasks or when they have to work in hazardous environments, it is incumbent on the employer to ensure the safety of their employees and others. They must therefore decide *how* the job is to be done in order to ensure that their employees are kept safe.

What is a permit to work or permit to enter system?

A permit to work or a permit to enter is a formal system of checks that records that the safe system of work that has been developed for the process is implemented.

The 'permit' process usually applies to hazardous areas, and is commonly used for

- entering and working in confined spaces
- working of electrical plant
- working on railways or traffic routes
- working in chemical plants

- working in hazardous environment
- hot works
- working near or over water
- working near overhead power lines.

It is a means of communication between site management, supervisors and those carrying out the hazardous work.

Essential sections of a permit to work or a permit to enter are

- clear identification of who may authorise particular jobs
- limitations in respect of anyone's authority
- clear guidance as to who is responsible for determining the safety procedures to be followed
- clear guidance as to what safety precautions are necessary
- details of emergency procedures
- information that must be relayed to site operatives
- specification of instructions and the training and competency requirements
- the condition for which the permit is relevant
- the duration of the permit
- the 'hand back' procedure
- monitoring and review procedures.

There are no set forms to use – employers should devise their own so that it suits the situation they want to control. Examples of permit to work forms are included in the appendix to this chapter.

What is a hot works permit?

'Hot work' is any work using open flames or sources of heat that could ignite materials or vapours/explosive atmospheres in the workplace or area. Examples are

- welding
- burning
- brazing
- propane soldering
- oxyacetylene cutting
- grinding ferrous metals
- use of blow torches for paint stripping and so on.

Hot works permits are permit to work documents issued to a contractor for a specific job that has to use hot work techniques.

Hot works permits enable a safe system of work to operate during the works, and ensure that all the hazards and risks associated with the task have been identified before the job starts and that the necessary control measures have been put in place.

Individuals will be held responsible for ensuring that safety measures are followed.

Permits are issued for defined periods only and will need to be returned and signed off by a competent person.

Principal contractors should implement hot works permits across their site and ensure that all contractors follow the procedures.

Hot works permits are particularly important during refurbishment works, when combustible materials are more likely to be present within the building.

The requirement for hot works permits on the site should be covered in the site induction procedures.

There is no set format for hot works permits, as they should reflect the situation being controlled. An example template is included in the appendix to this chapter.

Case study

Three men employed by a small plumbing and drainage company were called to unblock a sewer. The men lifted the manhole cover to find that the drains were about 3 m down. The step irons looked OK, and the first operative, the 17-year-old nephew of the company owner, descended into the sewer to unblock it with his rods.

When he got to the bottom of the shaft he collapsed.

His two colleagues panicked and descended into the sewer to rescue him, fearing that he might drown. Both men were overcome but managed to shout before passing out.

Fortunately, the site foreman had arrived and realised something serious was wrong. He called the emergency services.

The 17-year-old operative was dead by the time he was rescued. The other two operatives died 2 days later in hospital. The cause was hydrogen sulfide gas poisoning – a deadly gas that becomes more odourless and more lethal in higher concentration.

The company owner was prosecuted for health and safety offences – for failing to have a safe system of work to protect his employees whilst they were at work.

At the very least, the Health and Safety Executive said, there should have been a safe system of work operated by a permit to enter or work system.

There should have been gas monitoring/detection equipment used, the sewer sludge agitated to release any build-up of toxic gas, emergency procedures, breathing apparatus, training and so much more.

Appendix to Chapter 26

Permit to work

Site address: ______________________________

Site agent: ______________________________

Brief description of the work and location:
Sequence of work and control measures:
Supervision arrangements:
Individual responsible for controls and monitoring performance:
Plant and equipment to be used and operator training requirements:

Occupational health assessments (risk, noise, COSHH, etc.):
Measures to ensure safety of third parties:
Environmental controls:
First aid and personal protective equipment requirements:
Emergency procedures:

Permit issued by: ______________________________

Permit issued to: ______________________________

Permit valid until: ______________________________

Permit to work

Permit title:	Permit reference No:
Principal Contractor:	Sub-contractor:
Job location on the site:	
Description of works and limitations:	
Equipment/plant to be used:	
Hazards identified:	
Precautionary actions required (including emergency arrangements):	
Protective equipment required (including personal protective equipment):	
Person in control of work activity (name):	
Permit issued to (name and signature): Company:	

<table>
<tr><td colspan="3">Confirming that they understand the work to be done, the hazards involved and the precautions required. Confirming that permit information has been explained to all permit users and that all conducting works are competent to do so</td></tr>
<tr><td>Permit valid from (time):</td><td colspan="2">To (time):</td></tr>
<tr><td colspan="3">Valid on (date):</td></tr>
<tr><td rowspan="2">Time extension to:</td><td colspan="2">Authorised by:</td></tr>
<tr><td colspan="2">Signed:</td></tr>
<tr><td colspan="3">Shift handover signed by changeover operative:</td></tr>
<tr><td colspan="3">Confirming that checks have been made, that plant/works area remains safe to be worked on/within. Confirming that the new permit user is fully aware of the hazards and precautions</td></tr>
<tr><td>Issued by (name):</td><td colspan="2">Date of issue:</td></tr>
<tr><td colspan="3">Confirming that isolations and all precautions have been made, except where these can only be made during the work activity</td></tr>
<tr><td colspan="3">Permit closure and return</td></tr>
<tr><td colspan="3">Closed by:</td></tr>
<tr><td>Signed:</td><td>Date:</td><td>Time:</td></tr>
<tr><td colspan="3">Cancellation of permit
Cancelled by:</td></tr>
<tr><td>Signed:</td><td>Date:</td><td>Time:</td></tr>
<tr><td colspan="3">Certifying that plant has been tested and satisfactorily re-commissioned/works area is safe</td></tr>
</table>

Hot works permit

For all operation involving flame, welding and hot cutting

This permit is valid only for the job described and the timescales provided

Description of work

Location of work

Building	Floor	Room	Location		
Date required (maximum duration 1 day)		Valid from (time)		To	

Contact details (method of contact)

Mobile phone no.		Site phone no.		Main office phone no.	
Estates project officer		Estates help desk		Security	

Potential hazards	☐ Oxygen enrichment	☐ Oxygen depletion	☐ Toxic gas	☐ Explosive gas
	☐ Biohazard	☐ Poor lighting	☐ Heat	☐ Noise
	☐ Tripping/falling/striking objects			
	☐ Other – provide details			

Control measures	☐ Hazard/equipment isolated ☐ Department staff informed ☐ Protective equipment required – specify:		
Other identified hazards		Controls measures	

Mandatory safety requirements

	Actioned
All areas to be checked and combustibles removed or protected before commencement of work	
All areas to be screened, protected, roped off as necessary and warnings signs displayed	
All systems associated with the work to be isolated, inclusive of smoke alarms	
Assistant to standby with fire extinguisher suitable for task (must be competent in its use)	
Client/building facilities manager notified	
Area to be checked/inspected for combustion **1 hour** after completion of work	

Person entering work area

Permit issued by		Date		Time	
Permit received by		Date			

Permit cancellation

Name		Date		Time	

CDM 2015 Questions and Answers: A practical approach
ISBN 978-0-7277-6032-6

http://dx.doi.org/10.1680/cdmqa.60326.405

Chapter 27
Working at heights

What legislation covers working at heights?
The Work at Height Regulations 2005 cover all work at height activities.

What do the Work at Height Regulations 2005 require employers to do?
The Work at Height Regulations 2005 not only apply to employers but also to 'duty holders'. Duty holders will often be employers but may also be any person who controls the way that work at height is undertaken (e.g. clients commissioning construction work, managing agents and building owners).

The regulations require duty holders to ensure

- where possible, work at height is avoided
- all work at height is properly planned and organised
- those involved in work at height are competent
- the risks from work at height are assessed and appropriate work equipment is selected and used
- the risks from fragile materials/surfaces are properly controlled
- equipment for work at height is properly inspected and maintained.

What is work at height?
Work at height is work in any place, including a place at, above or below ground level, where a person could be injured if they fell from that place. Access and egress to a place of work can also be work at height.

Examples of work at height are

- working on a flat roof
- working from a ladder
- working near or adjacent to fragile materials
- working at ground level adjacent to an open excavation

- working anywhere where there is a risk of falling
- erecting and working from scaffolding.

When is a risk assessment required for working at height?

Every time someone is to undertake a task where there is a risk that they could fall and injure themselves, a risk assessment must be completed.

The risk assessment approach required is no different than for all other risk assessments completed by the employer or duty holder.

The risk assessment approach is to

1. identify the hazard
2. decide who might be harmed and how
3. evaluate the risks and decide whether the existing precautions are adequate or whether more needs to be done i.e. identify the control measures necessary to reduce the risks of injuries
4. record your findings
5. review the assessment.

What type of issues should be addressed in the risk assessment?

Some examples of what to consider are

- the work activity
- the equipment to be used
- the duration of the work
- the location where the work activity is due to take place (i.e. the presence of hazards such as overhead power lines, open excavations, underground services, vehicle access, people still at work, and public access)
- the working environment (e.g. weather conditions, lighting, type of ground and slopes)
- the condition and stability of existing work surfaces
- physical capabilities of the workers (e.g. vertigo sufferers and pregnancy).

What is the most important aspect of working at height that must be considered?

The most important aspect of working at height that an employer should consider is whether the work at height needs to be done at all.

The first principle of safe working at height is not to work at height. Consider whether there are alternative ways to do things: for example, if light fittings are high up over, say, a stairwell, and access to them is unsafe due to height and risk of falls, consider whether the light fittings could be put onto winch-downable mechanisms so that maintenance and cleaning can be carried out at ground level.

Work at height should only be carried out if there is no alternative safer way of carrying out the task.

The principal designer and design team should consider work at height requirements when formulating their designs and, if following the principles of prevention, should look to design out, as far as possible, the need to work at height.

People carrying out work at height have to be competent. What does this mean?

Competency is the experience, knowledge and appropriate qualifications that enable a worker to identify the risks arising from a situation and the measures needed to be taken to eliminate or reduce them.

Anyone working at height needs to be trained in the hazards and risks and in what control measures are being put in place to eliminate or reduce the risks of injury. Those undertaking work at height activity must be trained in the selected system of work and on the use of any particular equipment chosen (how to use mobile elevating platforms, tower scaffolds, etc.).

Managers and supervisors must check that anyone carrying out work at height is trained and competent to do so.

Is it now illegal to use ladders and step ladders?

The Work at Height Regulations 2005 do not ban the use of ladders or step ladders but do require consideration to be given to their use.

Ladders and step ladders should only be considered where the use of other, more suitable work equipment (e.g. mobile elevating platforms or tower scaffolds) are not appropriate.

Where ladders and step ladders are used, they should only be used for light work of short duration.

A suitable and sufficient risk assessment needs to be carried out.

What types of work equipment is suitable for work at heights?

The following types of equipment can be considered as options to reduce the hazards and risks of working at height:

- a work restraint system
- a work positioning system
- a rope access and positioning system
- a personal fall arrest system.

What is a work restraint system?

A work restraint system is a fall prevention system that relies upon personal protective equipment (i.e. a harness and lanyard) being adjusted to or set to a fixed length that physically prevents the person from getting to a place where they could fall.

Such a system requires strict supervision, as there may be a risk that the operative will unfix their lanyard to reach more distant work locations.

What is a work positioning system?

A work positioning system is a personal fall arrest system that includes a harness connected to a reliable anchor point to support the user in tension or suspension in such a way that a fall is prevented or restricted. Examples include a boatman's chair and the system used by people working on telephone lines.

All work positioning systems must be provided with a back-up system in case the primary support fails.

What are rope access and positioning systems?

This system is similar to abseiling, and involves the use of two ropes each secured to different anchors. One rope is connected to a harness and the other acts as a safety back-up rope. Rope access is often used to access cliff faces or the sides of tall buildings when other systems are not feasible.

What is a personal fall arrest system?

A personal fall arrest system is a fall protection system that uses a harness connected to a reliable anchor to arrest and restrict the fall and prevent the user from hitting the ground. It usually has an energy-absorbing device to limit the impact of gravity forces on the body.

What steps need to be taken to prevent objects falling from heights?

The Work at Height Regulations 2005 require the likelihood of falling objects to be considered and eliminated or reduced wherever possible.

Toe boards, solid guards, debris nets and so on should be considered.

Good housekeeping will reduce the likelihood of objects falling, as tools, materials and so on will be put away so as not to cause trip hazards and be in walkways where they can be inadvertently kicked over the edge of platforms.

If necessary, tools could be tied onto harnesses or belts, especially if they have to be put down frequently.

Does the 2.0 m rule regarding falling from heights still apply?

No. The 2.0 m rule was repealed by the Work at Height Regulations 2005, which apply to all work at heights, no matter how far the distance of the potential fall. Injuries occur from low-distance falls as well as high-distance falls.

If there is a risk of a person falling *any* distance, whether off a height or into a trench or pit, then the Work at Height Regulations 2005 require the employer, the self-employed, employees and anyone who controls the way work at height is undertaken to eliminate, reduce or control the risk of personal injury.

What are the safety requirements for using mobile elevating work platforms?

Mobile elevating work platforms, often known as MEWPs or 'cherry pickers', are a much safer way of working at height than using ladders.

The hierarchy of risk control states that if a hazard cannot be eliminated, a less hazardous approach should be considered. Working at heights is a hazard, and using a ladder is less risky, but using a mobile elevating work platform is even less risky than a ladder. This latter option should always be considered.

Safety procedures for using mobile elevating work platforms include

- only trained operatives to use the equipment
- equipment to be well maintained and checked
- equipment to be suitable for the job
- the ground to be level so that the equipment is not at risk of falling over
- access ways to be kept clear
- overhead obstructions, especially overhead power cables, should be assessed and work programmed to avoid contact
- safe working loads of the platform to be considered so that they are not overloaded with operatives and materials
- operatives to wear fall arrest equipment (e.g. a harness and lanyard)
- weather conditions to be considered, especially wind speeds.

Case study

The problem

Two operatives fell to the ground from the basket of a 'cherry picker' when the basket suddenly tipped down. They both sustained severe injuries, and were off work for many weeks.

As with many employees and the self-employed in the construction industry, they received no wages during their recuperation, and both endured financial hardship. The employer lost two experienced operatives and had to hire in people from an agency at a much greater cost.

What probably happened?

The investigation showed that it was likely that the equipment was not being properly maintained and that a fastening bolt connecting one of the rods on the basket was either missing or had worked its way loose during the operation of the cherry picker.

Neither of the two operatives were wearing safety harnesses, and neither had received any proper training in how to use the equipment or the hazards and risks associated with it. They also had not been trained in how to conduct a simple visual check of the equipment.

The solutions

The employer or whoever authorised the cherry picker to be used should have ensured that the equipment was safe to use and that it had had a suitable inspection by a competent person within the last 6 months – as required by the Lifting Operations and Lifting Equipment Regulations 1998.

The employer or person in control should have requested copies of their inspection certificates for the equipment.

An inspection of the cherry picker would have identified faults with the fixings.

The operatives should have had training in using the cherry picker, and then they would have been declared competent to use the equipment. The Management of Health and Safety at Work Regulations 1999 require all persons to be competent to undertake work activities.

A comprehensive risk assessment should have been carried out, identifying the hazards and risks associated with using the equipment. Were potential tipping hazards identified, was the weather and wind speed considered, was there a plan to reduce any hazards and so on?

Where hazards cannot be eliminated completely, leaving the possibility of accident and injury, there should be an emergency plan in place. What steps would be taken in an emergency, who would do what, how quickly would people be rescued?

What consideration was given to personal protective equipment (i.e. a safety harness and lanyard)? Although both operatives were inside the basket, which had guard rails, there was a risk that if the basket tipped up they could fall out. They would obviously then fall from a height, and the Working at Height Regulations 2005 requires that to be prevented where possible.

What were the costs?

The company could have been fined an unlimited amount for breaching the Working at Height Regulations 2005, up to £20 000 for failing to have a safe system of work and up to £5000 for failing to have a risk assessment, failing to provide personal protective equipment and so on.

Also, as the contract was for a large public body that has to be seen to be employing competent contractors, the company was at risk of being removed from the approved contractor list, and this would have prevented them obtaining further work.

Had the company instigated the correct health and safety procedures, it would have had to have spent approximately £300 per operative – for a safety harness and lanyard and a 1 day training course, and also about £200 to have had the equipment inspected by a competent person.

When the Health and Safety Executive inspects a site, what will it be looking for in respect of mobile elevating work platforms?

The Health and Safety Executive (HSE) inspector will be looking to enforce the Work at Height Regulations 2005 and the general provisions of the Health and Safety at Work etc. Act 1974.

HSE inspectors have an inspection checklist for a wide range of job tasks and for enforcing numerous pieces of legislation.

With regard to MEWPs, they will be looking to ensure that equipment and procedures meet the following:

- The MEWP should be safe plant (i.e. correct type for the job, provided with adequate guardrails and toe boards).
- The MEWP should be thoroughly examined in the last 6 months, properly maintained and inspected.
- The MEWP should be used on a safe site (i.e. firm, level ground that is free from slopes, holes, etc.) and can sustain the loads imposed by the MEWP, the operatives and material on it and so on.
- The MEWP should be segregated from site traffic to avoid collisions.
- The MEWP should have a safe operator who is trained and experienced (i.e. competent to use the equipment).

- The MEWP should have fall protection equipment provided.
- There should be no evidence of inappropriate use and unsafe practices (climbing over mid-rails, leaning over the edge to the extent that there is potential to fall, etc.).
- There should be appropriate anchorage for any fall arrest equipment.

Inspectors are encouraged to take a robust approach to enforcement, and their guidance advises them to serve prohibition notices for a number of contraventions.

Prohibition notices under the Health and Safety at Work etc. Act 1974 would be served for the following:

- no segregation of the use of a MEWP from other vehicles or pedestrians (e.g. street lighting works taking place with no control over traffic)
- any situation where there is a risk of vehicle impact
- any evidence of operating a MEWP at reckless speeds
- any evidence that other vehicles are being driven at speed in the vicinity of a MEWP
- evidence of poor ground conditions
- evidence of any likelihood of a MEWP overturning due to slopes, uneven ground and so on
- evidence of failure to consider safety issues from manhole covers, ducts, suspended floors and so on (i.e. whether they will take the weight of a MEWP)
- any evidence of collapse of any structure
- any evidence of the wheels or outriggers a MEWP sinking into the ground or not being properly supported
- unsafe practices being followed
- evidence that fall protection procedures are not in place
- evidence of inadequate anchorage points for any fall arrest system
- evidence that a MEWP had not been properly maintained and interpreted in the previous 6 months and some evidence that competent parts were faulty.

Breaches of prohibition notices carry fines of up to £20 000 per offence and up to 12 months' imprisonment. If a prohibition notice is being ignored and a fatal or major injury occurs, there would be an unlimited fine and a risk of 2 years' imprisonment, as the case is likely to be heard in the Crown Court. In addition to fines, fee for intervention charges may apply.

What are the requirements for guardrails?

Guardrails must be substantial and properly fixed so that they are secure. Falling against a guardrail that gives way is no protection at all.

Guardrails must

- have a top rail at a height of at least 950 mm from a level surface
- have an intermediate guardrail at an approximate height of 470 mm
- have a toe board of approximately 150 mm in height
- be continuous around an opening
- be clearly visible
- not be breached by openings or missing rails and so on (e.g. at lift/hoist openings).

Guardrails must be inspected regularly by a competent person, and any defects repaired immediately. Hazard warning signs must be displayed if necessary, but any defect to a guardrail renders the rail unsafe, and so work should be prohibited in the area until the guardrail has been repaired.

Do guardrails at the old height of 910 mm have to be changed to meet the new height of 950 mm specified in the Work at Height Regulations 2005?

No, not necessarily. Where existing handrails that measure 910 mm high are fixed in place, they can remain at that height until changed. However, any gap between the top rail and the intermediate rail must not be greater than 470 mm.

The responsibility for deciding on whether the exiting guardrail height should be changed will rest with the designer, and they must complete a design risk assessment.

If the area protected by the guardrails is to have more plant and equipment placed within it, or if more frequent access for maintenance will be required, it will be necessary to consider the effectiveness of the existing guardrails and whether, in light of new circumstances, the rail heights should be raised.

Anything that substantially changes an existing situation should be looked at as a new design, and the most up-to-date requirements must be applied.

When would it be reasonable to use fall arrest systems?

A fall arrest system is type of personal protective equipment that does not prevent someone from falling but instead *arrests* their descent if they did, so that they do not fall all the way to the ground.

Fall arrest systems usually consist of

- a lanyard or rope
- a harness for the body

- hooks and couplings to connect the restraint ropes and so on to the body harness
- hooks or couplings to connect the lanyard or rope to the securing point (e.g. an eye bolt or a permanent rail).

Fall arrest systems should only be used when fall prevention is impossible (e.g. when erecting scaffolding).

The risk assessment should state whether fall arrest systems are suitable control measures to reduce the risk of injury from working at heights.

Fall arrest systems may be the only safety precaution available if work is of a short duration at heights.

What needs to be considered when using fall arrest systems?

Harnesses and lanyards are made of artificial fibres, and as such are subject to degradation by sunlight, inclement weather, chemicals and so on.

It is important to carry out detailed daily checks of the equipment prior to use. A visual and tactile inspection is necessary – look and feel for faults in the lanyard or harness. Choose an area in good light.

If there is the slightest doubt about the condition of any part of the fall arrest system, do not use it. Label it as 'defective' and put it aside for further inspection.

Faults can be identified by

- discoloration
- tears
- fraying
- nicks
- grittiness
- rust on metal catches
- excessive scratching.

Harnesses do *not* prevent a fall. The person falling will not fall to the ground but will instead be suspended in mid-air or only fall to the next level down. There is a significant risk of injury even if using a fall arrest system, because people could be harmed by impact to the body when the lanyard goes taunt or when they strike against parts of the structure during the fall.

Consider the use of an energy-absorbing device fitted to energy-absorbing lanyards so as to reduce the risk of injury from impact loads.

Minimise the 'freefall' distance. Keep the anchor as high as possible so as to reduce fall distances.

Ensure that anchor points are suitable and have been checked by a competent person. Any anchor points that a fall arrest system is attached to is 'lifting equipment' under the Lifting Operations and Lifting Equipment Regulations 1998, and *must* be inspected by a competent person every 6 months.

Ensure that the lanyard is not too long, so that it will arrest the fall before the person hits the ground.

Anyone who needs to attach themselves to a fall arrest system needs to be able to do so before they have entered the area from which they need to be protected.

Develop a system of running lines and second lanyards so that a person can unclip and clip themselves in a continuous process without being put at risk of falling.

Ensure that everyone wearing a harness and lanyard knows how to wear it properly, check it, adjust it and so on. Everyone should carry out their own visual pre-use checks and must themselves be satisfied about its conditions.

Never force someone to wear a harness or lanyard when they believe it to be defective.

What needs to be considered in respect of temporary suspended access cradles and platforms?

Temporary suspended access cradles and platforms are *often* used for window cleaning and external façade maintenance and cleaning. They are usually fixed at anchor points on the top of buildings, and the cradle passed over the edge to 'hand down' the building façade.

Many accidents happen when using suspended access cradles, mostly due to

- unsafe access to and from the cradle (e.g. stepping over a parapet wall)
- insufficient or poorly secured counterweights and holding systems
- failure of the cradle platform or components such as pins and bolts
- failure of winches, climbing devices, safety gear and ropes as a result of poor maintenance
- poor erection and dismantling techniques.

Equipment must be chosen and installed by a competent person.

A comprehensive risk assessment is required.

Consideration must be given to

- the type of weather
- wind speeds
- access into the cradle
- egress from the cradle
- the safe working loads of cradles, to include the number of operatives and the type and number of equipment to be used and therefore kept in the cradle
- the protection of ropes or suspension cables over parapet walls and so on, so as to prevent friction and possible fraying
- the risk of overturning
- the likelihood of suspension cable collapse, and the use of an alternative back-up cable and so on
- guarding of the cradle
- the operating instructions
- the training and experience of the operatives.

An emergency plan must also have been devised so that if there is a failure of the system or an injury to operatives in the cradle, emergency measures can be taken for rescue and so on.

Access into the cradle should preferably be from ground level.

All access cradles should be raised when not in use, so that unauthorised access is prevented, especially by children. Power should be switched off and isolated.

Access cradles must display suitable safety signs and have safe working loads (SWLs) displayed. SWLs should be equated to understandable information: for example, the number of people as opposed to the weight in kilograms.

What are the common safety rules that should be followed for working at heights?

Committing the following to memory and always following them will help to reduce injuries for falls significantly:

- Do not work at height unless it is absolutely essential.
- Make sure any working platform is secure and stable.
- Check that any working platform will support the weight of workers and the necessary equipment.

- Make sure any access equipment or working platform is stable and will not overturn.
- Do not erect equipment on uneven ground.
- Ensure that everyone has adequate working space. Overcrowded and restricted work areas contribute to accidents.
- Foot any ladders or access towers or secure them to a stable structure.
- Provide guardrails, and barriers to all edges, openings, drops and so on.
- Check everything frequently and do not take anything for granted.
- If fall arrest systems are used, check their condition regularly.

What other hazards are associated with working at heights?

Falling through fragile materials, especially on roofs, is a common and frequent hazard on construction sites and in maintenance works.

Fragile materials can include

- asbestos sheeting
- asbestos cement
- fibreglass
- plastics
- steel sheets
- glass sheets.

The above materials may become brittle over the years when exposed to sunlight. Steel may rust. However, any material could give way without warning, and such an eventuality must always be considered.

Roof openings and roof lights are common hazards that contribute to people falling. Roof lights often become brittle, but they may look in perfect condition.

Structural engineers should be consulted if there is any doubt about the ability of any structure to carry and load.

What safety procedures need to be followed for working on fragile materials?

Generally, the first principle of risk assessment is eliminate the hazard. So, do *not* work on fragile materials unless there is absolutely no alternative.

Fragile materials may be disguised by loose coverings, moss and so on, so nothing should be taken for granted.

Work on fragile roofs has to be carefully planned to prevent falls through the roof.

Working platforms to spread the load are essential.

Purpose-built roof ladders are essential – these hook over the ridge tiles and should be secured at the base. A review should take place to ensure that any tiles or parts of the structure used to secure roof access ladders are stable and secure.

If there is no possibility of working off platforms, then fall arrest systems, running lines and so on must be considered.

Safety nets are also a possibility – although they do not stop people falling, they may reduce the possibility of serious injury.

Where there are localised areas of fragile materials (e.g. roof lights), these can be protected by adequate guardrails.

Risk assessments *must* be completed for all works undertaken at heights.

Other safety considerations are

- access to the place of work
- weather conditions
- wind conditions
- methods of getting materials to the place of work
- falling materials and objects
- emergency procedures and rescue plans.

What are the common safety procedures for using scaffolding?

All scaffolds, including mobile tower scaffolds, should be designed, erected, altered and dismantled by competent persons. All scaffolding work must be supervised by a trained and competent person.

Scaffolders must adopt safe systems of work when erecting scaffolds and must wear fall arrest equipment.

Scaffolds must be adequately braced to prevent them from collapsing. Consideration must also be given to wind loadings and materials and weight loadings.

Scaffold walkways should be at least four boards wide.

All scaffolds must be sheeted or netted to prevent objects falling.

Hoists, chutes and lifting tackle must be assessed to establish the effect it may have on the scaffold.

Do not take up boards, move handrails or remove tiles to gain access to work areas.

Changes to scaffolds should only be made when properly assessed, planned and undertaken by competent people.

Scaffold lifts must never be partially boarded. Boards must be secured or must overlap each other by at least 150 mm. Consideration must be given to trip hazards on the walkway.

Boards forming the walkways must be adequately supported by horizontal transoms. Spacings should be between 1.2 and 1.5 m. Inadequate support will cause the boards to bow and give way.

Safe means of access must be provide by way of secured vertical ladders or ladders placed at a suitable angle for ease of use. All ladders must be tied.

Scaffolding must be inspected

- before its first use (i.e. erection)
- after substantial alteration
- daily before use
- after any event likely to have affected stability (e.g. high winds)
- at regular intervals not exceeding every 7 days.

Suitable records of inspection must be kept.

Faults must be highlighted clearly or the scaffold put out of use until repaired – the only real action for a defective scaffold.

Any contractor using a 'communal' scaffold has a duty to inspect the scaffold and to be confident about its condition *before* any of their operators use the scaffold.

Is there a safe distance from the edge of a flat roof or raised platform when edge protection or a fall arrest system will not be needed?

It is possible to demarcate a 'safe area' on a flat roof or other raised work area in which operatives are to work, preventing them from gaining access to the edge.

There is no legal safe distance from an edge – it all depends on the circumstances and, of course, the risk assessment!

The commonly held view is that if limited work or infrequent work on a larger flat roof involves nobody going any closer than 2.0 m to an open edge, then full edge protection may not be required and a fall arrest system may not be necessary.

However, when considering the 2.0 m 'rule', you will need to consider the risk of a person tripping over something – such as a roof vent or a piece of equipment – or slipping on a wet, moss-covered or otherwise slippery surface, as the trip or slip could propel them very quickly to a distance greater than 2.0 m.

A demarcated safe area can be set up on the roof, outside of which nobody goes.

Demarcated areas should be

- operated by a clear 'roof permit' that clearly sets out the demarcated area
- limited to areas from which nobody could fall (including any possibility of slips and trips)
- indicated by an obvious physical barrier
- adequately signed with hazard warning notices advising of the safe system of work, prohibiting access beyond the barrier and so on
- subject to close supervision by a competent person so that the system of work put in place is followed.

It will not be acceptable to mark out the safe area with yellow paint to indicate a hazard, nor to run tape or bunting around the perimeter of the safe area.

The physical barriers will need to be substantial but may not need to be permanently fixed to the roof. Consideration must be given to any hazards that might be created by the barriers (instability in windy weather, being blown over, etc.).

Are safety nets legal?

Yes. There is no law that states that safety nets are illegal. However, safety nets do not *prevent* a fall – they arrest the descent and, as such, the Work at Height Regulations 2005 would not permit the use of safety nets unless all other options had been considered that would prevent a fall.

If a fall cannot be completely prevented, then the employer or other person in control of the work has to consider the hierarchy of risk control (see Chapter 5 on risk assessments) and substitute a lesser risk.

So, instead of potentially falling 20 m, a properly installed safety net might reduce the distance of fall to only 2 or 3 m. Injuries may still occur, but their severity would be reduced.

Safety nets also comply with the collective protective equipment requirement: that is, prioritising protective equipment solutions for multiple operatives over individual personal protection measures.

What are some of the key safety considerations when using safety nets?

The decision to use safety nets must be made by a competent person, and it is essential that they have considered all other options and that they have justifiable reasons as to why alternatives won't be suitable.

The competent person must compete the risk assessment for the use of safety nets.

If safety nets are to be used they must

- Be installed as close as possible below the roof surface so as to minimise the distance of fall.
- Be securely attached to a permanent part of the structure. All fixings should be assessed and inspected and, where necessary, structural engineers should be consulted to determine whether imposed loadings, weight limits and so on will be tolerated on the fixings and other load-bearing points.
- Be capable of securely withstanding the potential weight of all of the operatives who *could* fall into the netting. It may not be the case that just one operative falls – they could all fall at the same time, imposing considerable load on the netting and its fixings.
- Be installed and maintained by a competent person.
- Be regularly inspected by a competent person before work activity.

The safety netting may successfully do its job and arrest someone's fall from height – but the person cannot stay in the netting, and will need to be rescued. What emergency procedures have been considered? Do you have an emergency plan?

Rescue procedures need to be clearly written, and everyone needs to be trained on them. It will not be acceptable to think about the rescue after the event – a well thought out plan and procedure, and trained operatives could save someone's life.

What will an HSE inspector look for specifically in relation to safety nets if they were to undertake a routine visit?

The HSE inspector will be checking for the following:

- that nets are positioned as close as possible to the level at which persons are working so that any fall from height is minimised
- that there is sufficient clearance below the nets to avoid injury due to collision with an obstruction if a person fell into the net
- that the nets do not have any gaps at their edges through which a person could fall
- that the nets are securely anchored.

In addition to the above, the HSE inspectors will be checking

- the competency of the persons who erected the nets
- the competency of those carrying out the tasks
- the adequacy of risk assessments
- the availability of method statements or safe systems of work
- the emergency plans and procedures for rescue should a person fall into a net
- the safe erection and dismantling of the nets.

Ladders can be used for short-duration work. What are some of the safety precautions to consider?

Ladders should ideally be used as a means of access to a place of work and not as a place from which work takes place. However, ladders do need to be used for the latter as well as the former.

The following safety precautions should be followed when using a ladder:

- Ladders must be in good condition, and examined regularly for defects.
- A comprehensive management system must be in place to record checks, note remedial actions, record training and so on.
- Ladders must be secured to prevent slipping. Usually, ladders are tied at the top rung, but they can also be securely 'footed' and tied at lower levels.
- The ladder must be angled so as to minimise the risk of slipping outwards, and the 'rule of thumb' is 'one out for every four up'.
- Any access ladder must extend at least 1.0 m above the work platform so as to provide a handhold for people getting on or off the ladder.
- Persons working from the ladder must not overstretch or over-reach. The ladder must be positioned so that the work area can be reached safely (i.e. the ladder needs to be long and in the correct position).

- Operatives need to be able to hang on or hold on to the ladder. Work must be capable of being done with one hand so that the other can be used to hold on.
- Operatives must not work from the top rung of a ladder.
- The top of the ladder must rest against a solid surface – not against a gutter, window or other fragile or insecure material.
- Both feet of the ladder must be on a firm footing and not capable of slipping.
- If the ladder cannot be secured by tying, a second person must 'foot' the ladder.
- Landing areas or platforms must be provided if ladder runs are 9 m or more.
- Ladders must never be ended by gluing or re-fixing rungs or stiles.
- Defective ladders must be replaced.
- Ladders must be checked each time they are used – especially timber ladders.

Many sub-contractors use step ladders. What are some of the basic safety considerations for step ladders?

Common types of step ladders are described in Table 27.1, along with their advantages and disadvantages from the viewpoint of safety.

What are working platforms?

Under the Work at Height Regulations 2005, a working platform can now be any surface from which work is carried out, such as

- roofs
- floor
- platform on a scaffold
- a mobile elevating work platform
- the treads of a step ladder.

What are the requirements for working platforms?

Regulation 8 and Schedule 3 of the Work at Height Regulations 2005 set out the requirements for work platforms.

In particular Schedule 3 requires the following for *all* working platforms:

> **Condition of surfaces**
>
> **2.** Any surface upon which any supporting structure rests shall be stable, of sufficient strength and of suitable composition safely to support the supporting structure, the working platform and any loading to be placed on the working platform.
>
> **Stability of supporting structure**
>
> **3.** Any supporting structure shall—
>
> (a) be suitable and of sufficient strength and rigidity for the purpose for which it is being used;

Table 27.1 Types of ladders and their advantages and disadvantages

	Type	Advantages	Disadvantages
	Aluminium	■ Light ■ Strong ■ Robust ■ Low maintenance	■ Conducts electricity ■ Conducts heat ■ May fail suddenly after extreme heat exposure
	Wood	■ Does not conduct electricity when dry ■ Natural insulator against heat/cold ■ Show signs if heat is applied	■ Can dry and split – needs regular maintenance (clear varnish) ■ Heavy ■ Can fail suddenly
	Fibreglass	■ Does not conduct electricity ■ Withstands short exposure to high temperatures ■ Show signs if heat is applied	■ Heavy ■ Can crack and fail suddenly if overloaded ■ Can chip or crack under severe impact

(b) in the case of a wheeled structure, be prevented by appropriate devices from moving inadvertently during work at height;
(c) in other cases, be prevented from slipping by secure attachment to the bearing surface or to another structure, provision of an effective anti-slip device or by other means of equivalent effectiveness;
(d) be stable while being erected, used and dismantled; and
(e) when altered or modified, be so altered or modified as to ensure that it remains stable.

Stability of working platforms

4. A working platform shall—

(a) be suitable and of sufficient strength and rigidity for the purpose or purposes for which it is intended to be used or is being used;
(b) be so erected and used as to ensure that its components do not become accidentally displaced so as to endanger any person;
(c) when altered or modified, be so altered or modified as to ensure that it remains stable; and
(d) be dismantled in such a way as to prevent accidental displacement.

Safety on working platforms

5. A working platform shall—

(a) be of sufficient dimensions to permit the safe passage of persons and the safe use of any plant or materials required to be used and to provide a safe working area having regard to the work being carried out there;
(b) possess a suitable surface and, in particular, be so constructed that the surface of the working platform has no gap—
  (i) through which a person could fall;
  (ii) through which any material or object could fall and injure a person; or
  (iii) giving rise to other risk of injury to any person, unless measures have been taken to protect persons against such risk; and
(c) be so erected and used, and maintained in such condition, as to prevent, so far as is reasonably practicable—
  (i) the risk of slipping or tripping; or
  (ii) any person being caught between the working platform and any adjacent structure.

Loading

6. A working platform and any supporting structure shall not be loaded so as to give rise to a risk of collapse or to any deformation which could affect its safe use.

What is a supporting structure?

A supporting structure is: 'any structure used for the purpose of supporting a working platform and includes any plant used for that purpose'. As an example, mobile access equipment is a supporting structure, a fork lift truck will be a supporting structure and the 'tower' of a tower scaffold is a supporting structure.

If an upper floor is a working platform, then the walls and structure supporting the floor will be a supporting structure.

Roof maintenance (e.g. gutter cleaning)

Questions to ask include the following.

Height
How high is the job from the ground?

Surface
What surface will the access equipment rest on (e.g. a wall, cladding or a pitched roof)?

Ground
What is the ground condition under the area where access equipment might need to be set up? For example, is it sloping, muddy or uneven? The access equipment you use must be suitable for the ground conditions – stable, level and not liable to fall or collapse.

If the employee falls, what will they fall onto? What will be the severity of injury?

Weather
Is it raining hard, or very windy? Is it icy, snowing or generally wet?

Task
What tools will the employee use to clean the gutters? How will they manage the debris? How will they get the tools to where they need them?

Types of maintenance
When looking at what the employee needs to do the job, think about the following:

- From the ground:
  - Can the work be done from the ground?
- From the roof:
  - Is the roof above the gutter accessible?
  - Is the roof strong enough to work from?
  - Does the roof have guardrails or other equipment that will prevent a fall? If not, can they be installed?

- From a platform:
  - Can the work be done from a MEWP or tower scaffold?
- Fall protection:
  - Is personal fall protection required to allow safe access?
- From a ladder:
  - Is the work low risk and of short duration?
  - Will ladder reach the area?
  - Can the ladder be secured safely?
  - Can the ladder be used safely?

Hazards: what can go wrong
Could the access equipment

- slip?
- fall/topple over?
- block means of escape from the building?
- block means of access?
- obstruct footpaths and other routes?
- be knocked into by passers-by?
- be too big or awkward to be erected safely?

Could the task to be undertaken

- cause debris to fall onto passers-by?
- cause tools to fall onto passers-by?
- obstruct others carrying out different work nearby?
- interfere with any safety equipment?

Could the operative undertaking the task

- be incompetent/inexperienced?
- lack training?
- misunderstand the hazards?
- take short cuts or rush the job?
- not know how to use the tools or equipment?

Tips for safe working
- Deal safely with the gutter debris. It is best to bag it in small quantities and lower it slowly to the ground – if there is a risk of falling debris, make sure no-one comes into the area below the work.
- Take frequent breaks, especially when working from a ladder – do not work from a ladder for longer than 30 min at a time.
- If using a ladder, keep three points of contact wherever possible.
- Make sure that the people who will be doing the job have the right skills, experience and training to use the equipment safely and have been consulted about the right equipment to use.

- If hiring access equipment, the hirer must make sure they know how to install and dismantle it safely – ask the supplier for instructions or assistance if needed.

Case studies

Fall through a roof

A maintenance worker plunged 10 m to his death after falling through an asbestos roof, an inquest heard. He went onto the roof of a building to repair panels that had been damaged in a break-in, after being told by his boss to 'sort it'.

A MEWP should have been used

A scaffolding company was fined heavily after a court heard how a scaffolder was killed when he fell from a third-storey window ledge to the street below.

The scaffolder was removing a large hoarding from the front of a building. He climbed through the window on the third floor in order to gain access to a narrow ledge so that he could stand on it to reach the hoarding brackets.

He lost his balance and fell 12 m to his death.

The company was prosecuted for

- failing to have a safe system of work
- failing to carry out risk assessments
- failing to provide employees with adequate information, instruction and training.

The HSE inspector said that a MEWP should have been used. The procedure adopted was inherently unsafe.

Improper use of a ladder

An inquest was told that a store assistant who was working in a retail store died from head injuries after falling from a ladder. He was using the ladder to remove a bike part from an upper shelf in the store room when he fell from it.

A colleague found him on the concrete floor on all fours with blood coming from his head. The store assistant's injuries were severe, and were described in court:

> He had an abrasion and swelling on his head and there was fresh blood leaking from his right ear. He had a brain scan and suffered a fit while in the machine. The results of the scan showed serious head injuries, two skull fractures, bleeding around brain tissue and significant bruising in two parts of the brain.

The inquest jury returned a verdict of accidental death.

The HSE inspector said he had investigated the scene, and the ladder being used lacked proper feet and had been used at an incorrect angle.

Appendix to Chapter 27

Work at height checklist

1. Identification of work at height activities

			List activities
Are all the work at height activities identified? (Include routine and non-routine activities)	**Yes**	**No**	
			Who?
Are the employees/contractors/public at risk from work at height identified?	**Yes**	**No**	
Do your employees/operatives work at heights			**List activities**
Constantly >4 hours per day?	**Yes**	**No**	
Daily?	**Yes**	**No**	
Weekly?	**Yes**	**No**	
Monthly, or every few months?	**Yes**	**No**	
Rarely, once or twice per year?	**Yes**	**No**	

2. Hazard checklist

During work at height activities are the following hazards controlled?

				Comment
Operative falling from height	**Yes**	**No**	**N/A**	
Operative falling through a fragile roof	**Yes**	**No**	**N/A**	
Objects falling on workers/members of the public	**Yes**	**No**	**N/A**	
Slipping/collapsing access equipment	**Yes**	**No**	**N/A**	
Slips/trips on elevated access ways	**Yes**	**No**	**N/A**	

3. Control measures checklist

Planning

				Comments
Is working at heights avoided where possible?	**Yes**	**No**	**N/A**	
Is work at heights properly planned and organised?	**Yes**	**No**	**N/A**	
Has the work area been surveyed for hazards (e.g. overhead power lines, unsuitable ground conditions or moving machinery parts)?	**Yes**	**No**	**N/A**	
During work at height is there always a competent person present who can suspend work if conditions change (e.g. weather, traffic or the scope of the work)?	**Yes**	**No**	**N/A**	

Fragile roofs

Are fragile roofs identified and labelled, especially where fragility is disguised?	**Yes**	**No**	**N/A**

Are warning signs fixed on the approach to fragile roofs?	**Yes**	**No**	**N/A**	
Is unauthorised access to roofs/roof spaces prevented?	**Yes**	**No**	**N/A**	

Ladders/access equipment

Are ladders only used for light work of short duration?	**Yes**	**No**	**N/A**	
Are portable ladders inclined at 75° (ratio 1 : 4) and secured top and bottom?	**Yes**	**No**	**N/A**	
Are work platforms capable of supporting workers, equipment and materials? Are the safe working loads indicated and inspected every 6 months?	**Yes**	**No**	**N/A**	
Are work platforms with edge protection (e.g. tower scaffolds, cherry pickers and scaffolding) always used?	**Yes**	**No**	**N/A**	
Are work platforms sufficiently wide to walk on without risk of stumbling or losing balance?	**Yes**	**No**	**N/A**	
Are properly constructed safety platforms always fitted if fork lift trucks are used to gain access at heights?	**Yes**	**No**	**N/A**	
Are portable ladders, steps and all other access equipment regularly inspected?	**Yes**	**No**	**N/A**	
Have operators been trained not to over-reach on ladders or work platforms?	**Yes**	**No**	**N/A**	
Is a written log maintained and access equipment removed from service if broken/damaged?	**Yes**	**No**	**N/A**	
Is routine monitoring carried out to ensure that makeshift ladders or other makeshift access equipment is not being used?	**Yes**	**No**	**N/A**	**Comment**

Open edges

Are open vertical drops identified and properly guarded when open?	**Yes**	**No**	**N/A**	
Are handrails present on the open side of staircases?	**Yes**	**No**	**N/A**	
Are open edges (e.g. on mezzanine levels) provided with appropriate guardrails? (The use of chains, ropes, etc., is not suitable)	**Yes**	**No**	**N/A**	

Are safe working methods used during delivery/removal of goods to mezzanine areas?	**Yes No N/A**
Are measures always taken to prevent falling objects injuring others as a result of working at heights?	**Yes No N/A**
Is fixed edge protection provided to prevent falls where working at heights cannot be avoided?	**Yes No N/A**

Training and information

Have all operatives who work at heights been assessed as competent or closely supervised by someone who is competent, and is there evidence of suitable training?	**Yes No N/A**

Protective equipment

Where the risk of falls cannot be eliminated, are safety harnesses, fall arrest devices or safety nets used as required? (Safety nets protect those working in the space below, fall arrest equipment and safety harnesses require specialist training as well as statutory inspection every 6 months)	**Yes No N/A**
Are anchorage points for safety harnesses and ladders provided, used and regularly inspected?	**Yes No N/A**

If you answered **No** to any of the questions, record the action required below

	Target date
Additional control measures to be implemented:	
Additional control measures to be implemented:	

	Target date
Additional control measures to be implemented:	
Additional control measures to be implemented:	
Additional control measures to be implemented:	
Additional control measures to be implemented:	

Work at height inspection form

Name of premises:	Address:
Contact name:	Telephone number:
1. Location of the work area inspected:	
2. Description of any work equipment inspected:	
3. Date and time of inspection:	
4. Details of any matter identified that could give rise to the risk to the health or safety of any person:	
5. Details of any action taken as a result of any matter identified in Section 4.	
6. Details of any further action considered necessary:	
Name of person making report:	Position of person making report:

Ladder and step ladder checklist

Date:				
Checked area	**Condition good**	**Defects**	**Comments**	**Signature**
Ladder checklist **Ladder identification number:**				
Missing steps of rungs				
Loose steps or rungs (considered loose if they can be moved at all by hand)				
Loose nails, screws, bolts or other parts				
Cracked, split or very worn or broken stiles, braces, steps or rungs				
Slivers on stiles, rungs or steps				

Damaged or worn non-slip bases				
Twisted or distorted stiles				
Identification disc missing or illegible				
Extension ladder checklist **Ladder identification number:**				
Loose, broken or missing extension locks				
Defective locks that do not seat properly when ladder is expanded				
Rusted or corroded metal parts				

Checked area	Condition good	Defects	Comments	Signature
Worn or badly deteriorated cords				
Damaged, missing or seized pulleys				
Step Ladder Checklist **Ladder Identification Number:**				
Side strain (wobbly)				
Loose or bent hinge spreaders				
Stop on hinge spreader broken				

Broken, split or work steps				
Loose hinges				

CDM 2015 Questions and Answers: A practical approach
ISBN 978-0-7277-6032-6

http://dx.doi.org/10.1680/cdmqa.60326.439

Chapter 28
Lifting operations and lifting equipment

What is the legislation that governs the legal requirement for lifting and lifting equipment?

The Lifting Operations and Lifting Equipment Regulations (LOLER) 1998 contain all the legal requirements for the operating of safe lifting process and the use and maintenance of safe equipment.

The CDM Regulations do not contain specific requirements for lifting, but they do require contractors, designers, clients and others to be aware of all legal requirements relating to the health and safety of a project and to comply with the legislation as necessary.

The Manual Handling Operations Regulations 1992 also contain requirements for lifting, and the principal contractor and contractors should be aware of these in respect of lifting individual weights (blocks, sacks, materials, etc.).

Lifting equipment used at work is also subject to the general provisions of the Provision and Use of Work Equipment Regulations 1998.

What actually is lifting equipment?

Lifting equipment is defined as any equipment whose principal purpose is to lift or lower loads, including attachments used for anchoring, fixing and supporting it.

Lifting equipment can include

- cranes
- fork lift trucks
- jacks
- mobile elevating work platforms
- passenger lifts

- vehicle tail lifts
- ropes and pulleys
- hoists
- dumb waiters
- vehicle inspection platforms.

Lifting equipment also includes all lifting accessories, such as

- chains
- ropes
- slings
- shackles
- eye bolts
- harnesses
- lanyards
- running lines.

When do the LOLER 1998 apply?

The LOLER 1998 apply to all work premises, including all construction sites. They apply to all work situations that are subject to the Health and Safety at Work etc. Act 1974.

The principal contractor (or if one has not been appointed, the contractor) must be familiar with the LOLER 1998, and should ensure that the regulations are applied as appropriate to the construction site activities.

Should any contractor, or indeed any person, not be familiar with the LOLER 1998, they should ask the competent person they have appointed under the Management of Health and Safety at Work Regulations 1999 to advise them on how to apply the regulations.

What are the principal requirements of the LOLER 1998?

The LOLER 1998 require that the employer or person in control of the premises addresses the following:

- the suitability of lifting equipment
- the strength and stability of lifting equipment
- the position and installation of lifting equipment
- the marking of lifting equipment
- the organisation of lifting operations
- the examination and inspection of lifting equipment
- the adequacy of equipment for lifting people.

Contracts managers and site agents will need to ensure that either they, or the sub-contractor providing and using the equipment, have addressed the key requirements.

What needs to be considered in respect of the suitability of lifting equipment?

The following need to be considered:

- ergonomic risks when using the lifting equipment
- the suitability of the lifting equipment material
- safe means of access/egress to/from the lifting equipment
- the need to minimise the risks of slips, trips and falls from any part of the lifting equipment
- protection for operators of the lifting equipment, especially in inclement weather
- wind speeds and how high winds could affect either the equipment or the lifting operation.

There are many choices of lifting equipment available, and for complex lifting tasks it will usually be advisable to seek competent advice. Some equipment will not be suitable for certain lifting tasks, and choosing the wrong equipment could put people at risk of severe injury.

Ask the person choosing the equipment why they have chosen that type, have they considered all of the options, have they compiled risk assessments and so on.

What needs to be considered in respect of the strength and stability of lifting equipment?

Equipment must have adequate strength for the proposed job/use, with an appropriate factor of safety against failure.

Equipment must also have adequate stability and not be prone to overturning.

Any equipment with pneumatic tyres must be checked to ensure the correct inflation of the tyres – underinflated tyres can cause equipment to topple over.

Spreader plates may be necessary or other types of stabilising equipment. Conditions need to be checked to ensure that the stabilisers/spreaders can be fully extended and used correctly.

What are some of the key points to consider regarding the position and installation of lifting equipment?

Lifting equipment must be positioned or installed in such a way that the risk of a person being struck or a load moving in an uncontrolled manner is minimised.

If a load needs to be lifted over people, this should be carefully planned, and the position of the lifting equipment carefully considered so as to minimise the duration of the lift (also, the site agent should consider excluding all pedestrians and other vehicles from the area underneath the lifting span).

Equipment cannot be placed in any position where there is a likelihood that people could be crushed: if it were to topple over, where would it fall and onto whom?

Any path of travel of a load or the equipment itself should be protected by a suitable enclosure.

Lifting equipment in use should not come into contact with other lifting equipment (e.g. proper planning and siting of multiple tower cranes on a site, or the location of hoists not to be a risk of being entangled with netting).

Access and egress points onto and from lifting equipment should be protected with suitable gates.

Interlocking devices should be fitted.

What is the safe working load (SWL) of a lift or lifting equipment?

All lifts and lifting equipment must display the SWL. This is the maximum weight that the equipment is designed to take safely.

Ignoring the SWL could cause equipment to be overloaded and plummet to the ground or lower levels. Accidents happen because the SWL has been ignored.

Where there is a risk that the SWL will be ignored, the equipment should be fitted with interlocks or capacity limiters that prevent it being used in the overloaded state.

It is essential to check the safe working load of slings, shackles, hoists, eye bolts, harnesses, lifting beams and so on.

If in doubt about the capacity of a lift or lifting equipment to take the load, seek advice from the manufacturer.

Where time is of the essence, reduce the load significantly and increase the number of times the loads have to be lifted – 'little and often' may take longer but will reduce the risk of equipment collapse.

The organisation of lifting operations seems to be critical to the safe operation of a construction site. What exactly does this organisation entail?

Lifting operations need to be

- properly planned
- appropriately supervised
- carried out in a safe manner.

The person planning the operation should have adequate practical and theoretical knowledge and experience of planning such operations. They *must* be competent.

It is good practice to write the plan of lifting operations down as a record and *aide-mémoire*.

The plan must address

- the risks identified
- the resources required
- the procedures to be followed
- those involved and their responsibilities
- any other lifts going on in the area and how they will interface.

Proper planning is a combination of

- Initial planning:
  - Is the equipment suitable?
  - What is to be lifted?
  - What are the weight, shape, the centre of gravity and lifting points?
  - What is the travel distance?
  - What is the frequency?
  - What are the weather conditions?
  - What is the experience of operatives?
- Appropriate consultation:
  - Who will do the task?
  - When does it need to be done?
  - By when?
  - Who might be affected?
- Preparing method statements:
  - How?
  - Who?

  - What?
  - When?
  - What could go wrong?
- Contingency planning:
  - What will be done in emergencies?
  - What will be done in the case of an accident?
- Information, instruction and training:
  - Evidence of training.
  - Evidence of experience.
  - Sharing of information on the site-specific hazards (e.g. overhead power lines and uneven ground).
  - Itemising preferred sequences of work.
  - Detailing equipment to be used.
  - Sharing the method statement.

Will the principal contractor have to supervise the lifting operation at all times in order to comply with the LOLER 1998?

No. However, all lifting operations will need to be supervised by a competent person. This may not necessarily be the principal contractor (or, if one has not been appointed, the contractor) but could be the specialist contractor engaged to carry out the lifting operation.

The principal contractor must be aware of the lifting activity to be undertaken, and should have a copy of the written lifting plan.

The principal contractor is not expected to be an expert in lifting operations but is expected to be competent enough to assess whether due regard has been paid to health and safety and compliance with the LOLER 1998.

The principal contractor must ensure that someone competent is supervising the lifting operations. It would be good practice to record this on the written plan.

The principal contractor should review the operation once it is underway and ensure that the work plan/method statement is being followed. This will be especially necessary where the principal contractor is responsible for the safety of others and the lifting operation involves loads being swung over people's heads.

The principal contractor will have the authority to stop the lifting operation – or the use of lifting equipment if it is deemed unsafe, or if, for instance, the weather conditions have deteriorated.

What is required regarding inspection and testing of lifting equipment?

The LOLER 1998 require that lifting equipment is thoroughly examined by a competent person to ensure that it is suitable for use.

Thorough examination should be carried out at regular intervals, and at least

- every 6 months if it is used for lifting people or is an accessory used for lifting people
- every 12 months if it is used for lifting goods.

Thorough examination may be required more frequently than the above if specified by the competent person.

Records of all thorough examinations must be kept.

The principal contractor (or, if one has not been appointed, the contractor) should check the records of thorough examination of all lifting equipment being brought onto the site.

Regular inspection of lifting equipment is not the same as thorough examination, and can be carried out much more frequently.

All users of lifting equipment should be trained to carry out visual checks and routine tests of their equipment before using it.

Lanyards, harnesses, ropes, slings and so on should all be regularly checked for

- fraying strands
- cut surfaces
- contamination with chemicals
- defective karibiners and similar items
- overstretching
- faulty mechanisms.

Any equipment found to be faulty must be put out of use, labelled accordingly and returned to the supplier for disposal/repair.

The records of lifting equipment can be disposed of as soon as the equipment ceases to be used or operated.

Any reports for lifting accessories (eye bolts, harnesses, slings, etc.) must be kept for a minimum of *2 years* after the report was made.

Reports on passenger lifts or permanent lifting equipment within the building must be kept as long as the equipment is used and until the next formal report is issued. Reports should be kept on the premises where the equipment is installed.

It is good practice to keep all records of inspection and testing, as this will show a 'due diligence' defence should things go wrong. If you can demonstrate to the court that you were aware of your duties and kept records and that any accident was not through any negligence of your own, you may be able to show that you had done everything reasonably practicable to avoid the commission of an offence.

Case study 1

Incident

A vehicle crane operator was killed after the lifting equipment on his vehicle came into contact with overhead power lines while he delivered materials.

What would have prevented the incident?

- Proper planning of the job.
- Identifying overhead lines.
- A site-specific risk assessment.
- Reviewing the equipment to be used.
- Ensuring the competency of the operator.
- On-site supervision.

Case study 2

Incident

A man sustained life-threatening injuries after being thrown from the platform of a mobile elevating work platform as he worked in the street to repair the fabric of residential buildings. It is alleged that a passing vehicle struck the platform support as it passed.

What would have prevented the incident?

- Proper planning of the job.
- Using more appropriate lifting equipment, which may have had a different type of stabiliser.
- Using more appropriate barriers and controlling passing traffic.
- The operator wearing fall arrest equipment (harness and lanyard).
- Adequate supervision.

Case study 3

Incident

A man died in a vehicle that was picked up and crushed by a hydraulic grab machine at a car breaker's yard. The man was a member of the public visiting the site, and it is thought he entered the vehicle just as it was being lifted.

What would have prevented the incident?

- Proper supervision and control of all work procedures.
- Strict controls on public access to areas.
- Prohibition of any lifting activity in the area where people are present
- Better grab operator training.
- Giving visitors high-visibility vests/jackets.
- Better planning of operations.
- Risk assessments and method statements.

Appendix to Chapter 28

Lifting equipment/operations checklist

		Yes	No
1	Do you carry out any lifting operations or have any lifting equipment on site that is subject to the Lifting Equipment and Lifting Operations Regulations 1998?		
2	Do you ensure that equipment is:		
	■ Suitable?		
	■ Of adequate strength?		
	■ Stable?		
	■ Suitable for lifting people and meets any specific requirements for this task?		
3	Is all lifting equipment and lifting accessories marked with the safe working load?		
	If not individually marked, is there adequate signage close to the equipment to advise the operatives of the safe working load?		
4	Are all lifting operations properly planned, with a written plan available?		
5	Are all lifting operations properly supervised by competent persons?		
6	Have you checked all the thorough examination records for the lifting equipment?		
	Are they the correct ones for the equipment?		
7	Are the operatives trained, instructed and informed about the lifting operations, the site-specific hazards, any acceptable risks, etc.?		
8	Are the operatives trained to carry out visual inspections and to carry out function tests of the equipment they are to use?		
	Have spot checks been undertaken?		

9	Have you instigated a system of immediate reporting of defective equipment and subsequent remedial actions?		
10	Have you kept records for the appropriate time and will relevant information be passed to the building user (e.g. permanent passenger lift thorough examination reports or eye bolt reports when used for fall arrest systems)?		

CDM 2015 Questions and Answers: A practical approach
ISBN 978-0-7277-6032-6

http://dx.doi.org/10.1680/cdmqa.60326.451

Chapter 29
Manual handling

What are the main hazards and risks associated with manual handling activities?

Manual handling activities were responsible for approximately 1 million lost working days during the year 2013–2014. and over 25% of all reportable injuries involve manual handling of some sort.

'Bad backs' are common, and are not always caused by lifting exceptionally heavy weights. Often, it is an accumulation of inappropriate lifting techniques, bad posture and poor systems of work. Sometimes, bending down to pick up a hand-held tool will be sufficient to cause a muscular spasm, slipped disc or other injury.

The main hazards are

- lifting too heavy a weight
- lifting awkward weights
- tripping while carrying
- pushing, pulling and shoving loads
- repetitive lifting
- repetitive actions
- dropping weights/loads.

The hazards will often result in the following injuries:

- slipped discs
- pulled muscles
- strained muscles
- torn muscles
- torn ligaments
- cracked bones (e.g. ribs)
- repetitive strain injuries
- impact injuries
- hernias.

What are the legal requirements regarding manual handling activities?

The Manual Handling Operations Regulations 1992 set down the legal duties surrounding manual handling activities at work.

Employers are required to complete risk assessments for all manual handling activities carried on at work, and the significant findings must be recorded in writing.

The responsibility under the Manual Handling Operations Regulations 1992 for the employer is to eliminate or reduce the amount of manual handling undertaken by employees.

The risk of injury must be reduced to the lowest level possible.

What are some of the practical steps that can be taken to reduce manual handling on construction sites?

The first duty to reduce manual handling activities on construction sites sits with designers under the CDM Regulations.

Designers have a duty to design out hazards associated with their designs both in relation to the future use of the building and to how it will be constructed.

The principal designer should ensure that both they and the other designers have considered their duties and provided design risk assessments for the principal contractor.

Designers should consider

- the weight of materials they specify
- the size of the materials they specify
- the shape of materials they specify
- the type of materials they specify
- the sequence of construction.

The principal designer should encourage all designers to reduce the likelihood of injury from manual handling activities. Consideration should also be given as to how materials will be delivered and moved around the site: for example, carrying 3 m × 2 m boards up ten flights of stairs may be excessive manual handling, and the designer should have considered smaller boards or hoist access into the building.

Materials should be available in the lightest weights possible. Bags, blocks and so on should weigh 25 kg or less.

Once the principal contractor (or, if one has not been appointed, the contractor) is in receipt of any design risk assessments and the pre-construction health and safety pack (as this may outline the philosophy of the design in respect of manual handling), the site agent should carry out a risk assessment of the materials needed on site, the sizes specified in the design, access, and movement of materials around the site and where they need to be used. Any concerns should be raised, if appropriate, with the principal designer, or the client's independent safety advisors if appointed, who should then liaise with the design team and seek solutions.

Manual handling must be eliminated where possible, and in any event reduced to acceptable levels. Small improvements in manual handling will have a significant long-term impact, and must not be ignored.

Mechanical aids, lifts, hoists, trolleys and so on eliminate most of the manual handling activity, and the risk assessment should consider these options.

Plan what equipment is needed throughout the site.

Avoid double handling of materials – it increases the risk of injury.

Reduce the height to which materials need to be lifted by positioning loads at height by mechanical aids. Scissor lifts and scissor tables could assist.

Reduce the distance materials need to be carried – consider the delivery plan and where materials will be stored, where they will be needed, how often they need to be replenished and so on.

Encourage the sharing or lifting of loads that are too heavy for one person or which are of an awkward shape. Two-people lifts will (usually) be safer than one-person lifts.

What are the key steps in a manual handling risk assessment?

There are four key elements to undertaking a manual handling risk assessment, namely

- consider the task
- consider the load
- consider the working environment
- consider the individual capability.

The tasks

Do they include the following?

- Holding or manipulating loads at a distance from the trunk.
- Unsatisfactory bodily movement or posture.
- Twisting the trunk.
- Stooping.
- Reaching upwards.
- Excessive movement of loads, especially
  - excessive lifting or lowering distances
  - excessive carrying distances.
- Excessive pushing or pulling of loads.
- A risk of sudden movement of loads.
- Frequent or prolonged physical effort.
- Insufficient rest or recovery periods.
- A rate of work imposed by a process.

The loads

- Are they heavy?
- Are they bulky or unwieldy?
- Are they difficult to grasp?
- Are they unstable or with contents likely to shift?
- Do they contain sharp edges, protruding nails, screws and so on?
- Are they hot from process?

The working environment

- Are there space constraints due to small work areas, restricted heights and so that will prevent good lifting postures?
- Are there uneven, slippery, defective or unstable floors?
- Are there changes in the floor level either via steps or ramps?
- Are there extreme temperature changes that could affect physical exertion?
- Are there ventilation/air changes that could raise dust, cause gusts of wind and so on?
- Is there overcrowding of work spaces?
- Are there physical obstructions?
- Is there poor lighting?

Individual capacity

- Does the job require unusual strength?
- Does the job require special knowledge of the load?
- Does the job create any other hazards to the individual carrying the load?

Control measures

The information collected in the stages above will form the basis of assessing the hazards and risks from the manual handling task to be undertaken. The likelihood of injury will range from high to low, depending on the interrelationship of the above factors.

Control measures then need to be identified to reduce the risk of injury: for example,

- remove obstructions
- increase lighting
- reduce load weight
- use mechanical aid
- provide training on lifting techniques.

When the control measures are implemented, the risk of injury will reduce.

The risk assessment will need to be reviewed regularly for the same manual handling process.

Can manual handling be done safely?

There are several key rules for safe manual handling. Information should be freely available on safe lifting techniques in the mess room, site office and so on so that operatives can be reminded of the correct way to lift objects.

The fundamentals of safe lifting are

- take a secure grip
- use the proper feet position – feet apart with the leading foot pointing in the direction you intend to go
- adopt a position with bent knees but a *straight* back
- keep arms close to the body
- keep the head and chin tucked in
- keep the load close to the body
- use the body weight where possible
- push up for the lift, using the thigh muscles.

Operatives should also consider the environment that they are to lift in, and pre-plan any interim rest positions, how to change direction and so on. The principal contractor should oversee manual handling activities that are site-wide.

What topics should be covered in a manual handling toolbox talk?

An effective toolbox talk agenda will cover

- what checks to carry out before starting manual handling
- how to judge your capability
- environmental conditions

- wearing of personal protective equipment (e.g. gloves and safety boots)
- carrying out a trial lift first
- getting help (it should not be seen as being a 'wimp')
- good handling techniques
- reducing the weights of objects
- using mechanical aids
- checks to carry out during the manual handling tasks (e.g. obstructions in the travel route).

What other activities need to be considered to successfully manage manual handling on the site?

Constrained postures

With stretched arms or bent postures, the muscles have to do extra work in order to maintain the posture. In a constrained posture the muscles can produce less force than in a more extended, comfortable one. This means that muscles will get tired faster in awkward postures, even when the work activity does not demand high muscle forces. Also, the mechanical load on the spine and joints is higher in these postures than in comfortable ones.

Kneeling and squatting result in high loads to the knees. In a kneeling posture, the front part of the knee can be irritated due to contact with the hard floor, pavement or ground. If working when kneeling cannot be avoided by technical changes, the operators should be provided with kneepads for protection.

Working above head level

Working with the arms extended above head level is common in several tasks such as assembling electricity or ventilation channels in the ceiling, or painting the ceiling. While working with the arms stretched upwards, the small shoulder muscles have to do extra work in order to hold the weight of the arms. The load is extremely high if the operator also holds a tool or load in a hand far from the shoulder. To see the work being done, the operator also has to bend the neck backwards, which stresses the neck. These situations result in extra risk for shoulder and neck disorders.

Repetitive work

Repetitive tasks require the same movements to be repeated several times in a minute. Construction work includes repetitive manual tasks such as

- hammering
- drilling
- driving screws
- sawing

- painting with brushes
- cutting sheet metal with scissors
- loading and unloading small pieces (e.g. tiles or bricks) to be transported from intermediate storage locations to the final assembly site.

Doing these tasks repeatedly or for a long period of time can result in work-related upper-limb disorders – more commonly called repetitive strain injury (RSI).

The harm to muscles, nerves and tendons that can be caused by repetitive movements is acerbated by their combination with the large muscle forces used for gripping. Grasping heavy objects between the thumb and the fingers needs more force if a pinch grip is used (for thin objects) or too broad (for large objects). The risk is increased if the wrist cannot be held straight during these tasks.

Providing ergonomically designed hand tools can reduce this risk. Electric or pneumatic tools can also reduce repetitive forceful movements. These tools can, however, bring new problems, such as vibration, or more gripping force can be needed to hold the powered tool if it is too heavy or incorrectly designed.

Static work

Static muscle work means the continuous contraction of some muscles to maintain the posture or to hold the force level constant. If static contraction is held for an extended period, the circulation in the muscles will be disturbed, and this can result in disorders.

With high muscle load, fatigue will force the operator to take a rest. With lower loads the level of fatigue is not so evident, and the operator can spend too long a time in the same posture.

Operators in the cabins of modern machines (lorries, caterpillar trucks, etc.) can easily spend too long in the same posture without any clear indications of fatigue. Therefore, it is important when undertaking these tasks to stop regularly and move the body around to avoid disorders.

Local compression of tools and surfaces

All sharp edges or hard surfaces can harm the body if the contact compression lasts for too long. Using the hand as a hammer can cause local injuries to the hand. The symptoms may not appear during the task but several hours later. This bad habit can be continued without any early warnings of danger.

Musculoskeletal injuries are common in the construction industry, and everyone involved in a project – clients, principal designers, designers and all contractors and workers – have a part to play in reducing the number of incidents. The principles of prevention can be applied successfully to all activities requiring any kind of manual handling, and such an approach should form the basis of the risk assessments.

Improvements in manual handling: evaluation

There are many ways to improve manual handling on a construction site and when carrying out refurbishment works. However, sometimes what seems to be an improvement in one area creates additional hazards or risks in another.

The following questions might help to ensure that one hazard is not being substituted for another.

Will the improvement

- reduce or eliminate most or all of the identified risk factors?
- add any new risks that have not been previously identified?
- be affordable: that is, is there a simpler, less expensive option that would be equally effective?
- affect productivity or effectiveness?
- affect product or service quality?
- provide a temporary or permanent fix?
- be accepted by site operatives?
- be fully implemented, including any training, within a reasonable time?
- affect any specific trades more than others, thus affecting the work rate and so on?

Will the equipment

- reach far enough to cover the work area?
- handle the weight and shape of the product?
- re-orient the load as needed?
- be easy to load/unload?
- require much force or energy to push, steer or stop it at the destination, given the typical ground conditions of a construction site?
- be heavy or large and create its own manual handling issues?
- handle the load in a safe and controlled manner?
- hold the load securely and well balanced?
- allow too much movement from any chains or cables if used?
- allow an adequate field of vision for the operator?
- slow operatives down too much, thus causing them to take shortcuts and create greater risks?
- interface with existing equipment and structures?
- obstruct the movement of people, materials or equipment around the rest of the site?
- need an additional power supply beyond the capacity of the system already in place?

Appendix to Chapter 29

Manual handling assessment form

Activity:					
Questions to consider (If the answer to a question is 'yes', place a tick against it and then consider the level of risk)	**Level of risk** (tick as appropriate)				**Possible remedial action**
	Yes	Low	Med.	High	
The tasks – do they involve ■ holding loads away from body? ■ twisting? ■ stooping? ■ reaching upwards? ■ large vertical movement? ■ long carrying distances? ■ strenuous pushing or pulling? ■ unpredictable movement of loads? ■ repetitive handling? ■ insufficient rest or recovery? ■ a work rate imposed by a process?					
The loads – are they ■ heavy? ■ bulky/unwieldy? ■ difficult to grasp? ■ unstable/unpredictable? ■ intrinsically harmful (e.g. sharp/hot?)					
The working environment – are there ■ constraints on posture? ■ poor floors? ■ variations in levels? ■ hot/cold/humid conditions? ■ strong air movements? ■ poor lighting conditions?					

Individual capability – does the job ■ require unusual capability? ■ create hazards for those with a health problem? ■ create hazards for those who are pregnant? ■ call for special information/training?					
Other factors: ■ is movement or posture hindered by clothing or personal protective equipment?					
Deciding the level of risk will inevitably call for judgement: use the information contained within the policy document to assist you.					

CDM 2015 Questions and Answers: A practical approach
ISBN 978-0-7277-6032-6

http://dx.doi.org/10.1680/cdmqa.60326.461

Chapter 30
Personal protective equipment

What are the legal requirements with regard to personal protective equipment?

The Personal Protective Equipment at Work (PPE) Regulations 1992 apply in all work environments.

Employers have a duty to provide to the employees, *without charge*, all necessary PPE that has been identified as necessary within the risk assessment.

PPE must be suitable and sufficient for use and appropriate for the risks it has been chosen to reduce. It must suit the worker for whom it is intended, and afford adequate protection.

PPE must be comfortable to wear and fit the user properly. If more than one type of PPE is to be used, all items must be compatible: for example, goggles and a face mask must be suitable to be worn together.

Employers must ensure that employees have been given adequate information, instruction and training about how to wear PPE, why it has to be worn, its benefits, how to check it for defects, maintenance procedures and so on.

Suitable accommodation must be made available for the storage of PPE if it is impracticable for employees (and others) to store it on the site.

Employers could be prosecuted for failing to supply appropriate PPE or for failing to maintain it in a suitable condition for use.

What are the responsibilities of the site agent with regard to issuing PPE?

The duty to issue PPE rests with the employer or the self-employed person and not with the site agent or principal contractor.

Where the site agent is the employer's on-site manager, then the employer has a duty to ensure that their *own* employees are correctly issued with PPE.

The principal contractor (or, where one is not appointed, the contractor) will determine the site rules for a construction site, and in the rules it can stipulate the minimum requirement for PPE.

The site agent must ensure that their site rules are being followed – they have the duty to monitor and review safety on site.

What are the rules regarding the wearing of hard hats on site

Generally, the principal contractor (or, where one is not appointed, the contractor) will determine whether the site will be a hard-hat site or not, using the criteria set out in the PPE Regulations 1992.

The Construction (Head Protection) Regulations 1989 were repealed during 2014 as part of the UK government's initiative to reduce red tape, as it was felt that there were sufficient legal duties included in the PPE Regulations 1992.

The site agent, on behalf of the principal contractor, or indeed any other employer, must determine the need for head protection in the same way that they determine the need for any other protective equipment.

Rules about who should wear head protection, and where and when, must be in writing and clearly displayed on the site.

Usually, all entry points, mess rooms and canteens display safety signs stating that 'hard hats must be worn'.

Wear there is 'no foreseeable risk of head injury, other than by falling' (PPE Regulations 1992), hard hats do *not* need to be worn. In effect, the employer or site agent needs to complete a risk assessment that identifies the hazards and risks of working on the site.

If a site agent relaxes the hard hat rule, there should be a risk assessment that identifies that the risk of head injury is minimal.

If may be possible to relax the rule for hard hats for parts of the site, although this is not recommended, as operatives may be confused as to where and when to wear their hard hat.

Many fatalities and severe injuries are caused because employees and others do not wear protective hats. Brain damage is often irreversible, and although someone may be physically very able, the damage done to the brain by way of a head injury can result in severe learning disabilities for the individual.

As a sub-contractor on the site, can I decide that the area in which my employees are working does not have to be a hard hat area?

An as employer in control of your own site, you can decide when hard hats need to be worn, but as a sub-contractor on a multi-occupied site you will have to follow the site rules and wear head protection.

Under the CDM Regulations, the principal contractor (or, where one is not appointed, the contractor) must make site rules with regard to the implementation and management of site safety. As a contractor, you have a duty to co-operate with the principal contractor and to follow the site rules.

There would be no harm in discussing the situation with the site agent with a view to seeking authority to relax the rules for head protection in your work area.

There may of course be other works being undertaken that you are not aware of, and these could pose head injury hazards to your employees. It is the principal contractor's duty to co-ordinate and manage the hazards and risks of other contractors on a multi-occupied site.

Is the site agent responsible for issuing PPE to site visitors?

It depends on the visitors. Usually, a site agent will provide PPE for the client and the client's representatives, as it would be unreasonable to expect the client to provide the appropriate equipment. Hard hats are absolutely essential if there is a risk of falling objects or of hitting one's head on scaffold tubes, ceiling projections and so on.

It may not be reasonable for the site agent to provide PPE for

- architects
- designers
- quantity surveyors
- building services consultants
- project managers

unless it is *specialist* PPE that is site specific. All of the above will be employees of professional service firms or will be self-employed. The duty to provide PPE falls to the *employer*. They should provide their own!

What PPE would usually be expected on a main construction site?

The usual PPE comprises

- hard hats
- safety boots/shoes
- high-visibility vests/jackets
- gloves
- goggles/eye protectors.

It may be necessary to provide overalls and protective clothing as well.

It may also be necessary to provide hearing protection.

Should a risk assessment be completed in order to determine what PPE is required?

All work activities where there is a risk of injury to employees or others should be subjected to a full risk assessment.

Under the Management of Health and Safety at Work Regulations 1999, employers are responsible for completing risk assessments and for recording their significant findings. Employers should therefore conduct risk assessments for any site-based activity, and issue appropriate PPE.

The site agent should consider the hazards and risks in communal areas on the site, and set down the findings in the risk assessment. From this, the site rules will be drawn up that stipulate what PPE is to be worn when and by whom.

The site agent may need to issue PPE to persons who are exposed to risk from hazardous substances from the activities of others.

CDM 2015 Questions and Answers: A practical approach
ISBN 978-0-7277-6032-6

http://dx.doi.org/10.1680/cdmqa.60326.465

Chapter 31
Noise

What is noise at work?

Noise at work comes in many different forms, including from machinery, music and factory processes. Noise at work can damage hearing and in some cases lead to deafness, depending on how loud the noise is and how long a person is exposed to it. All employers have a duty under health and safety law to reduce the risk of hearing damage to their employees by controlling exposure to noise.

Noise can also be a safety hazard at work by interfering with communication and warning sirens, making them harder to hear.

How is noise measured?

Noise is measured in decibels (abbreviated to dB). Sometimes you may see it written as dB(A) – this is known as an A weighting, and is an average of the noise level. The decibel is a logarithmic unit used to describe a ratio. Because of the logarithmic effect of the decibel scale, a small increase such as 3 dB can actually mean that the level of noise has doubled, so what might seem like a small increase in the noise level can in fact be a very significant one.

What do the Control of Noise at Work Regulations 2005 require?

The Control of Noise at Work Regulations 2005 place duties on all employers whose employees are or may be exposed to noise while at work. The legislation sets out action levels and states what action should be taken at each level by the employer to control noise levels (Table 31.1).

The action levels are the noise exposure levels at which employers are required to take certain steps to reduce the harmful effects of noise to their employees. The action levels shown in Table 31.1 are a daily or weekly average of noise exposure.

Do the Control of Noise at Work Regulations 2005 apply to all workplaces?

The Control of Noise at Work Regulations 2005 apply to all premises and workplaces where noise may affect persons at work, and apply to all construction sites.

Table 31.1 Control of Noise at Work Regulations 2005: sound levels and actions required

Exposure action level		Action required if the level is exceeded
Lower exposure action level	80 dB(A)	■ Carry out a risk assessment ■ Make suitable ear protection available ■ Implement a maintenance programme for the ear protection ■ Implement a training programme
Upper exposure action level	85 dB(A)	■ Reduce the noise at the source ■ Implement ear protection zones ■ Provide ear protection, which must be used by employees (the use of hearing protection is mandatory if the noise cannot be controlled by any other measure) ■ Provide health surveillance for employees

The regulations set noise levels as shown in Table 31.1 at which action must be taken to control the noise at work.

How do I determine if I have a noise problem at work?

There are several factors that need to be taken into consideration when determining if noise is a problem at work. These are

- how loud the noise is
- how long people are exposed to the noise.

As a simple guide, you will need to do something about the noise if any of the following apply:

- the noise interferes with the day-to-day work activities for most of the day (e.g. a busy street or a vacuum cleaner being used non-stop)
- employees have to raise their voices to carry out a normal conversation.
- noisy tools or machinery are used for more than half an hour throughout the working day
- employees work in a noisy industry (e.g. construction, road repair, engineering, canning, production, manufacture, foundry, or paper or board making)
- there is noise in the workplace due to machinery impacts (e.g. hammering, pressing, forging, pneumatic equipment, and explosive sources).

What are some typical noise levels associated with construction?

Sound pressure: dB(A)	Situation
140	Peak action level: immediate irreversible damage Jet at 30 m
130	Threshold of pain Pneumatic breaker (unsilenced) at 1 m
120	Pneumatic digger 600 hp scraper at 2 m
110	Rock drill Diesel hammer driving sheet steel at 10 m
100	Scrabbling 7 hp road roller on concrete at 10 m
95	Concrete pouring
85	Second action level (Control of Noise at Work Regulations 2005) Drilling/grinding concrete
80	First action level (Control of Noise at Work Regulations 2005) Scaffold dismantling at 10 m 8 hp diesel hoist at 10 m
70	5 hp power float at 7 m
60	Typical office
50	Living room

What do I do if I have a noise problem?

If any of the above apply, then an assessment of the risks will need to be carried out to decide whether further action is needed. This is known as a risk assessment. The aim of the risk assessment is to provide you with the information so that a decision can be made on what needs to be done to ensure the health and safety of employees who are exposed to noise. In some cases, measurements of noise may not be necessary: it is about collecting as much information as possible.

The noise risk assessment should contain the following information:

- Who is at risk from noise.
- Who may be affected.
- An estimate of the employees' exposure to noise, compared with the lower and upper exposure action levels in the Control of Noise at Work Regulations 2005 (see Table 31.1).
- What needs to be done to comply with the law. These are often called noise control measures, and may include the provision of hearing protection, such as ear

defenders or ear plugs. If noise control measures are required, then the type should be included in the risk assessment.
- Details of any employees who need to be provided with health surveillance and whether any particular employees are at risk, because of the nature of their work

There is no right or wrong way to complete a risk assessment. The law requires only that it is 'suitable and sufficient'.

A risk assessment must contain suitable information to be useful to an employee to understand what hazards they may be exposed to when carrying out the task.

How do I estimate an employee's exposure to noise?

The key is to ensure that the estimate of the employees exposure is a true reflection of the work that they do, and should therefore take into account the following:

- the work that they do and may be doing in the future
- the way in which they do the work
- how it might vary from one day to the next.

The best way to collect this information is to speak to the employees concerned, as the estimate must be based on correct information. The suppliers of machinery will also be able to provide data sheets that will show the noise level associated with the use of a particular piece of machinery.

The information gathered must be recorded in a risk assessment. The risk assessment should set out what you have done and what you are going to do to ensure that employees' exposure to noise is controlled. There should also be a timetable showing when the measures will be implemented, along with who will be responsible for the work.

Once a risk assessment has been carried out, the information should be used to determine if a noise assessment is required.

In order for employers to be able to ensure that they are controlling their employees' exposure to loud noise, they need to know which employees are at risk, and what that level of risk is.

For an employer whose employees are exposed to noise while at work, they may have to implement measures to control their employees' exposure to that noise, if the level of noise is deemed to exceed the action levels stated in the Control of Noise at Work Regulations 2005.

The best way of doing this is to have a noise assessment carried out. The Control of Noise at Work Regulations 2005 require the employer to carry out an adequate noise assessment, which will help to provide the information necessary to control noise in the workplace. In addition, the noise assessment assists in determining the most suitable hearing protection to provide for employees, if ear protection zones are required and where they should be. The noise assessment as a document helps employers in their compliance with duties relating to controlling noise exposure.

What is the difference between a noise risk assessment and noise assessment?

A noise risk assessment is the first step in the process of determining whether employees are exposed to noise at work. The risk assessment process should be used to gather as much information as possible about the type of work that is carried out, who is at risk from the work, what the estimated level of exposure is, how employees may be affected and what is or will be done to reduce the noise that employees may be exposed to.

If the risk assessment suggests that there is a noise problem, a competent person may need to be employed to measure the noise and determine the representative daily or weekly personal noise exposure: this is called a noise assessment. The noise assessment includes measurement of the sound pressure level at the different places the employee works and for the different tasks carried out during the day. The average noise levels are calculated from these values and the time spent in each place or at each task. Information on getting started with a noise risk assessment can be found in the Health and Safety Executive (HSE) leaflet *Noise at Work: A Brief Guide to Controlling the Risks* (INDG362(rev2), and more detail is provided in another HSE publication, *Controlling Noise at Work: Guidance on the Control of Noise at Work Regulations 2005* (L108).

Is it a legal requirement to carry out a noise assessment?

In order for employers to be able to ensure that they are controlling their employees' exposure to loud noise, they need to know which employees are at risk, and what that the level of risk is.

An employer whose employees are exposed to noise while at work may have to implement measures to control their employees' exposure to that noise, if the level of noise is deemed to exceed the action levels stated in the Control of Noise at Work Regulations 2005.

The best way of doing this is to have a noise assessment carried out.

Is there a standard format for a noise assessment?

No, although there is certain information that should be contained in the noise assessment.

The noise assessment must contain information on the likely exposure to noise and at what level, the expected duration of exposure and the control measures necessary to reduce the risk of hearing loss.

The HSE publishes a range of guidance documents to help employers. *Controlling Noise at Work* (L108) is the comprehensive guidance document for the Regulations.

Do I need to employ a consultant to carry out my noise assessment?

No, not necessarily. However, the measurement of noise and all the many factors that need to be taken into consideration when carrying out a noise assessment can be complex. If you feel unable to carry out the task yourself or you do not have anyone at your workplace capable of carrying them out, then you should employ a noise consultant to carry out the noise assessment.

Specialists will have the necessary noise-measuring equipment to carry out the assessment, and will be able to provide you with recommendations on how to tackle any noise issues in the workplace.

Consultants can provide a wealth of knowledge and experience, and will be able to make the whole process simple and painless.

Remember

The rules of thumb provided as to whether a noise risk exists should only be used as a guide, and if it is likely that a noise hazard exists in the workplace, then the employer will more than likely need to have a noise assessment carried out by a competent person.

By law, employers are required to assess and identify measures to eliminate or reduce risks from exposure to noise, in order that employees' hearing can be protected.

Where the risks are low, the actions taken may be simple and inexpensive, but where the risks are high, they should be managed using a prioritised noise control action plan.

Remember to review what is being done on a regular basis to ensure that if there have been any changes to the work process or job they are not affecting the noise exposure of employees.

What are the health effects of noise at work?

Noise at work can cause hearing damage, which may be temporary or permanent. This damage can lead to a partial loss of hearing or total deafness, depending on its extent.

Some people often experience temporary deafness after leaving a noisy place (e.g. after leaving a noisy nightclub or bar). Although this type of hearing loss usually results in the hearing recovering within a few hours, it should not be ignored. Temporary hearing loss is a sign that if the person continues to be exposed to the noise, their hearing could be permanently damaged.

Permanent hearing damage can be caused immediately by sudden, extremely loud, explosive noises (e.g. from guns or cartridge-operated machines), but is more usually a gradual process. It may only be when damage caused by noise over the years combines with hearing loss due to ageing that people realise how deaf they have become.

Hearing loss is not the only problem. People may develop tinnitus (ringing, whistling, buzzing or humming in the ears).

Anyone of any age can suffer from hearing damage.

What is a low-noise purchasing policy?

A low-noise purchasing policy basically means that when new equipment is bought or hired, the employer will ensure that the quietest equipment or machinery is selected. The cost of introducing noise reduction measures is often reduced quite significantly if quiet equipment can be introduced into the workplace.

What should be included in the low-noise purchasing policy?

The following tips may be helpful:

- Consider at an early stage how new or replacement machinery could reduce noise levels in the workplace – set a target to reduce the noise levels if possible.
- Ensure a realistic noise output level is specified for all new machinery, and check that tenderers and suppliers are aware of their legal duties.
- Ask suppliers about the likely noise levels under the particular conditions in which the machinery will be operated, as well as under standard test conditions. (Noise output data will only ever be a guide, as many factors affect the noise levels experienced by employees.)
- Only buy or hire from suppliers that can demonstrate a low-noise design, with noise control as a standard part of the machine.
- Keep a record of the decision process that was followed during the buying or hiring of new machinery, to help show that legal duties to reduce workplace noise have been met.

What are manufacturers and suppliers of machinery required to do?

Under the Health and Safety at Work etc. Act 1974 and the Supply of Machinery (Safety) Regulations 1992 (as amended), a supplier of machinery must do the following:

- provide equipment that is safe and without risk to health, with the necessary information to ensure it will be used correctly
- design and construct machinery so that the noise produced is as low as possible
- provide information about the noise the machine produces under actual working conditions.

New machinery must be provided with

- a 'declaration of conformity', to show that it meets essential health and safety requirements
- a 'CE' (Conformité Européenne) mark
- instructions for safe installation, use and maintenance
- information on the risks from noise at workstations, including
  - the A-weighted sound pressure level, where this exceeds 70 dB
  - the maximum C-weighted instantaneous sound pressure level, where this exceeds 130 dB
  - the sound power (a measure of the total sound energy) emitted by the machinery, where the A-weighted sound pressure level exceeds 85 dB
- a description of the operating conditions under which the machinery has been tested.

When should hearing protection be used?

If noise cannot be controlled by other methods, such as new machinery, acoustic screening, provision of anti-vibration mounts to machinery or a change in working patterns, then extra protection for employees will be needed. In addition, it may be needed as a short-term measure while other methods of control are being implemented.

Hearing protection should not be used as an alternative to controlling the noise by technical or organisational methods. It should only be used where there is no alternative way of protecting employees from noise exposure.

What are the general requirements for hearing protection?

- If employees ask for hearing protection, it should be provided for them.
- Hearing protection must be made available for employees to use when the lower action level of 80 dB(A) is exceeded.
- Hearing protection must be used by employees when the upper action level of 85 dB(A) is exceeded.

- Employers must provide training and information in the correct use of hearing protection.
- Employers must ensure that any hearing protection that is provided is properly used and maintained.

What should maintenance of hearing protection involve?

Any hearing protection that is provided for employees should be checked and maintained on a regular basis to ensure it remains in good working order. As a minimum, the employer should check that

- it is in good condition and clean
- the seals on ear muffs are not damaged
- the tension on headbands of ear defenders is still good and fits well when worn
- employees have not made any modifications to the hearing protection that may mean that it does not provide the correct level of protection
- earplugs are soft, pliable and clean.

What else can be done?

It is best practice to include the wearing, maintenance and care of hearing protection in the company safety policy.

Managers and supervisors should be encouraged to set a good example by ensuring that they wear hearing protection in areas where it is required.

The hearing protection used must give enough protection. As a guide, it should give sufficient protection to reduce noise levels to below 85 dB.

Employers should ensure that the hearing protection is suitable for the working environment, and consideration should be given to how hygienic and comfortable it is.

If other personal protective equipment is worn by the employee, such as dust masks or hard hats, the employer should consider how the hearing protection will fit in with this other protection. For example, the wearing of ear defenders may be difficult if the employee also has to wear a hard hat: in this case, a hard hat incorporating a set of ear defenders would be required.

It is important to ensure that the hearing protection provided is not

- designed to cut out too much noise, as this can cause isolation and lead to employees not wanting to wear them
- compulsory to wear, unless the law requires it.

What information, instruction and training do employees need?

Employees need to understand the risks that they may be exposed to from noise in the workplace. If employees are exposed to the lower action level of 80 dB, as a minimum employers should tell them

- the estimated noise exposure and the risk to hearing
- what is being done to control the risks and exposure (this may include the use of acoustic screening, increased maintenance of noisy pieces of equipment, changes in work patterns, or provision of hearing protection)
- where they can get hearing protection from
- who in the company will be responsible for the provision and maintenance of the hearing protection and who they should report defects to
- the correct way to use the hearing protection, how to look after it, how it should be stored and the area where it needs to be used
- details of any health surveillance programme that may be in place.

What is health surveillance for noise?

Health surveillance for hearing usually means

- regular hearing checks in controlled conditions by a trained professional
- telling employees about the results of their hearing checks
- keeping health records
- ensuring employees are examined by a doctor where hearing damage is identified.

Ideally, health surveillance should be started before employees are exposed to noise. This helps to give a baseline, and can be used to determine any changes in noise exposure and the effects those changes may be having.

What is the purpose of health surveillance?

Health surveillance is designed to provide the employer with an early warning system as to when employees might be suffering from the initial signs of hearing damage. It gives the employer an opportunity to do something to prevent the damage getting worse, and ensures that the control measures that are in place are working.

It is important that employees understand the aim and importance of health surveillance, and that it is there to protect their hearing.

When does health surveillance need to be provided?

If employees are regularly exposed to the upper exposure action level or are at risk for any reason (e.g. they already suffer from hearing loss or are particularly sensitive to hearing damage), the employer is required to provide health surveillance in the form of hearing checks for those employees.

How often should checks be carried out on employees?

After the initial check, a programme of health surveillance should be implemented with a regular check carried out annually for the first 2 years of employment, and then at 3 year intervals thereafter. However, this may need to be more frequent if any problems with hearing are detected or where the risk of hearing damage is high.

Who should carry out the hearing checks?

The hearing checks need to be carried out by someone who has the appropriate training. The health surveillance programme needs to be under the control of an occupational health professional (e.g. a doctor or a nurse with appropriate training and experience). The employer is responsible for making sure that the health surveillance is carried out properly.

How can health surveillance be arranged?

Larger companies may have access to in-house occupational health services that may be able to carry out the programme. Where there are no facilities in house, an external contractor will need to be used. Further information about occupational health services may be available from trade associations or from the local healthcare services.

What should be done with the results of health surveillance?

The results should be used to make sure that employees' hearing is protected. The records of the health surveillance should be kept and used to provide advice for each employee.

Recommendations should be given by the person carrying out the surveillance, and these should be acted upon by the employer. The information may need to be used to amend any risk assessments that have been carried out.

Controlling noise on construction sites: a good planning guide

A client should include noise control requirements for both occupational and environmental noise early in the planning stage for a new project. The desired noise control requirements may be included in a client specification list in the tender document. This can help to avoid unexpected and often very expensive noise controls during the construction phase. It allows tenderers to plan how to overcome noise problems in advance.

The client's specifications may include

- specified noise exposure levels during the construction phase, as per legislative requirements or company policy
- the use of quiet/silenced equipment
- the adoption of quiet alternative techniques
- the use of noise control measures such as silencers, barriers and enclosures

- the erection of warning signs identifying noise hazard areas
- time restrictions
- the provision of personal hearing protectors and training.

The tenderer's proposal should cover all the client's specifications. The tenderer should prepare a noise control policy and a noise control plan to be included in the site-specific safety management plan.

The noise control plan may be a set of actions required to achieve the noise control policy and to reduce noise exposure. It may also include information on how the company is planning to meet its obligations, such as

- a list of equipment to be used – with noise levels at the operator position and/or at 1 m
- the methods undertaken to lower noise exposure (e.g. maintenance, barriers and enclosures)
- restricted hours, the rotation of workers in noisy places, and special time arrangements such as noisy work done after hours
- the identification of noisy equipment and processes by signs
- the site induction for employees and contractors to include noise levels, noise controls, and the correct use and maintenance of personal hearing protectors
- the selection and provision of appropriate personal hearing protectors
- audiometric tests.

The main contractor should plan to co-ordinate sub-contractors so that the activities of one do not unnecessarily expose employees of another to noise hazards. It is good practice to nominate one person as the noise co-ordinator for all noisy activities. Site planning should include

- the preparation of guidance for workers on hazards and the methods to reduce noise
- the preparation of schedules of noisy plant and exposure estimates for each phase of work
- laying out the site to separate noisy activities from quieter ones
- scheduling noisy activities to take place when the minimum number of nearby workers are present (out-of-hours noise needs to be carefully planned to avoid neighbourhood annoyance)
- rostering workers to minimise exposure times
- ensuring that workers are well trained, instructed and supervised in noise matters and responsibilities, including the correct use and maintenance of personal hearing protectors.

Once the construction work is in progress, it is essential to monitor the implementation of the noise control plan. This could be carried out by the client or the main contractor, and could include the following:

- checking if equipment brought onto the site complies with specifications (this could be done by obtaining information available from suppliers or by noise assessments)
- reducing noise from identified noise sources by exchanging equipment and/or processes for a quieter alternative or by engineering control methods to quieten existing ones
- ensuring that all plant is properly maintained (e.g. all noise control measures such as silencers and enclosures are intact)
- monitoring work schedules to check that noisy work is carried out as specified, away from other workers, outside hours and so on
- monitoring if noisy areas are identified and well marked so that employees and contractors can avoid entering them unnecessarily
- monitoring whether training and hearing tests have been carried out and if personal hearing protectors are adequate and are being worn and maintained correctly
- ensuring that the cause of any hearing loss shown up by audiometry is investigated
- utilising safety toolbox meetings to provide feedback on the effectiveness of noise control measures and personal hearing protectors to workers, employers and contractors
- posting on safety notice boards the results of the noise assessments conducted and additional noise information.

CDM 2015 Questions and Answers: A practical approach
ISBN 978-0-7277-6032-6

http://dx.doi.org/10.1680/cdmqa.60326.479

Chapter 32
Fire safety on construction sites

What causes a fire to start?

A fire needs the following to start:

- fuel
- ignition
- oxygen.

These three component parts are often referred to as the 'triangle of fire' (Figure 32.1). If one or more of the component parts of a fire are eliminated, a fire will not start or continue.

Figure 32.1 The 'triangle of fire'

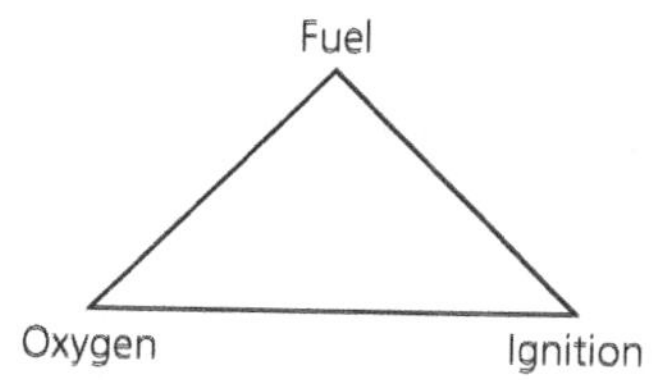

As the elements of fire are not successfully managed on all construction sites, the industry suffers approximately 4000 fires a year, costing millions of pounds in destroyed buildings, materials, plant and equipment and delayed projects.

Often, insurance cover for buildings and construction work is difficult to obtain without a proper fire safety management plan in place.

What is the Regulatory Reform (Fire Safety) Order 2005 (RRO)?

The RRO contains the legal requirements for managing fire safety in premises. It places duties on a responsible person to ensure that people are protected from fire and that adequate means of escape are available from all premises.

What type of premises does the RRO apply to?

The RRO applies to virtually all premises, and covers nearly every type of building, including open spaces:

- offices
- shops
- factories
- care homes
- hospitals
- churches and places of worship
- village and community halls
- schools, colleges and universities
- pubs, clubs and restaurants
- sports centres
- tents and marquees
- hotels, hostels
- warehouses
- shopping centres
- construction sites.

The RRO does not apply to private dwellings (including individual flats in a block or house). (Note: fire safety requirements for houses in multiple occupation are contained in Housing Act 2004.)

Who is legally responsible for complying with the RRO?

The RRO refers to a 'responsible person' as being the person with responsibility for complying with the order.

In the workplace, the responsible person will be the employer or any other person who may have control of any part of the premises (e.g. the owner or occupier or, in the case of a construction site, the principal contractor).

In all other premises, the person or persons in control of the premises will be responsible for complying with the RRO.

What happens if there is more than one person in control of premises?

Where there is more than one responsible person in any type of premises (e.g. a multi-occupied complex), all of the persons must take reasonable steps to co-operate with one another. If there are managing agents responsible for common parts, then they will be the responsible person for those areas.

The requirements of the RRO are in fact imposed on *any* person who has, to any extent, control of premises so far as the requirements of the order relate to matters within their control.

What are the main requirements or rules under the RRO?

The responsible person must

- carry out a fire risk assessment, identifying any possible dangers and risks of fire
- consider who may be especially at risk
- reduce the risk of fire within the premises as far as is reasonably practicable
- provide general fire precautions to deal with any residual fire risk
- implement special measures to control the risks from flammable materials, explosive materials or other hazardous substances
- create an emergency plan for fire safety
- record significant findings in writing
- review and monitor fire safety arrangements.

In addition, the responsible person must ensure that everybody who could be affected by their business undertaking receives information or training and instruction.

The responsible person must also prepare an emergency plan for fire safety arrangements.

Any equipment and facilities (fire-fighting aids, lifts, etc.) provided for the use and safety of fire fighters must be adequately maintained.

Does the duty to manage fire safety relate only to employees?

The RRO refers to all 'relevant persons'. This includes everyone – employees, public, customers, contractors, visitors, students, patients and so on.

The responsible person must consider the fire safety needs of those at special risk e.g. disabled people, young people, people not familiar with the building.

The RRO introduces clear responsibility for the 'responsible person' to consider the needs of the public in respect of fire safety and not just employees.

What are the main requirements of the RRO?

The main requirement of the RRO is for the responsible person to carry out a fire risk assessment. The fire risk assessment must focus on the safety in cases of fire of all 'relevant persons'.

In addition to a fire risk assessment, the responsible person must

- appoint one or more competent persons to assist with fire safety
- provide employees with clear and relevant information on the risks to them identified by the fire risk assessment, about measures taken to prevent fires and, generally, how they will be protected in the event that a fire occurs
- consult with employees about fire safety matters
- take special steps regarding the employment of young people and inform their parents about the fire measures that have been taken
- inform all persons who are not employees about fire safety measures in the premises (e.g. contractors and temporary workers)
- co-operate with other responsible persons and co-ordinate fire safety matters in multi-occupied buildings
- inform the employer of any temporary workers from an outside organisation with clear and relevant information about how the safety of their employees will be protected in the event of fire
- consider any special precautions needed for managing fire safety from any dangerous, flammable or hazardous substances, including explosives
- establish a suitable means for contacting the emergency services and providing them with relevant information about dangerous substances
- provide all relevant persons, where reasonable, with suitable information, instruction and training on fire safety
- provide and maintain suitable fire-fighting equipment, fire detection equipment, fire-warning systems, emergency lighting and so on as is deemed necessary in the premises by the fire risk assessment
- advise employees that they have a duty to co-operate with their employer so as to enable them to comply with their statutory duties.

Can anyone be appointed as the competent person?

Yes, provided they have enough knowledge and experience to understand the basic principles of fire safety and to be able to implement the measures contained in the RRO.

There are no formal qualifications to become a competent person, but as the onus is on the responsible person to appoint someone – or more than one person if necessary – to provide safety assistance, the responsible person will need to be able to demonstrate that the person so appointed has suitable credentials to take on the role.

The type of person who could be classed as competent persons could be a

- fire safety consultant
- fire engineer
- health and safety manager/officer – provided that have fire safety training
- employer – if they have attended basic fire safety training

- unit/department manager – with fire safety training
- health and safety consultant.

The competent person is there to help the responsible person fulfil their statutory duties.

What are the principles of prevention in respect of fire safety?

The RRO requires all responsible persons to undertake fire risk assessments of their premises, operations and any other business activity that could expose persons to risks to their safety.

If the risk assessment indicates that preventative and protective measures are required, then those measures must be in accordance with the requirements of Schedule 1, Part 3 of the RRO, the 'principles of prevention':

(*a*) avoiding risks;
(*b*) evaluating the risks which cannot be avoided;
(*c*) combating risks at source;
(*d*) adapting to technical progress;
(*e*) replacing the dangerous with the non-dangerous or less dangerous;
(*f*) developing a coherent overall protection policy which covers technology, organisation of work and the influence of factors relating to the working environment;
(*g*) giving collective protective measures priority over individual protective measures;
(*h*) giving appropriate instructions to employees.

The principles of prevention listed above are more detailed than in other health and safety legislation.

The Health and Safety Executive (HSE) inspector has indicated that my fire risk assessment is not 'suitable and sufficient'. What should I do about it?

If possible, discuss it with the inspector. Sometimes they don't like the format of a risk assessment rather than the content!

The inspector may be concerned that you have not considered all aspects of fire safety and that your risk assessment is missing a vital element. Generally, the inspector will tell you why they are not satisfied with your risk assessment.

If you feel that you have covered everything and that the inspector is being over-zealous, then ask to discuss the matter with a senior officer in the HSE.

Be prepared and review your risk assessment together with your competent person. Or seek professional advice.

A fire risk assessments must address the major fire safety hazards associated with your workplace – it does not have to be perfect.

What is the legal requirement for a fire risk assessment?

The RRO requires the 'responsible person' to make a suitable and sufficient assessment of the risks to which relevant persons are exposed in respect of fire.

Any premises in which persons are employed or to which others have access must be subject to a fire risk assessment.

The principles of risk assessment to be followed are the same as those as listed in the Management of Health and Safety at Work Regulations 1999.

Where there are five or more employees, the significant findings of the risk assessment must be recorded in writing.

What actually is a fire risk assessment?

A fire risk assessment is in effect an audit of a workplace and work activities in order to establish how likely a fire is to start, where it would be, how severe it might be, who it would affect and how people would get out of the building in an emergency.

A fire risk assessment should be concerned with *life safety* and not with fire engineering matters.

A fire risk assessment is a structured way of looking at the hazards and risks associated with fire and the products of fire (e.g. smoke).

Like all risk assessments, a fire risk assessment follows *five key steps*, namely

1. identify the hazards
2. identify the people and the location of people at significant risk from a fire
3. evaluate the risks
4. record the findings and the actions taken
5. keep the assessment under review.

So, a fire risk assessment is a written record that shows that the likelihood of a fire occurring in a workplace has been assessed, identifies those who could be harmed and how, and lists the steps needed to reduce the likelihood of a fire (and therefore its harmful consequences) for the workplace.

Definition

- *Risk assessment* – the overall process of estimating the magnitude of risk and deciding whether or not the risk is tolerable or acceptable.
- *Risk* – the combination of the likelihood and consequence of a specified hazardous event occurring.
- *Hazard* – a source or a situation with a potential to harm in terms of human injury or ill health, damage to property, damage to the environment or a combination of these.
- Hazard identification – the process of recognising that a hazard exists and defining its characteristics.

What are the five steps to a fire risk assessment?

Step 1: identify the hazards

Sources of ignition

Sources of ignition can be identified on a site by looking for possible sources of heat that could get hot enough to ignite the material in the vicinity. Such sources of heat/ignition could be

- a smoker's materials discarded due to smoking illicitly on the site
- naked flames (e.g. fires and blow lamps)
- electrical, gas or oil-fired heaters
- hot work processes (e.g. welding and gas cutting)
- cooking, especially frying in any mess rooms
- faulty or misused electrical appliances, including plugs and extension leads
- lighting equipment, especially halogen lamps
- hot surfaces and obstructions of ventilation grill (e.g. radiators)
- poorly maintained equipment that causes friction or sparks
- static electricity
- flammable materials
- arson.

Look out for any evidence that things or items have suffered scorching or overheating (blackened plugs and sockets, burn marks, cigarette burns, scorch marks, etc.).

Check each area of the premises systematically:

- work areas and site offices
- the staff kitchen and mess rooms
- store rooms and chemical stores
- plant rooms and motor rooms
- refuse areas

- materials storage areas
- external areas.

Sources of fuel

Anything (generally) that burns is fuel for a fire. Fuel can also be invisible in the form of vapours, fumes and so on given off from other less flammable materials.

Look for anything on the site that is in sufficient quantity to burn reasonably easily, or to cause a fire to spread to more easily accessible fuels.

Fuels to look out for are

- wood, paper and cardboard
- flammable chemicals (e.g. cleaning materials)
- flammable liquids (e.g. cleaning substances and liquid petroleum gas (LPG))
- flammable liquids and solvents (e.g. white spirit, petrol and methylated spirit)
- paints, varnishes, thinners and so on
- furniture, fixtures and fittings
- textiles
- ceiling tiles and polystyrene products
- waste materials and general rubbish
- gases (e.g. leaks from LPG cylinders).

Consider also the construction of the premises – have any materials been used that would burn more easily than other types? For example, hardboard, chipboard and blockboard burn more easily than plasterboard.

Identifying sources of oxygen

Oxygen is all round us in the air that we breathe, but sometimes other sources of oxygen are present that accelerate the speed at which a fire ignites (e.g. oxygen cylinders for welding).

The more turbulent the air the more likely the spread of fire will be: for example, opening a door brings a 'whoosh' of air into a room, and a fire will be fanned and intensified. Mechanical ventilation also moves air around in greater volumes and more quickly.

Also, while ventilation systems move oxygen around at greater volumes, they will also transport smoke and toxic fumes around a building.

Step 2: identify who could be harmed

It is essential to identify who will be at risk from a fire and where they will be when a fire starts. The law requires employers to ensure the safety of their employees and others (e.g. sub-contractors and delivery personnel).

Would anyone be affected by a fire in an area that is isolated? Could everyone respond to an alarm, or evacuate? Will contractors working in plant rooms or on the roof, in excavations, lift shafts and so on be adversely affected by a fire? Could they be trapped or not hear alarms?

Who might be affected by smoke travelling through the building? Smoke often contains toxic fumes especially if volatile substances are in use around the site.

Step 3: evaluate the risks arising from the hazards

What will happen if there is a fire? Does it matter whether it is a minor or major fire? Remember: small fires can grow rapidly to infernos, and fires spread quickly across construction sites because of the volume of combustible material.

A fire is often likely to start because

- people get careless with heat-generating equipment
- people purposely set light to things
- cooking equipment and canopies catch fire due to grease build-up
- people put combustible material near flames/ignition sources
- equipment is faulty because it has not been maintained
- electrical sockets are overloaded and 110 V power supplies on the site are abused.

Will people die in a fire from

- flames
- heat
- smoke
- toxic fumes?

Will people get trapped in the building or in work areas?

Will people know that there is a fire – will they be able to hear the alarm and will they be able to get out?

Step 3 of the risk assessment is about looking at what *control measures* are in place to help control the risk or reduce the risk of harm from a fire or what measures will be put in place.

Remember: fire safety is about *life safety*. Get people out fast and protect their lives. Property is replaceable.

The fire risk assessment will need to record the fire precautions that have or will be put in place: that is,

- What emergency exits are present and are they adequate and in the correct place?
- Are the emergency exits easily identified and are they unobstructed?
- Is there fire-fighting equipment?
- How is the fire alarm raised?
- Where do people go when they leave the building – to an assembly point?
- Are the signs for fire safety adequate?
- Who will check the building and take charge of an incident (i.e. has a fire warden been appointed)?
- Are fire doors kept closed?
- Are ignition sources controlled and fuel sources managed?

Considering all the fire safety precautions for the premises, is there anything more that needs to be done?

Are the staff trained in what to do in an emergency? Can they use fire extinguishers? Are fire drills held? Is equipment serviced and checked (emergency lights, fire alarm bells, etc.).

Step 4: record the findings and the action taken

Complete a fire risk assessment form and keep it safe. Make sure that the information is shared with everyone on the site.

When contractors come to the site, make sure that *their* fire safety plans are discussed and that they are told what the fire precaution procedures are on the site.

Step 5: keep the assessment under review

A fire risk assessment needs to be reviewed regularly – about every 6 months or so and whenever something has changed (e.g. alterations to the layout of the premises).

Who must I consider when preparing my risk assessment?

Responsible persons must consider the following people as being at risk in the event of a fire:

- employees and those on temporary or agency contracts
- employees whose mobility, sight or hearing might be impaired
- employees with learning difficulties or mental illness
- other persons in the premises if the premises are multi-occupied
- anyone occupying remote areas of the premises
- visitors and members of the public, including contractors
- anyone who may sleep on the premises
- anyone with any special needs or disabilities.

Does a fire risk assessment have to consider members of the public?

A fire risk assessment must be carried out by the responsible person and must consider the risks to the safety of *relevant persons* (i.e. all persons who are, or could be, lawfully on the premises). This will include members of the public.

The RRO has made the inclusion of all persons, including the public, a legal requirement when completing risk assessments.

The fire risk assessment for the construction site must include any risk to adjoining premises and the effect any fire could have on the occupants of adjoining buildings.

Who can carry out a fire risk assessment?

The RRO states that the person who carries out a fire risk assessment must be *competent* to do so. They do not necessarily have to have had formal training, but it is essential that they understand about the risks of fire in the construction industry.

Competency is not defined specifically in the RRO, but is generally taken to mean having a level of knowledge and experience that is relevant to the task in hand.

A fire risk assessment is a logical, practical review of the likelihood of a fire starting in the premises and the consequences of such a fire. Someone who has good knowledge of the work activities and the layout of the site, together with some knowledge of what causes a fire, would be best placed to carry out a fire risk assessment.

What are the general requirements for fire safety on construction sites

Planning for fire safety must include making available adequate resources, in terms of time, materials and money for the provision and maintenance of suitable and sufficient general fire precautions for the duration of the project.

A management ethos should be adopted such that suitable procedures and standards are laid down and adopted by all parties concerned in the project regarding the prevention of fires.

The fire prevention and protection measures to be adopted must be set out in the fire risk assessments undertaken in compliance with the RRO.

The principal contractor will require sub-contractors to observe their duties relevant to national fire safety legislation to ensure that the fire risk and potential for damage have been properly assessed and are kept to a minimum during construction.

The fire risk assessments in respect of all construction sites must be reviewed periodically in compliance with the legislation. This is important and necessary due to the rapidly changing nature of the hazards on all construction sites.

Fire safety must be recognised as an integral part of the management strategy for every construction site. In each case, therefore, a designated person must be responsible for the fire safety management system and inspections.

Because of the increased risk to personnel on high-rise construction sites and other locations where the means of escape may be compromised in the event of a fire, these sites should be identified in the fire risk assessment, and in these instances the individual who is responsible for the fire safety management system and inspections should be permanently based on the site.

What are some of the basic fire safety measures to include in the fire safety plan?

The fire safety plan should be developed as an integral part of the construction phase health and safety plan or the overall work plan for the project.

The fire safety plan should include all aspects of fire safety that affect the multi-occupied parts of the building and any overlap with the client's undertaking.

The basics are

- fire exit routes from the site
- fire-fighting equipment
- methods to raise the alarm
- lighting
- signage
- fire protection measures to prevent spread
- safe systems of work
- prohibiting smoking on site
- appointment of fire wardens
- regular checks
- emergency procedures.

Fire exit routes

Where possible, there should always be more than one exit route from a place of work. If the travel distance is more than 45 m to an exit, there must be two or more exits. This travel distance will vary depending on the risk rating of the site.

If the number of exits cannot be improved, then the risk rating of the site for fire must be reduced.

Fire exit routes must be unobstructed, clearly defined, of adequate size and width, and not locked.

Doors leading onto fire exit routes should open via a push bar in the direction of travel.

No fire exit route must lead back into the building or site.

All fire exit routes must lead to a place of safety.

The fire exit routes must in themselves be protected from fire by fire-protected enclosures or doors. Doors must be kept shut.

Exit routes using ladders (e.g. on scaffolding) need to be especially assessed as part of the site-specific risk assessment.

Fire exit routes must be clearly visible from all parts of the work area. Exit signs which meet the Health and Safety (Safety Signs and Signals) Regulations 1996 must be displayed.

If the lighting is poor, photo-luminescent signs should be used.

All fire signage must display pictograms. Text can also be used alongside the pictogram, as can directional arrows.

Fire-fighting equipment

Suitable fire extinguishers need to be placed in appropriate locations around the site, and always at fire points near the fire exit routes.

Multi-purpose foam or powder extinguishers are suitable, but so too are water and carbon dioxide. The fire risk assessment should determine which type is required.

Fire-fighting equipment should be visible, properly signed, inspected weekly and ready to use if needed. Operatives should not need to climb over materials, move plant and so on to use the extinguishers.

Either a designated number of operatives in each work area or all operatives should be trained in how to use the fire-fighting equipment.

Regular re-assessment of the working area is needed to ensure that the location of the fire extinguisher points are appropriate.

Methods of raising the alarm

A fully integrated alarm system would be beneficial on all sites, activated by break glass points and linked to an alarm control panel in the site office. However, this is not always possible, and alternatives are permissible such as

- hand bells
- klaxons
- sirens
- hooters.

The alarm in use on the site should be clearly identified, and all operatives *must* receive training in fire alarm procedures.

Fire alarm points must be clearly visible, easily accessible and so on.

If alarms cannot be heard in all areas of the site, there must be a procedure for fire wardens to warn other floor fire wardens and others.

On small sites, a simple shouting of 'Fire! Fire!' may be all that is needed.

If the site is multi-occupied and without employers (e.g. a major department store refurbishment), then the construction site alarm system must integrate with that of other employers so that total building evacuation is occasioned as necessary.

Lighting

Emergency lighting is not necessarily required on all construction sites, but if there is a risk of power failure and no natural daylight to the areas of work, emergency lighting will be essential.

A simple system of torches may suffice.

All emergency exit routes must be adequately lit at all times.

Emergency back-up lighting should activate when the main power supply fails.

Regular checks of emergency lighting will be necessary.

Signage

Signage enables people to be guided to safe places – to emergency exit routes, safe places (e.g. refuges) or assembly points.

Signs, where possible, should be photo-luminescent.

Signs must be visible from all work areas, non-confusing, of large enough size and accurate in the information they portray (e.g. they must not lead to a dead end as a fire exit route).

Fire protection measures to prevent fire spread

Generally, it is best to try to consider floors and different areas as compartments, with fire-protected, closed doors, stopping and fire protection to voids and ducts, and so on

Fire and smoke, including toxic fumes, spread rapidly. Compartmentation constrains it to one area.

Safe systems of work

Any work activity that looks as if it could increase the risk of a fire starting *must* be controlled by a permit to work or hot works permit system.

Hot works should be prevented whenever possible. Controls need to be localised (e.g. additional fire extinguishers, regular checks and additional fire wardens).

Combustible materials, flammable gases and so on should be removed. Flashover should be considered.

Only trained operatives should carry out hot works or use flammable materials and so on.

Smoking on site

There is no other fire safety procedure acceptable other than to ban it completely within all areas of the site.

Legislation is in force that prohibits smoking in enclosed or substantially enclosed public and work areas, and so smoking will be illegal inside any building being refurbished or within a new structure as it is completed and enclosed, and also within any mess rooms and changing rooms.

Smoking may be permitted in outside areas, but it would be good practice to clearly define in which areas smoking is permitted and to make suitable arrangements for fire safety to be maintained in these areas.

Smoking should ideally be prohibited until operatives are off site, as this is the most effective way to control the risk of fire from smoking material and carelessness.

Appointment of fire wardens

Each floor or work area should have an appointed fire warden (i.e. a person who is trained to know what to do in the event of a fire and to evacuate their area, raise the alarm and so on).

There should be enough fire wardens to cover for absences. Fire wardens should receive regular training.

Regular checks

Daily and weekly fire safety checks are advisable on all sites. Checks are *always* necessary after hot works, and usually approximately 1 h after the end of hot works so that any smouldering materials can be identified.

Records of fire safety checks should be kept for the duration of the project. Remember: the responsible person needs to demonstrate that you they what they are doing.

Emergency procedures

These must be specific for each site and written down clearly. Emergency procedures must be displayed in prominent positions.

They should include

- the type of fire alarm
- how to raise the alarm
- how to evacuate the site
- the location of assembly points
- the names of the fire wardens
- any highly hazardous areas
- storage details for flammable materials
- procedures for visitors to the site
- the names and telephone numbers of the local emergency services
- liaison procedures for the emergency services when they arrive on the site.

Tips: basic fire safety measures

- Plan fire safety before works start, and consider all aspects of the site and adjoining buildings and work operations.
- Reduce the number and amount of combustible materials.
- Reduce the number of ignition sources.
- Keep fire exit routes clear.
- Display adequate and suitable fire signage.
- Train operatives in emergency procedures.
- Keep fire extinguishers on site, in suitable locations and of the correct type.
- Put in emergency lighting if possible.
- Clearly describe the fire alarm raising procedure.
- Get everyone out rather than fight the fire.

Does the site agent need to consider site security against arson when preparing the fire safety plan?

Arson protection should be a feature of the site fire safety plan, and must be addressed as part of the fire risk assessment for the site.

Buildings must be suitably protected against theft and deliberate fire raising in accordance with the fire risk assessment.

The most effective method of deterring trespassers, as well as helping to prevent malicious fire, is to ensure, as far as reasonably possible, that the site is secured against unauthorised entry. This may be achieved by erecting a suitable hoarding around the perimeter of the site or securing all access points such as windows and doors on refurbishment sites.

Flammable liquid stores, LPG cylinder storage and combustible material stores must be fenced or otherwise suitably protected.

Illumination of the site is an effective deterrent to unauthorised access and is recommended.

The installation of CCTV cameras should be seriously considered on sites where arson, vandalism or theft may occur.

The recruitment of security personnel should be considered on all sites, especially for employment on site outside of normal working hours.

A permanent security presence must be provided on sites where the construction is predominantly of combustible materials.

The installation of intruder alarms in temporary buildings and temporary accommodation is strongly encouraged.

What should be considered for temporary buildings and temporary accommodation

The site fire safety plan must include a suitable and sufficient fire risk assessment for all temporary buildings and temporary accommodation. The assessment should be reviewed on a regular basis..

Temporary buildings should be separated from the building under construction or refurbishment and other permanent buildings to provide a fire break, which, where possible, should be at least 10 m, and 20 m in the case of sites where the structure under construction is predominantly of combustible materials. Similarly, rows of temporary buildings should be separated in a similar manner to provide a reasonable fire break.

Temporary buildings need to be considered individually for potential fire risks. Mess rooms where cooking facilities are used could be considered higher risk than materials stores, but stores containing flammable/combustible substances will be an even greater fire risk. Control measures to prevent an outbreak of fire or the spread of the fire could be different for each category of building.

Glazing must be fixed shut and provide a degree of fire resistance as determined by a risk assessment where the glass is within 20 m of another building or structure.

Often, temporary buildings are raised above the ground level, and the space beneath becomes the storage area for all sorts of items and rubbish. Such areas must be enclosed to prevent accumulation of rubbish, while still allowing under-floor ventilation. No combustible materials should be stored under a temporary building.

Temporary buildings must not be constructed within a building or structure under construction or refurbishment (e.g. no timber sheds within the enclosed construction site area) – all temporary buildings must be placed outside in the open.

Automatic fire detection systems should be installed in temporary accommodation and in temporary buildings when these are

- within 6 m of a building under construction or refurbishment
- used for the storage of flammable liquids or compressed gases
- used for the drying of clothes.

Temporary buildings or temporary accommodation where cooking (including the use of toasters) is undertaken should have automatic fire detection installed, as these areas are often high risk for fires due to abuse and carelessness with the cooking equipment.

Heaters and cooking equipment should not be based on the use of LPG or other gases, as not only is there fire risk but also health and safety hazards from oxygen depletion and carbon monoxide gas caused by poor ventilation.

Heaters for use in temporary buildings and temporary accommodation must preferably be of an electrical nature, and

- have enclosed elements
- be fixed in position, preferably above the floor level
- be fitted with securely fixed metal guards
- be thermostatically controlled
- be maintained in a sound condition.

Carelessly drying of clothes causes fires. Coat stands and drying racks must be firmly positioned at a safe distance from heaters.

All heaters and cooking appliances must be properly installed, with adequate ventilation provided. Microwave ovens are preferable to proprietary electric or gas cookers.

Consideration should be given to the installation of automatic sprinkler systems in temporary buildings on sites where construction is predominantly of combustible materials.

What is good practice regarding the site storage of flammable liquids and gas cylinders?

Ideally, flammable liquids and gas cylinders should not be stored on site but instead use the 'just in time' concept, where only the amount needed to fulfil a day's activities is brought on to site.

The fire safety plan and the fire risk assessment must consider all aspects of storage and use, and appropriate control measures must be implemented.

Flammable liquids, compressed gases and/or LPG must not be stored together.

Containers of flammable liquids, gas cylinders and LPG cylinders should preferably be stored in open compounds that are securely fenced, shaded from the sun and remote from pits, drains and low-lying areas.

Any electrical fittings (e.g. lights and switches) within these stores must be suitable for an environment where a flammable or explosive atmosphere may be present, and be selected and installed by competent persons.

Adequate numbers of extinguishers appropriate to the hazard should be sited at the entrances to storage areas.

The HSE and the Fire Protection Association, as well as numerous insurance companies, all offer advice and guidance on the safe storage of combustible and flammable materials and substances.

Flammable liquids

There are many products used on a construction site that are classified as flammable (e.g. solvents, thinners and adhesives). Often, it is the vapours given off by a solid or liquid product that is the fire risk, and the fire risk increases if lids are left off tins and so on.

Stores of flammable liquids must be situated on an impervious base and surrounded by a bund sufficient to contain the maximum contents of the largest drum stored plus 10%. The bund must not be allowed to accumulate water or waste material.

Where it is necessary to store flammable liquids inside a building under construction or renovation, the quantity so stored must be the minimum necessary and no more than a day's supply. The containers must be kept in a store, cupboard or bin that is of fire-resistant construction.

Ideally, storage areas should be sited at least 10 m from permanent and temporary buildings and 20 m from structures fabricated predominantly from combustible materials. Containers must not be stored within 4 m of any building or boundary fence unless the boundary is a wall with at least 30 min of fire resistance and is at least 2 m high. In the latter case, containers and drums should be at least 1 m below the top of the wall.

Different products should be stored in separate areas (i.e. no mixing and matching of products) because often there can be a reaction between two or more products/ substances, which in turn can increase the risk of fire or explosion.

Gas cylinders

All gas cylinders should be stored outside unless they are in use. As soon as they are no longer in use, they must be returned to outside storage compounds.

The floors of LPG and other cylinder stores should be paved or compacted level with a suitable hard standing provided for the delivery and dispatch of cylinders. The area must be kept clear of all combustible materials, weeds and rubbish.

The provision of automatic flammable gas detection equipment should be considered for enclosed storage locations.

All permanent LPG and natural supplies and their connections to gas/LPG fuelled appliances must be installed by a competent gas fitter.

The gas supply to appliances should be by fixed piping or armoured flexible tubing, and the gas cylinders sited externally to any building (e.g. the mess room).

Gas cylinders should be located outside buildings and be secured and protected from unauthorised interference.

Acetylene

Acetylene is commonly used for welding activities, and is serious hazard on construction sites and elsewhere. It is a flammable gas that at elevated temperatures and pressures, or following impact of the cylinder, becomes unstable and liable to spontaneous decomposition. As a result, acetylene in cylinders, once suspected to be unstable, constitutes a serious fire hazard.

Wherever possible, following the principles of prevention, the use of acetylene on a construction site should be avoided and suitable alternatives sourced. Where the use of the gas is unavoidable, its presence must be minimised.

Spare cylinders must not be kept on site.

The cylinders must be removed from the workplace and returned to the storage area as soon as the work requiring their use is complete. The cylinders should be removed from the site as soon as they are no longer needed.

Gas cylinders should always be adequately supported, preferably by mounting on purpose-built trolleys. Equipment should be hired from responsible and authorised dealers/suppliers.

Equipment, hoses and flashback arrestors used with oxyacetylene and similar equipment should be in good condition, set up in accordance with the manufacturer's instruction and be subject to a visual inspection by a competent person before each period of use.

Gas welding and cutting procedures should only be carried out by trained personnel.

The use of acetylene cutting and welding equipment must be subject to a hot works permit once fitting out work has commenced on site and in all that which are being refurbished.

What steps should be taken to manage hot works?

Works involving naked flames and heat-generating equipment are major causes of fires on construction sites, and such activities must be addressed in the fire safety plan and the site fire risk assessment. Adequate control measures are essential, and operatives must be fully informed of the safety precautions to be taken to minimise risk. Hot works should be a last resort.

Alternative methods to hot work should be adopted wherever possible. Where hot work cannot be avoided, the guidance set out in the HSE's guide Fire Safety on Construction Sites (HSG168) and in the Confederation of Fire Protection Associations in Europe's *Fire Safety Basics for Hot Work Operatives* (Guideline 12), should be followed.

Hot work should only be undertaken by suitably trained staff, as this will reduce the likelihood of uncontrolled events.

When there is no alternative to hot work, the hot work should be undertaken, where possible, in a dedicated area away from the main area of construction work or storage of materials.

A 'permit to work' system must be adopted where hot work is being undertaken unless there is no risk of damage to any surrounding property.

Roofing works often require the installation of bitumen felt, lead flashings and so on, and hot works are an integral part of the job. Tar boilers are common for heating up the bitumen to enable it to be applied to the roof boarding or substrate. Tar boilers need to be risk assessed to ensure that they will not be creating hazards to users and those in the vicinity. Hot works permits are essential to ensure that all hazards are identified and control measures implemented.

Tar boilers must be sited on stable ground and in locations which do not pose risks (e.g. separated from combustible materials).

Suitable fire-fighting equipment must be available, and a fire watch instigated to manage any residual risks.

Is there any fire risk from the protective coverings used during the final fit-out stages?

Protective coverings are a common feature during the fit-out stages where final fixtures, such as doors, handrails, floor coverings and panels, need to be protected against damage. Such coverings can be a substantial contribution to the overall fire load in circumstances where ignition sources are common. Particular risks occur where protective coverings are used to protect floor surfaces and features in fire escape stairways. The risk can be reduced by using protective materials that are flame retardant and that comply with the Loss Prevention Standard (LPS) 1207, and satisfy flame-retardant criteria. These materials are usually labelled clearly as being flame retardant or as meeting the LPS.

Fire-retardant materials can still burn, and, therefore, their use will increase the fire load on the site. It will be essential to review the Fire Risk Assessment and amend it accordingly to address additional control measures.

Risks arising from protective coverings can be reduced by

- installing vulnerable features needing protection as late as possible in the fit-out stage
- ensuring that the coverings are to flame-retardant specifications wherever possible – this may require liaison with suppliers of vulnerable items and/or protective coverings.

What subjects need to be covered in relation to fire safety during any site induction sessions?

Employers must provide adequate fire safety training for their staff. The type of training should be based on the particular features of the site, including any fire precautions required by adjacent buildings, the client's activities or other occupiers, and should

- take account of the findings of the fire risk assessment
- explain the emergency procedures
- take account of the work activity and explain the duties and responsibilities of staff
- take place during normal site working hours and be repeated periodically, especially when new operatives or sub-contractors come on to the site
- be easily understood by all operatives and other people who may be on site, using material translated into different languages if necessary or by using pictograms and signs that are universally comprehended.

Any induction information should include the following

- what to do on discovering a fire
- how to raise the alarm and what happens then
- what to do upon hearing the fire alarm
- the procedures for alerting contractors and visitors including, where appropriate, directing them to exits
- the location of the assembly points and the procedure for reporting to the fire wardens
- the location of fire-fighting equipment
- the arrangements for calling the fire and rescue service
- the procedure for the reporting of incidents and any near misses.

Who enforces the RRO on construction sites?

The HSE is responsible for the enforcement of fire safety legislation within the curtilage of the construction site. This includes the construction site itself and any buildings provided in support of the construction activity, such as office accommodation and welfare facilities (including, where provided, sleeping accommodation). Where premises provided in support of the construction activity are outside and separate from the curtilage of the construction site (e.g. by a road), the local fire and rescue authority has enforcement responsibility.

On any site contained within, or that forms part of, another premises, where it either remains or becomes occupied by persons other than those working on construction, general fire precautions enforcement for the premises containing the construction site rests with the enforcer for those premises, normally the fire and rescue authority (i.e. they are responsible for the general fire precautions of the construction site as well). This is to ensure that if a fire on the construction site can affect persons within the other premises or vice versa, then the general fire precautions need to be co-ordinated.

An exception to this is where the construction site is separated from the remaining premises by an unbreached partition, such as a fire wall affording an appropriate period of fire resistance. In such cases the HSE is the enforcing authority.

If an HSE inspector is not satisfied with my fire safety arrangements, what action can they take?

Authorised officers can take various actions under the RRO and the most common ones are the service of two types of formal notice:

- *Enforcement notice* (article 30 of the RRO). An enforcement notice is issued where the responsible person has failed to comply with the RRO, and details corrective measures that they are legally obliged to complete within a set timescale, to comply with the law.
- *Prohibition notice* (article 31 of the RRO). A prohibition notice is issued where the use of the premises may constitute an imminent risk of death or serious injury to the persons using them. This may be a restriction of use (e.g. imposing a maximum number of persons allowed in the premises) or a prohibition of a specific use of all or part of the premises (e.g. prohibiting the use of specific floors or rooms for sleeping accommodation).

The issue of a prohibition notice under the RRF is the most serious enforcement option available to either an HSE inspector or a fire officer other than prosecution.

Inspectors can decide that breaches are so severe that direct prosecution is the only remedy available to them.

All formal and informal action (e.g. an advisory letter) is likely to carry a fee for intervention cost, and the responsible person will receive an invoice to cover the costs of the visit and any subsequent actions carried out by the inspector.

CDM Questions and Answers: A Practical Approach
ISBN 978-0-7277-6032-6

http://dx.doi.org/10.1680/cdmqa.60326.503

Bibliography

Confederation of Fire Protection Associations in Europe (CFPAE) (2012) *Fire Safety Basics for Hot Work Operatives* (Guideline 12), Copenhagen: CFPA.

Health and Safety Executive (HSE) (2005) *Controlling Noise at Work: Guidance on the Control of Noise at Work Regulations 2005* (L108), HSE, Sudbury, UK.

HSE (2010) *Fire Safety in Construction: Guidance for Clients, Designers and Those Managing and Carrying Out Construction Work Involving Significant Fire Risks* (HSG168), HSE, Sudbury, UK.

HSE (2012) *Noise at Work: A Brief Guide to Controlling the Risks* (INDG362(rev2)), http://www.hse.gov.uk/pubns/indg362.pdf, HSE, Sudbury, UK.

HSE (2013) L74 – *First Aid at Work: The Health and Safety (First-Aid) Regulations* 1981. HSE, Sudbury, UK.

HSE (2014) *Five Steps to Risk Assessment* (INDG 163(rev4)), http://www.hse.gov.uk/pubns/indg163.pdf, HSE, Sudbury, UK.

HSE (2014) *Avoiding Danger from Underground Services*, (HSG47, Third edition), http://www.hse.gov.uk/pubns/books/hsg47.htm, HSE, Sudbury, UK.

HSE (2015) F10 – Notification of construction project, Construction (Design and Management) Regulations 2015, www.hse.gov.uk/forms/notification/f10.htm, HSE, Sudbury, UK.

HSE (2015) *Managing Health and Safety in Construction, Construction (Design and Management) Regulations 2015. Guidance on Regulations L153*, HSE, Sudbury, UK.

HM Government (1992) *Approved Code of Practice for the Workplace (Health, Safety and Welfare) Regulations 1992 and the British Standard BS 6465 2006*, Norwich: The Stationery Office.

HM Government (2005) *Control of Noise at Work Regulations 2005: Noise Assessments – Getting What You Need*, Norwich: The Stationery Office.

CDM 2015 Questions and Answers: A practical approach
ISBN 978-0-7277-6032-6

http://dx.doi.org/10.1680/cdmqa.60326.505

Index

Page locators in *italics* refer to figures separate from the corresponding text.

Printed and bound by CPI Group (UK) Ltd, Croydon, CR0 4YY

16/09/2026

14963400-0001